Higher Superstition

Higher Superstition

The Academic Left and Its Quarrels with Science

PAUL R. GROSS
University of Virginia

NORMAN LEVITT
Rutgers University

THE JOHNS HOPKINS UNIVERSITY PRESS
BALTIMORE AND LONDON

© 1994, 1998 The Johns Hopkins University Press
All rights reserved
Printed in the United States of America on acid-free paper

Johns Hopkins Paperbacks edition, 1998
9 8 7 6 5 4 3 2 1

The Johns Hopkins University Press
2715 North Charles Street
Baltimore, Maryland 21218-4319
The Johns Hopkins Press Ltd., London

Library of Congress Cataloging-in-Publication Data will be found
at the end of this book.

A catalog record for this book is available from the British Library.

To our wives, our children, and our students, with gratitude

Contents

Preface to the 1998 Edition   ix

Acknowledgments   xv

ONE The Academic Left and Science   1

TWO Some History and Politics: Natural Science and
Its Natural Enemies   16

THREE The Cultural Construction of Cultural Constructivism   42

FOUR The Realm of Idle Phrases: Postmodernism, Literary Theory,
and Cultural Criticism   71

FIVE Auspicating Gender   107

SIX The Gates of Eden   149

SEVEN The Schools of Indictment   179

EIGHT Why Do the People Imagine a Vain Thing?   215

NINE Does It Matter?   234

Notes   259

Supplementary Notes to the 1998 Edition   289

References   305

Index   319

Preface to the 1998 Edition

Like most scientists, engineers, and mathematicians, including those who founded the U.S. National Science Foundation and encouraged its social-science programs, we have always valued social analysis of science and its history. We still do. As for anti-science, that is a very old story in our culture. In its accustomed forms, it would hardly have stirred us to take the trouble to study and write about it. The writing of *Higher Superstition* was undertaken only when it became clear to us, from separate but remarkably similar experiences at our respective universities, that something new and unwelcome had found its way into the academic bloodstream and thence into lecture rooms, journals, books, and faculty chit-chat: the systematic disparagement of modern science. A public response was clearly needed. Even the silliest criticisms of science, dressed up as social analysis, hermeneutics, or emancipatory politics, were going largely unanswered. Neither scientists as individuals, nor scientific organizations, nor scholars within the disciplines whence issued the disparagement, showed any inclination (might it have been any *courage?*) to rebut the kinds of antiscientific nonsense and flawed scholarship we were encountering in the academy. We didn't know what sort of response the book we considered writing would evoke. In fact, we rather feared, since the argument would necessarily alternate between broad-brushing and technical detail and between polemic and analysis, that it would slip unremarked from new release to backlist to remainder without stirring much interest one way or the other.

The fear turned out to be groundless: it was dispelled by astonishing numbers of allies and opponents promptly having their say in conferences, seminars, journals, and on the internet. We found ourselves in the eye of a storm generated by the book. Predictably enough, in the flood of reviews, most of those from scientists were strongly positive, while most of the antagonism came from the "science-critique" academic subculture and its allies. Interestingly, however, some of the most hostile criticism appeared in journals affiliated with the scientific community, which lent their pages (as they have been doing for some time) to anxious apologists for the antiscientific, pseudo-sociological fads that are the subject of this book. It was lauded in some conservative journals of opinion despite our disclaimer of any wish to advance conservative political causes—or indeed *any* political program. Most of the indignant huffing and puffing came from the circled wagons of what we called, with repeated misgivings, the "academic left," which tried to encourage the perception that we are deep-dyed conservatives (closeted or otherwise) pursuing reactionary agenda, and protecting our fat research grants.

As it happened, the instinctive dislike of science that lurks historically in the conservative woodwork was mostly dormant when we wrote the book. Since then it seems to have begun scratching again, in the form of new denunciations of "Darwinism" (which is not to suggest that there is no anti-Darwinism on the left). If, therefore, we were writing this book ab ovo, the "academic right" would have to join the academic left in its subtitle and there would have to be a chapter on "Intelligent Design Theory." Indeed, "right" and "left" are curiously united on these issues, with leftist adherents of identity politics just as comfortable with the doctrine of special creation—in its "Native American" version—as are rightist defenders of biblical orthodoxy. We know at least one public incident in which a senior and greatly esteemed member of the science studies establishment, fashionably leftist in most respects, defended the right of fundamentalist school boards to mandate the teaching of creationism in science classrooms. Apparently, for her (a noted scholar not mentioned in our original text), the intrusion of religion into the science classroom is less troubling than the idea that natural science gets at truths about nature that transcend socio-cultural particulars.

For us, however, the greatest surprises have been pleasant ones. Chief among them was the international uproar occasioned by the publication of Alan Sokal's now-famous hoax, "Transgressing the Boundaries: The Transformative Hermeneutics of Quantum Gravity," in the trendy cultural studies journal *Social Text*. The ongoing saga of Sokal's pleasantry is instructive on several levels. The joke arose from Sokal's reading of our book. Originally, as a principled leftist, he suspected that we two might be conservatives as charged, advancing antiliberal agenda under the pretext of defending science. However, he ultimately found much of our argument persuasive. In fact, his own researches convinced him that we had in some respects understated the case. His dismay at the clear evidence that a once-vigorous intellectual tradition of radical dissent is slipping into irrationality prompted him to put aside physics for a few weeks in the fall of 1994 in order to compose his delightful parody. It was submitted to *Social Text*, in all apparent seriousness, early that winter.

Unknown to Sokal at the time, that publication, under the leadership of Prof. Andrew Ross (see Chapter 4), was preparing a special issue on what it dubbed "the science wars." The intention was to vindicate assorted poststructuralist, multicultural, and feminist critiques of science and to denounce their critics, most notably the depraved Gross and Levitt. Sokal's piece, with its seconding and fulsome praise of such intentions, was snapped up by the editors.

The tainted issue appeared in due course (May 1996), Sokal's revelation of the hoax appeared a few days later in *Lingua Franca,* and then all hell broke loose. Predictably, some conservatives crowed, citing the "Sokal Text" affair as further proof that left-wing sympathies equate to outright dementia (notwithstanding Sokal's own leftist views). But the reaction of a large number of left-intellectuals was more lasting and perhaps more significant for the academy. Sokal's hoax brought into the open a widespread reaction, years in the making, against the sesquipedalian posturings of postmodern theory and the futility of the identity politics that so often travels with it. Cutting-edge celebrities, long used to dictating the tone of political discussion in "progressive" circles, suddenly found themselves on the hot seat. As of this writing, the recriminations continue with no sign of abatement.

We had hoped to include in this new edition, as an appendix, the full text of Sokal's article, which cites verbatim and to exceedingly comic effect a goodly collection of postmodernist "authorities" on science and its philosophical import. Alas, their tender sensibilities intervened to frustrate the hope. Duke University Press, under the leadership of Prof. Stanley Fish, is the publisher of *Social Text* and thus owns the copyright to Sokal's piece. That scholarly organization responded to a Johns Hopkins University Press request for reprint permission with what can only be described as an exorbitant demand for royalties. The proffered justification was greed: they expected, it was explained, that the Johns Hopkins Press would make lots of money from such a deal, and they wanted their cut. The grapevine brings to our ears, however, credible rumors suggesting that spite, rather than lucre, might have been the chief reason for this dogfish-in-the-manger attitude. Whatever the case, we apologize to the reader. The text of Sokal's highly relevant put-on must be consulted elsewhere.

Interest stirred up by our book convinced us to try to extend the discussion by organizing a conference under the sponsorship of the New York Academy of Sciences. This conference, held in New York in the spring of 1995, was called "The Flight from Science and Reason"; several dozen scholars and writers—whom we do not blush to call distinguished—contributed to it. It had at least the virtue of demonstrating that misgivings about the spread of relativism and antirationalism, or, more broadly, the increasing loss of nerve within an intellectual community faced with the need to defend logic, evidence, and rational thought, are not the parochial concern of self-interested scientists. The wide spectrum of political views and academic disciplines represented confirmed that the proposition that there is a growing disdain for reason and science is not simply a bugbear of conservatives looking for a pretext to discredit the left. The conference proceedings, now reissued by the Johns Hopkins Press, incorporate a number of essays recruited subsequent to the actual meeting, expanding the range of topics and perspectives. Whatever else might be said about this conference, its breadth was, in our age of narrow specialization, astonishing (at least it astonished us). Where else, in a single volume, can one find informed debate about editorial practice in preparing editions of King Lear, on the one hand, and about the implications of the quantum-mechanical formalism for our view of physical reality, on the other?

By the time those proceedings were published, the "science wars" (we dislike the term, but the coinage is out there) were in full career, with the Sokal affair but one obvious stimulant. Radical science studies, with its do-it-yourself epistemologies, had long enjoyed a certain immunity from serious challenge or criticism. That immunity began to melt away as more and more scientists became aware of the breadth and depth of the misconceptions about science being propagated by constructivist historians, sociologists of scientific knowledge, and feminist epistemologists, among others. Benchmarks in this process included Nobelist M. F. Perutz's astringent New York Review of Books article on Gerald Geison's The Private Science of Louis Pasteur and the ensuing correspondence. The Sokal parody generated its own large literature, including essays by another Nobel laureate, physicist Steven Weinberg, also in the New York Review of Books, and philosopher Paul Boghossian, in the Times Literary Supplement. The science studies establishment seems divided about how to respond to such arguments. Heavy manifestos in defense of the absolute truth of relativism and the permanent ephemerality of scientific knowledge issue from some quarters; historian Paul Forman and literary theorist Barbara Herrnstein Smith have provided paradigmatic samples of the genre. On the other hand, there have been some frantic signals from once-stalwart adherents of the science-as-social-convention orthodoxy that the time has come to back away from the now-blunted cutting edge. Angry arguments for the defense, of the "nobody ever believed that" variety, erupt from time to time into print—M. Norton Wise has provided important type-specimens. These come to grief, however, as time and again something else pops up in print to prove that someone pretty famous in the science studies game does indeed believe exactly that.

The issues now reverberate in elite scholarly institutions. Stanford University, for instance, raised serious questions about its own science studies program. Even more striking, the Institute for Advanced Study, at Princeton University, perhaps the most prestige-laden academic research facility in the country, has been in the thick of the argument. Unknown to us, shortly before this volume was written, there was conflict when "anthropologist of science" Bruno Latour (see Chapter 3) was proposed as a permanent member by the Faculty of Social Science. The Institute's mathematicians and physicists, acquainting themselves with Latour's writings, raised the roof, and the nomination was withdrawn. Our own comments on Latour were written in ignorance of these events; in fact, owing to the institute's code of silence, nearly a year passed following the publication of *Higher Superstition* before rumors of the fracas were confirmed. Quite recently (within the past few weeks) the issue has flared anew in the same place. M. Norton Wise, a cultural historian of science at Princeton, was proposed for the institute position once denied Latour, and rejected. Readers of Wise's review of this book (in *Isis*) and of *The Flight from Science and Reason* (in *American Scientist*) will perhaps understand why we are not as disheartened by this event as the social-scientist commentators say we ought all to be. It is among the signals that scientists and mathematicians have begun, finally, to take notice of what some of the leading figures of science studies have been writing and teaching for nearly two decades.

Conferences and seminars throughout the United States and Western Europe, as well as a flood of new publications, now reflect the argument, or struggle: in that sense, the flippant coinage "science wars" can be justified. Many notable historians and sociologists of science have long held misgivings about the intellectual nihilism that offers itself as "cultural constructivism"; but they have been reluctant to challenge it for fear of gaining a reputation as sissies, too weak-kneed to play the exhilarating game of "epistemological chicken." Just as often, they have been cowed by fear of the academically fatal accusations: political conservatism, sexism, disdain for the Other. Moreover, in order to commit themselves publicly to professional conflict, honest scholars need time to do painstaking research, and they usually need to feel that they won't be just wasting breath. Now, however, some of them seem to be in a fighting mood. Highly touted works of the science studies avant garde have recently come under systematic scrutiny and have been found lacking, not only because of perverse philosophical assumptions but also for flaws in methodology and for serious historical inaccuracies. As evidence of this new mood, we cite *A House Built on Sand*, a compilation of such critiques soon to appear from Oxford University Press. If the publication of *Higher Superstition* played some catalytic role in these reactions, then we are pleased to have written it.

A word about changes and supplements to this second edition. We have corrected some typographical errors in the text and a few awkwardnesses in wording, without altering the underlying meaning. A few minor factual slips have been corrected. We have had second thoughts about some points, often at the prompting of well-informed readers; a new set of Supplementary Notes sets those forth. The notes also include evidence on matters whose further investigation has made our case stronger than it was when first examined. A few new, relevant facts complete the addenda.

During the preparation of this new edition, we have had deeply appreciated support for it and related projects: N. L., from the Open Society Institute's Individual Project Fellowships Program, and P. R. G., via a grant from the Esther A. and Joseph Klingenstein Fund to the New York Academy of Sciences. Finally, we thank Douglas Armato and the Johns Hopkins University Press staff for their expert assistance in bringing this paperback edition to timely printing.

Acknowledgments

The following friends and colleagues have helped, beyond the possibility of adequate thanks, in the writing of this book. All have read, or heard, or discussed with us at least some part of it. All had blame as well as praise to report—in varied proportions. In no case have we taken all of someone's advice; but in every case we have taken some of it. We owe special thanks to those correspondents who identified examples and sources we had overlooked, and who went out of their way to provide us with their own published and unpublished work and with other documents that might otherwise have eluded us. We are deeply grateful to the Johns Hopkins University Press, whose director and staff have from the start displayed the mixture of flexibility and intellectual seriousness that one does, and should, expect from a university press. Our gratitude to editors Marie Blanchard and Terry Schutz is boundless.

In this book we speak, at one point, of threats to the essential grace and comity of scholarship and the academic life. We mean what we say there; but the generous responses of our colleagues to various versions of our writing show clearly that threats are by no means equivalent, as yet, to major damage. None of those listed here bears the slightest responsibility for errors of fact, judgment, or style in what follows. We alone are culpable. On the other hand, a good many such errors are not here because the following people graciously gave some thought to our project: Tony Bahri, Felix Browder, Ralph Cohen, Eric Davidson, Robin Fox, Sheldon Goldstein, Wendy Gross, Antoni Kosinski, Steven Levitt, Martin Lewis, Eve Menger, R. L. Norman, Bernard Ortiz de Montellano, Philip Pauly, Christopher Phillips, R. K. Ramazani, Eugenie Scott, Keith Spurgeon, E. O. Wilson, and Joe Wisnovsky.

We consider it proper to note that our university affiliations are given for identification only. No position we take in this book is an official one of either institution or of any academic unit with which we are associated for teaching or administrative purposes.

Higher Superstition

CHAPTER ONE

The Academic Left and Science

Ever since puberty I have believed in the value of two things: kindness and clear thinking. At first these two remained more or less distinct; when I felt triumphant I believed most in clear thinking, and in the opposite mood I believed most in kindness. Gradually, the two have come more and more together in my feelings. I find that much unclear thought exists as an excuse for cruelty, and that much cruelty is prompted by superstitious beliefs.

BERTRAND RUSSELL, *AUTOBIOGRAPHY*

Muddleheadedness has always been the sovereign force in human affairs—a force far more potent than malevolence or nobility. It lubricates our hurtful impulses and ties our best intentions in knots. It blunts our wisdom, misdirects our compassion, clouds whatever insights into the human condition we manage to acquire. It is the chief artisan of the unintended consequences that constitute human history. To crusade against muddleheadedness, therefore, may be the most futile, and hence the most muddleheaded, quest of all. Inasmuch as that is the aim of this book, we concede that we may be as misguided as any of our subjects. Still, passivity in the end is more reprehensible than quixotry.

Sub specie aeternitatis, it hardly pays to subdivide muddleheadedness into different categories arising from different social and historical circumstances. An attempted taxonomy implies ingratitude toward the philosophers and poets who have demonstrated, in their varied and powerful ways, the ineluctability of folly. What one is muddled about may well be the consequence of one's specific relation to society, politics, and history. Muddleheadedness, as such, goes on forever. Humankind does not live, however, sub specie aeternitatis; neither of us has ever tried to do so and we don't think we would like it. And so we are tempted to make, this once, a categorization that we reject as in general illusory. We take on a species of muddleheadedness that abounds in a particular contemporary community—more precisely an interlinked net of communities—that has flourished in its present form for a decade or two and that might, although we doubt it, fade out of existence before another decade has passed.

We hope to be as clear in our thinking as Bertrand Russell would have wished, though it will be difficult to be as kind. Since much of what we write will appear, and in places may actually be, polemical, a certain gleefulness may be imputed to some of our observations. It would be sanctimonious to plead innocence, or to seek forgiveness in advance. But it should be noted that such glee as we may inadvertently exhibit is an ambiguous thing: in the last analysis, the subjects of our rancor are not enemies but friends. There is inescapable irony in that, but, we trust, no hypocrisy. Our chief hope in writing this is to convert friends (whose asseverations are for the moment our subject), or at least to persuade them to reflect. If we succeed only in gratifying their traditional foes, providing one more shaft to be launched against them, we shall have failed utterly.

The Academic Left

Our subject is the peculiarly troubled relationship between the natural sciences and a large and influential segment of the American academic community which, for convenience but with great misgiving, we call here "the academic left." The academic left cannot be said to have a well-defined theoretical position with respect to science—it is far too diverse and internally contentious for that—but there is a noteworthy uniformity of tone, and that tone is unambiguously hostile. To put it bluntly, the academic left dislikes science. Naturally enough, it dislikes some of the uses to which science is put by the political and economic forces controlling our society, especially in such areas as military hardware, surveillance of dissidents, destructive and environmentally unsound industrial processes, and the manipulation of mass consciousness through the technologies of popular culture.[1]

This is hardly surprising: such dislikes are widespread, and scientists themselves display them as much as anyone. Within the academic left, however, hostility extends to the social structures through which science is institutionalized, to the system of education by which professional scientists are produced, and to a mentality that is taken, rightly or wrongly, as characteristic of scientists. Most surprisingly, there is open hostility toward the *actual content* of scientific knowledge and toward the assumption, which one might have supposed universal among educated people, that scientific knowledge is reasonably reliable and rests on a sound methodology.

It is this last kind of hostility that scientists who are aware of it find most enigmatic. There is something medieval about it, in spite of the hypermodern language in which it is nowadays couched. It seems to represent a rejection of the strongest heritage of the Enlightenment. It seems to mock the idea that, on the whole, a civilization is capable of progressing from ignorance to insight, notwithstanding the benightedness of some of its members. We have the sense, encountering such attitudes, that irrationality is courted and proclaimed with pride. All the more shocking is the fact that the challenge comes from a quarter that views itself as fearlessly progressive—the veritable cutting edge of the cultural future. On the surface at least, the phenomenon is not a case of nostalgia. These critics of science do not repine for the traditional mores and devout certainties of a prescientific age. They accuse science itself of a reactionary obscurantism, and they revile it as an ideological prop of the present order, which many of them despise and hope to abolish.

We try to use the troubling term *academic left* with reasonable precision. This category is comprised, in the main, of humanists and social scientists; rarely do working natural scientists (who may nevertheless associate themselves with liberal or leftist ideas) show up within its ranks. The academic left is not completely defined by the spectrum of issues that form the benchmarks for the left/right dichotomy in American and world politics, although by reference to that standard set—race, women's rights, health care, disarmament, foreign policy—it unquestionably belongs on the left. Another set of beliefs—perhaps it is more accurate to call them attitudes—comes into play in an essential way, shaping this subculture. What defines it, as much as anything else, is a deep concern with cultural issues, and, in particular, a commitment to the idea that fundamental political change is urgently needed and can be achieved only through revolutionary processes rooted in a wholesale revision of cultural categories.

This apocalyptic break with things-as-they-are is supposed to displace a vast array of received cultural values and substitute an entirely novel ethos. From this perspective feminism, for example, means more than full juridical equality for women, more than income parity and equal access to careers, more than irrevocable "reproductive rights." It means, in fact, a complete overthrow of traditional gender categories, with all their conscious and unconscious postulates. By the same token, racial justice, on this view, does not mean peaceful assimilation of blacks into the dominant culture, but the forging of an entirely new culture, in which "black" (or "African") values—in social relations, economics, aesthetics, personal sensibilities—will have at least equal standing with "white" values. Similarly, environmentalism, as understood and preached on the academic left, extends far beyond concrete measures to eliminate pollution, or to avoid extinction of species and elimination of habitats. Rather, it envisions a transcendence of the values of Western industrial society and the restoration of an imagined prelapsarian harmony to humanity's relations with nature.

Most scientists are made aware of the academic left's critique only by fragmentary and sporadic contact. They may have heard feminist or environmentalist or multiculturalist criticisms of their disciplines; but they have given them—generally—scant attention. They may have encountered books and popular articles proclaiming epistemological revolution in science and everything else on the basis of postmodernist philosophical insight; but they have not been much moved. They think they know that species of proclamation. They have heard rumors of literary critics waxing sententious over the uncertainty principle, or Gödel's theorem; but if so, they have almost certainly written these efforts off as harmless, even charming, examples of the literary temperament.

We do not suggest that there is any reason for scientists to be *acutely* alarmed in the short run. The academic left's rebellion against science is unlikely to affect scientific practice and content; nor will it penetrate the attitudes of those who study the philosophical implications of science from a position of genuine familiarity. The danger, for the moment at least, is not to science itself. What *is* threatened is the capability of the larger culture, which embraces the mass media as well as the more serious processes of education, to interact fruitfully with the sciences, to draw insight from scientific advances, and, above all, to evaluate science intelligently. To the extent that the academic left's critique becomes the dominant mode of thinking about science on the part of nonscientists, that thinking will be distorted and dangerously irrelevant.

Sources of Indignation

Much of this critique is informed or inspired by what is usually called "postmodern" thought and its concomitant value system.[2] In turn, postmodernism is embedded and elaborated in the scholarly work of the academic left, notably in fields such as literary criticism, social history, and a new hybrid called "cultural studies." Postmodernism is grounded in the assumption that the ideological system sustaining the cultural and material practices of Western European civilization is bankrupt and on the point of collapse. It claims that the intellectual schemata of the Enlightenment have been abraded by history to the point that nothing but a skeleton remains, held together by unreflective habit, incapable of accommodating the creative impulses of the future.

Postmodernism, however, is but one of the strands from which the academic left weaves its indictment. Other notions both new and old enter into the cloth. The traditional Marxist view that what we think of as science is really "bourgeois" science, a superstructural manifestation of the capitalist order, recurs with predictable regularity, in its own right or refurbished as the doctrine of "cultural constructivism." The radical feminist view that science, like every other intellectual structure of modern society, is poisoned and corrupted by an ineradicable gender bias, is another vitally important element. An analogous accusation comes from multiculturalists, who view "Western" science as inherently inaccurate and incomplete by virtue of its failure to incorporate the full range of cultural perspectives. A certain strain of radical environmentalism condemns science as embodying the instrumentalism and alienation from direct experience of nature which are the twin sources of an eventual (or imminent) ecological doomsday.

These ideas are the chief elements alloyed to form the academic left's challenge to conventional scientific thinking. It must be noted, however, that there is no canonical way of combining them. Although we have been speaking of an academic left critique, it must be stressed—and we are compelled to stress it throughout the discussion to follow—that this is not a self-consistent body of doctrine. Rather, it is a congeries of different doctrines, with no well-defined center, each of which draws upon the notions we have cited in an idiosyncratic way, elaborating some of them with enthusiasm while leaving others in the background and rejecting still others completely. What enables them to coexist congenially, in spite of gross logical inconsistencies, is a shared sense of injury, resentment, and indignation against modern science.

Natural science is one of the last major features of Western life and thought to come systematically under the critical gaze of the academic left. The reason is obvious. In order to think critically about science, one must understand it at a reasonably deep level. This task, if honestly approached, requires much time and labor. In fact it is best started when one is young. It is scarcely compatible with the style of education and training that nurtures the average humanist, irrespective of his or her political inclinations. On the other hand, science, together with its immediate realization as technology, is—as much as anything can be—the single aspect of Western thought and social practice that defines the Western outlook and accounts for its special position in the world. Non-Western societies—Japan, to take the obvious example—can simultaneously succeed and maintain their identities only to the degree that they naturalize the science and technology of Western culture. Consequently, if one is predisposed to regard that Western position of privilege as wicked, for its prejudices and for its history of conquest, then one will inevitably regard Western science with suspicion and perhaps with contempt. Sooner or later any critique of Western values aspiring to be comprehensive must offer an analysis of natural science, preferably scathing.

It would seem to follow, then, that the last eight or ten years should have seen a flock of earnest humanists and social critics crowding into science and mathematics lecture rooms, the better to arm themselves for the fateful confrontation. This has not happened. A curious fact about the recent left-critique of science is the degree to which its instigators have overcome their former timidity or indifference toward the subject not by studying it in detail but rather by creating a repertoire of rationalizations for avoiding such study. Buoyed by a "stance" on science, they feel justified in bypassing the grubby necessities of actual scientific knowledge. This is not because any great number of science apostates has flocked to their banner, although a handful of figures with scientific credentials, as well as the occasional refugee from an unsatisfactory scientific career, can be found on the movement's fringes. The assumption that makes specific knowledge of science dispensable is that certain new-forged intellectual tools—feminist theory, postmodern philosophy, deconstruction, deep ecology—and, above all, the *moral authority* with which the academic left emphatically credits itself are in themselves sufficient to guarantee the validity of the critique.

Thus we encounter books that pontificate about the intellectual crisis of contemporary physics, whose authors have never troubled themselves with a simple problem in statics; essays that make knowing reference to chaos theory, from writers who could not recognize, much less solve, a first-order linear differential equation; tirades about the semiotic tyranny of DNA and molecular biology, from scholars who have never been inside a real laboratory, or asked how the drug they take lowers their blood pressure. We speak not only from an acquaintance with the literature, but from regular experience of lectures, seminars, and symposia, where speakers tend to say more directly what they think than they do in print, and whose reigning attitude is that the cultivation of an authentically postmodern cultural critique requires the avoidance of dialogue with anyone who happens to be a working professional scientist. Above all, it demands a principled refusal to learn the substance of the science one proposes to criticize.

It would appear alarmist, then, for us to get in a lather about what is, by our own account, a feckless enterprise. Life is short; the impulse to let asses bray is strong. It would certainly have saved us the trouble of writing this book had we honored it. We do not anticipate, after all, that radio astronomers will be hanged from the Very Large Array, or that topologists will be shown the instruments of torture and forced to recant the h-cobordism theorem. We expect little *early* change in the teaching and learning of science on the basis of these politicized critiques (although proposals in that direction, including some from people who should know better, pop up now with regularity). Nevertheless, we judge it worthwhile to analyze at some length the animus toward science currently expressed by the academic left. Its existence has to be read as the manifestation of a certain intellectual debility afflicting the contemporary university: one that will ultimately threaten it. One learns, after many years of observing university life and of experiencing most of its levels, that these institutions are fundamentally unlovable,[3] but also that their health has become incalculably important for the future of our descendants and, indeed, of our species.

Insularities

The academic left is embedded in a nearly inviolable insularity, which extends and intensifies that of traditional humanists. The classicists and historians of whom C. P. Snow spoke famously in *The Two Cultures and the Scientific Revolution* were excoriated for their self-satisfied ignorance of the most basic principles of science. Today we find ourselves, as scientists, confronting an ignorance even more profound—when it is not, in fact, simply displaced by a sea of misinformation. That ignorance is now conjoined with a startling eagerness to judge and condemn in the scientific realm. A respect for the larger intellectual community of which we are a part urges us to speak out against such an absurdity. This, we consider, is one of the duties of the scientific thinker, a duty commonly ignored.

Working scientists undeniably have the habit—more than a bit arrogant—of assuming that laymen can't get it right anyway, and that comical misunderstanding of what science is up to is inevitable. Why should yet another episode of the same old thing—they have been seeing it all their working lives—exercise them? We reply that the proliferation of distortions and exaggerations about science, of tall tales and imprecations, threatens to poison the intellectual cohesion necessary for a university to work as anything other than a collection of fiefdoms, trying to avoid each other's concerns—and students—as much as possible. We recognize that it is necessary for science patiently to abide social scrutiny, since science and its uses affect the prospects of the entire society. That kind of scrutiny is a serious enterprise, requiring painstaking attention to fact and a disinclination to extrapolate beyond the bounds of reasonable inference. In the current climate, such sane and indispensable scrutiny threatens to be displaced by myth-making of the most fanciful sort. The key function of these myths is to gratify the resentment and self-righteousness of those who propose them, and to serve as symbolic wish-fulfillment in a world that is notably indifferent to their politics.

On our own, we cannot hope to undo any significant fraction of the damage already done. Our wish is that thoughtful people in and out of the academy, but most particularly scientists, engineers, physicians, and the like, might become aware that this hostility to science exists, that it has coherence as a political project, if not, strictly speaking, as a body of doctrine, and that time and effort should be devoted to examining and refuting it. The countering arguments must be made with reason and patience, but also with a determination not to be put off by sloganeering or by the insinuation—which is certain to arise—that one is covertly inspired by a desire to keep the oppressed from having their say or winning the justice due them.

The academic left's critiques of science have come to exert a remarkable influence. The primary reason for their success is not that they put forward sound arguments, but rather that they resort constantly and shamelessly to *moral one-upmanship*. If you decry the feminist critique of science, you are guilty of trying to preserve science as an old-boy's network. If you take exception to eco-apocalyptic rhetoric, you are an agent, witting or otherwise, of the greed of capitalist-industrialist polluters. If you reject the convoluted cabalistic fantasies of postmodernism, you are not only sneered at for a dullard, but inevitably told that you are in the grip of a crumbling Western episteme, linked hopelessly to a failing white-male-European hegemony. This is not pleasant to encounter in debate; but it is very far from unanswerable. Be assured that it conceals fundamental weaknesses of fact and logic in the argument of the accuser.

We are treading now on the slippery territory of the "political correctness" debate; and we face the fact that this book will be read as yet another salvo in that dreary war. If we are to be candid, we must admit that nothing we offer will comfort those who describe the PC furor as a vicious invention of the political right. Their enemies, on the other hand, may find here a certain amount of ammunition, if they wish. It would be idle of us to lay claim to a prim neutrality. Nonetheless, we are not happy to be classed as reflex partisans of the right on any issue. There has been plenty of bad faith, dissimulation, sanctimony, and hypocrisy from *all* quarters. The academic right is all too eager to use the grotesqueries of the academic left as an excuse for walking away from deep and intractable problems. For its part, the left is ready, at the slightest hint of challenge, to play the martyr and to find fascism, racism, or "denial" in it, no matter how judicious and well reasoned the challenge may be.

We refuse on principle to take sides in the dispute over the literary canon, in the fights over affirmative action, in the question of whether it is well to have "studies" departments for subpopulations with a history of victimhood. It's not that we don't have opinions on those questions: we do; but they are simply not what this book is about. What we have to say is narrowly concerned with science and with misconceived attacks on science that grow out of a doctrinaire political position. The left has to take the blame, because that's where most (but certainly not all) of the silliness is coming from on this issue, at this time, although there has been an abundance of it in the past from the other side.[4] The campus right has had the good tactical sense to leave the matter alone, except to comment on the foibles of the left. It may well be that there are dead-of-night confabulations between leading anti-PC activists and, let us say, the Institute for Creation Research. If so, these have yet to come to light. In any case, we are not stalking-horses for social conservatism. If the academic left were to choose to abandon the most extravagant of its philosophical lucubrations, particularly those that lead to misguided assaults on natural science, the occasion for books like this one would disappear.

Distinctions

In this vein, let us reinforce some fussy but essential distinctions. When we use the phrase *academic left* we do *not* refer merely to academics with left-wing political views. There are plenty of such people with whom we have no quarrel. There are countless academics who do excellent and penetrating work, in appropriate fields, from a left-wing viewpoint. There are countless left-wing scientists—although we are stodgy enough to insist that there is no such thing as left-wing science. We are using *academic left* to designate those people whose doctrinal idiosyncrasies sustain the misreadings of science, its methods, and its conceptual foundations that have generated what nowadays passes for a politically progressive critique of it. If this terminological improvisation causes confusion, we apologize; but under the circumstances, it seemed the least arbitrary, if not the least inflammatory, choice. We hope that as our analysis progresses the necessary distinctions will become clearer to those who read it.

Of course, distinctions between "left" and "right," between the "academic left" as we have defined it and other points of view that generally belong on the left of the political spectrum, cannot be hard and fast. Notwithstanding that, many of the groups and individuals we discuss will be eager to denounce us as disguised reactionaries and as apostles of right-wing malignity. But the academic left makes a grave mistake when it believes that its ideological extravagances irritate no one but reactionaries and apologists for oppression. The saddest part of the situation is that professorial types who are, by any standard, well-meaning have developed a fatal facility for making enemies much faster than they make allies. One need look no farther for evidence of this than the declining public respect for them.

The most perplexing difficulty in responding with a critique of our own to the newly fashionable critique of science is the absence of a central body of doctrine that can be said to constitute the quintessence of that view. There is no intellectual core to the process by which the critics scrutinize and, for the most part, disparage science; thus there is no obvious target for a definitive rejoinder. If one examines some of the best-known and most highly praised assaults as they come to hand, one immediately notes that each goes off on its own tack, showing little correlation with the others in its choice of particular scientific practices to focus upon, its analytical methodology, or its ultimate conclusions beyond the most general. If one were to compare, say, a traditionally Marxist analysis, a hard-core postmodern epistemological critique, and an ecofeminist harangue from the Goddess-worshiping camp, there would be little to connect any two of them in terms of language or philosophy. The first would view science as a construct of capitalist social relations, misdirected by bourgeois idealism, or by some similar error. The second might concentrate on the instability of language and the indeterminacy of meaning, on the view that the "subversive" implications of these insights apply in full force to the scientific theories we construct, even those couched in the abstract and supposedly rigorous language of mathematics and physics. The third would, presumably, view science as a product of the patriarchal paradigm of dominance, control, and objectification, and would call for its transformation or even its abolition as a step on our path back to an edenic, nature-centered society.

The rank incompatibility of these views would seem, at first glance, to defy any attempt to characterize them as products of one ideological subculture. By claiming to do so, we lay ourselves open to the charge of taking an illegitimate polemical shortcut, of mere journalism, of ignoring crucial distinctions in order to condemn irritating opponents. This, however, would be to misread our case. What we assert is that the examples above, isolated from one another as they may be in strictly logical terms, are, in fact, extreme points in a contiguous ideological field. One can easily connect one to the other by a chain of intermediate examples, positions that synthesize something of this, a little of that, and just a bit of the other. This, moreover, would be more than just a theoretical exercise; the intermediate positions *exist* as components of the critique: they are being taught—increasingly—in university classes.

It is in fact not an exaggeration to say that the science criticism of the academic left is just such a game of mix-and-match. Each practitioner assembles his or her arsenal from favorite polemical bits and pieces—a little Marxism to emphasize the twinship of science with economic exploitation, a little feminism to arraign the sexism of scientific practice, a little deconstruction to subvert the traditional reading of scientific theory, perhaps a bit of Afrocentrism to undermine the notion that scientific achievement is inevitably linked to European cultural values. Proportions and emphases vary from text to text; but, as one becomes familiar with this body of theory, the underlying unities appear. They are more a matter of rhetorical style than of logical articulation. A strong sense of fellowship and common purpose unites the array. It is commonplace for one writer to cite, in flattering terms, a host of others for their bold, incisive, and devastating insights into the injustices and illusions of the official scientific culture. A's praise of B, like B's praise of A, takes little account of the fact that the respective views of A and B are fundamentally at odds. Still, the name of the game is solidarity. Differences are soft-pedaled in the interest of an overriding common purpose, which is to *demystify* science, to undermine its epistemic authority, and to valorize "ways of knowing" incompatible with it.

Although there is no true center, no foundational axiomatics, to the left-wing critiques of science, a few broad perspectives may be identified. Sociologists and social theorists, including quite a few Marxists, tend to produce what may be called "cultural constructivist" analyses, viewing scientific knowledge as historically and socially situated and encoding, in unacknowledged ways, prevailing social prejudices. The strongest and most aggressive versions of these theories view science as a wholly social product, a mere set of *conventions* generated by social practice. The critics whom, for convenience, we label as postmodern, attempt to exploit the linguistic and psychological theories grouped under that cognomen. A radical epistemological skepticism informs their commentaries on science, though rarely is it seen to impeach their own researches. For its part, academic feminism has generated a vast literature commenting on science from its own perspective. These critiques are probably the best known and most widely read of all, owing to the ubiquity of women's studies programs on American campuses and the reality of women's exclusion from science in the past. As well, the environmental movement has generated its own challenges to science, concentrating on the degree to which the orthodox practice of science maintains our alienation from nature while sustaining technologies that are shattering the fragile equilibria of planet Earth.

There are no firm and fast divisions among these approaches; they merge into one another; in most cases we find ourselves dealing with hybrids. What is common to all of them becomes clear at the rhetorical level. Modern science is seen, by virtually all of its critics, to be both a powerful instrument of the reigning order and an ideological guarantor of its legitimacy. It is stained by all the sins of the culture that engenders and nurtures it. Thus, whoever attacks it with a view to vindicating the oppressed, no matter how quixotic the methods, is seen to be fighting the good fight.

At the same time, in a largely unacknowledged fashion, other old scores are being settled. For decades certain assumptions about the epistemological ranking of various fields have prevailed, though rarely explicitly, among academic intellectuals. The rule of thumb has been that the hard scientists produce reliable knowledge, assembled into coherent theories. Historians, it is conceded, generate reliable factual knowledge (as long as they keep their methodological noses clean); but this is often contaminated by unprovable and bootless speculation. Economics has rigor of method; but its assumptions are serious, often fatal, oversimplifications of the real world. In the other social sciences impressionistic description and subjective hermeneutics rule, though they may come dressed in elaborate statistical costumes. The more theoretical the social scientists are, the less respect they get. Literary criticism, finally, has been looked upon as a species of highly elaborated connoisseurship, interesting and valuable, perhaps, but subjective beyond hope of redemption, and thus out of the running in the epistemological sweepstakes.

How justified or absurd this folklore may be is beyond the scope of our inquiry. The point is that it *has* been folklore for more than a century, no matter how much presidents, provosts, chancellors, and deans deny it. The resentment it provokes has been festering. Thus the fact that theoretical social scientists and professional literary critics—at least those on the left— are prominent among the current critics of science should not be a surprise. The recent critiques of science incarnate attempts to regain the high ground, to assert that the methods of social theory and literary analysis are equal in epistemic power to those of science.

Designing a Rejoinder

In view of the enormous range of left-wing criticism of science in terms of philosophical assumptions, historical focus, and working methodology, the only way to compose a coherent rejoinder within a reasonable space is to examine a range of specimens. Our study begins with a brief historical survey of the relation between the sciences and what may be called, without prejudice, emancipatory politics since the European Enlightenment. We point to certain intellectual and emotive roots of the current critique, some of them quite unacknowledged by recent writers who have demonstrably been influenced by them. We then devote a number of chapters to surveying the best-known critiques of science that have emanated from the academic left during the past few years. In aid of clarity, we group these as categories in accordance with the distinctions outlined above. We look first at the "sociological" or "anthropological" approach, that is, at the cultural constructivists. Next we examine the background and the specific practice of the new academic cult of postmodernism, reflecting upon its attempt to bring science within its empire. A separate chapter is devoted to the theory and practice of feminist science criticism. Finally, we examine the strange and vexatious mixture of good sense and folly that constitutes the radical-environmentalist approach to science, although, strictly speaking, in dealing with this topic we move a bit away from the theoretical atmosphere of the campus and become involved with public political and technological disputes of great moment.

Our approach in these sections is conspective and polemical. Nothing else will get attention. We synopsize and, as we see it, refute some of the representative work in each area. Our space for doing so is limited. Consequently, the synopses are brief. Simultaneously, we extract from the critic's work certain crucial arguments which, in our view, exemplify methodological weaknesses and expose the fallacies of the underlying viewpoint. Admittedly, this is rebuttal of a swift and selective sort. Nevertheless, given the volume and the diversity of left-wing science-criticism, we see no other way of giving a comprehensive idea of the range and variety of these attacks, while at the same time revealing what we see as their flaws.

A word about scholarly apparatus: Our experience and instincts, where scientific substance is at stake, call for presentation of necessary detail within the text, and for comprehensive citation of precedents. That is the style to which we are accustomed, and it is the one we would want to use in addressing one part of our hoped-for readership: professional scholars, including scientists, engineers, and physicians. Another of our hopes, however, is to reach a wider audience, whose members would be, if not put off, then at least distracted by the standard tools of scholarly communication. Compromise, clearly, is required. It takes the following form: In general, the facts of science are dealt with in text, but economically and—to the best of our ability—comprehensibly. Where there is necessary detail, it is relegated to endnotes. We have kept the number of those to the minimum consistent with conscience. Bibliographic citations cannot, therefore, be comprehensive, as the best scholarly practice would demand. We have taken pains to choose for citation representative writing and, where possible, especially among the cited scientific items, to select books and articles that are themselves bibliographically competent. The combination of endnotes for any chapter with the appropriate sources should provide for any reader who wants to pursue its subject, or check up on us, an adequate guide to the relevant literature.

The thinkers we examine are by no means obscure or peripheral to the academic left's assault upon science. Most of them are VIPs in academia and some are public figures as well. All have published widely read work on the subject (although occasionally we shall concentrate on a more obscure paper or a recent, as-yet-unpublished lecture). To exemplify cultural constructivism, we have chosen sociologists and historians of science: Stanley Aronowitz, Bruno Latour, Steven Shapin and Simon Schaffer. For postmodernism, we have settled upon the philosopher Steven Best, the "cultural critic" Andrew Ross, and the literary critic N. Katherine Hayles. The feminist theorists we consider include some of the best known: Sandra Harding, Donna Haraway, Evelyn Fox Keller, Helen Longino. As for the radical environmentalist attack on science, we concentrate on academics like Carolyn Merchant, but also on theorists—Jeremy Rifkin and Dave Foreman to take the most obvious examples—who may not be "academics" in the narrow sense, but whose writings and activist crusades have gained credence and widespread support from the academy, and whose language and intellectual temperament seem to echo closely the styles and prejudices of the most prominent contemporary critics of science.

We follow up these argumentative chapters with a brief survey of the effect of antiscientific rhetoric from the academic left on a few well-known (if not particularly powerful) dissident movements. Specifically, we take a look at the activism that has sprung up within the gay community in response to the AIDS epidemic, at the animal rights movement, and at the movement for Afrocentrism (representing here "multiculturalism")[5] in education. In reference to the last, we are particularly concerned with the active championing of an Afrocentric curriculum, as it affects the teaching of the history of science as well as of science itself.

In the final chapters, we claim the right, as does everyone else, to philosophize a bit, going beyond the confines of polemical necessity. We speculate on the deeper psychological, cultural and social roots of left-wing hostility to science, viewing it as a phenomenon to be explained with the help of some cultural constructivism of our own. We consider the importance of the phenomenon, its implications for education, for women's participation in science, and for effective environmental advocacy, among other serious issues. We argue that, although the criticisms we have examined amount, individually and collectively, to very little in strictly intellectual terms, it is nonetheless important for scientists and fair-minded intellectuals—and this includes many left-wing thinkers—to take them very seriously. It is not without historical precedent that incoherent or simply incomprehensible opinions have had great and pernicious social effect.

Our greatest hope is to stimulate awareness and debate. If other commentators sympathetic to our views arise to outdo us in polemical efficiency and pertinacity, we shall have done science a needed service. If the subjects of our analyses rise to the occasion and sharpen their own analyses to the point that arguments like ours become irrelevant, we shall have learned something important about science and the social organism that begets it. And, of course, we will be delighted either way, or both.

CHAPTER TWO

Some History and Politics: Natural Science and Its Natural Enemies

---

Having in past days perused Signor Galileo Galilei's book entitled The Assayer, *I have come to consider a doctrine already taught by certain ancient philosophers and effectively rejected by Aristotle, but renewed by the same Signor Galilei. And having decided to compare it with the true and undoubted Rule of revealed doctrines, I have found that in the Light of that Lantern which by the exercise and merit of our faith shines out indeed in murky places, and which more securely and more certainly than any natural evidence illuminates us, this doctrine appears false or even . . . very difficult and dangerous.*

DOCUMENT G3, ARCHIVE OF THE SACRED CONGREGATION
FOR THE DOCTRINE OF THE FAITH

What is the advantage . . . of again digging up a controversy that was a dialog among the deaf, having no beneficent (but only a delaying) effect on the history of modern science? Certainly the theories of physics, unlike theology, did not obtain results from the century-old struggle, only obstacles.

But there was an effect and that history will make us appreciate it. It was the effect of making us conquer the autonomy of research and reason from which we benefit today. And one might appreciate the fact that it did not descend to earth from the heaven of Plato's ideas, but was conquered at great cost in the seventeenth century, like every other human freedom. It is a common good, which must be safeguarded.

PIETRO REDONDI, IN *GALILEO HERETIC*

Fresh from the horrors of the Thirty Years' War, late seventeenth-century Europe produced a generation of intellectual giants whose collective accomplishment was to set in motion an epistemological enterprise that has continued to flourish over the past three hundred years, an effort that accelerates and expands continually in its scope, precision, and reliability. The true scientific revolution instituted by Galileo, Kepler, Newton, Halley, Harvey,

Boyle, Leibniz, and others is to be found, not in their particular discoveries about the world, stupendous as these were, but rather in the creation, almost in passing, of a methodology and a worldview capable of expanding, modifying and generalizing these discoveries indefinitely. It was, moreover, a methodology that almost unwittingly set aside the metaphysical assumptions of a dozen centuries, under which a description of the physical world would have been incomprehensible had it stood apart from a vision of transcendent divine order on the Christian model. That Newton, say, or Leibniz sought in all sincerity to affirm some version of this divine order through his scientific work is almost beside the point. The implicit logic of their work turned out to be of immensely greater importance than the explicit pious intentions of those who achieved it.

In its ineluctable dynamic, the science of the turn of the eighteenth century could not be contained within the shell of any theological system. It was, in important ways, already fully modern. Open-endedness is the vital principle at stake here. It constitutes the lifeblood of ongoing science. Newton said it best: an "ocean of truth" lies undiscovered before us. Unless we are unlucky, this always will be the case.

Having escaped most of the constraints of systematic theology, the new science was hardly to be contained within the ideological matrix of the society and political system within which it arose. The birth of Western science as a powerful, systematic, and ever-expanding set of interlinked disciplines very nearly coincides with the birth of its prestige as a uniquely reliable and accurate way of describing the phenomenal world. Consequently, philosophers and political thinkers of all shades of opinion attempted eagerly to conscript that prestige on behalf of their own favored ideas. Newton, along with his contemporary Locke, is often thought of as the tutelary figure of the Glorious Revolution and of the gentlemanly class, devoted equally to the pursuit of mercantile wealth and to an Anglican faith forever irreconcilable with Catholicism, that forged this constitutional and dynastic upheaval—this despite Newton's own peculiarly heterodox Protestantism. A faction within the Established Church came to be called "Newtonian" for its stout insistence that, just as Newton's singularly English genius revealed the eternal regularity of God's law made manifest in celestial mechanics, so too did the Church of England and the social system with which it was intertwined reveal the intentions of the Deity for the rightful ordering of human affairs.

But the partisans of social stability could scarcely maintain a monopoly over the totemic power of Newtonian physics. A deeper understanding of science itself, as well as an entirely different set of speculations as to what it might imply for the human political order, emerged throughout Western

Europe. This is symbolized by the growth of Freemasonry as a sort of philosophical "shadow government," by the attempts of the Encyclopedists to systematize and codify the full range of human knowledge, by the development of political economy as a fruitful intellectual enterprise.[1] The relationship between science as such and these various tendencies was by no means fixed according to any particular stereotype. It is, however, certain that science—in particular Newtonian physics and its related mathematics—held sway as a privileged model and inspiration, the very emblem of the power of the human intellect to probe beneath surface appearances, to rectify vulgar prejudices, and to exile habits of thought more ancient than accurate.

In the sphere of social thought, the success of physics inspired emulation in the form of analyses of society seeking general principles that might be made to yield a deep understanding of the dynamics of history, politics, and economic activity. The urge to prescribe, as well as to describe and predict, ran strong in these attempts, in a manner quite uncharacteristic of physics itself; but the boldness, indeed the arrogance, required to set forth schemes for the radical improvement of the human condition and for the rapid cure of its ancient ills reflects an intellectual self-assurance that derives largely from contemplation of the well-confirmed triumphs of eighteenth-century mathematical science. If the few simple axioms adumbrated by the *Principia* could be induced to yield precise accounts of the orbits of planets and comets, of the eccentricity of the earth and the precession of its axis, of the pattern of oceanic tides, why should there not be an equally elegant, comprehensive, and reliable systemization of the study of human affairs?

The obligations of hindsight impel us to look on most of these attempts as failures variously fatuous, quixotic, or disastrous, whose culmination is to be found in the self-defeating utopianism of the French Revolution. Admirers of Adam Smith's economics, or of the abiding wisdom of the American Constitution, will, on the other hand, discern triumphs amidst the scores of false starts and blind alleys. Our own position is that even on the most optimistic view such triumphs are drenched in irony and soured by an unending stream of historical misfortunes. These disputes, however, are not central to our point. What we wish to emphasize, rather, is that the underlying strategy that guides the intellectual enterprises of Smith, Diderot, Locke, Gibbon, Herder, Hume, Jefferson, and (what was until recently) a pantheon of others remains as an ongoing tradition that is unlikely to disappear within the imaginable future. This is simply, in its most naked form, the strategy of taking the social order, per se, as the object of one's critical investigations, seeing it as describable, in large measure, on the basis of discoverable first principles. It is to be implemented by combining careful and exhaustive attention to solid empirical fact with the construction of a more or less rigorous deductive model.

At their best, such theories yield chains of propositions which themselves may be variously regarded as confirmed insights into the social organism, or as tentative hypotheses to be tested in the hard world of experience, as a trial of the soundness of the fundamental postulates of the theory. That this accords, if only in a very rough sense, with the epistemological model already set in place by the physical sciences is, we think, so obvious as to need no further argument.

It is important to attend to another aspect of such Enlightenment—for that is what we are describing—social thought. It seems to us that what a broad spectrum of thinkers have in common is their determination to regard the social position of individuals as resulting neither from the decrees of a transcendent divinity nor from the processes of an optimal social mechanism. *Rank, wealth, and power are seen as contingent facts, rather than as the emblems of an innate or achieved social perfection.* Whatever their differences, none of these philosophers cry along with Pope and Handel "Whatever Is, Is Right." Rather, schemes and prescriptions abound for the reconstitution of the social organism to bring it into alignment with the dictates of reason and nature. Furthermore, the ills and malfunctions of the existing order are almost always located in the undeniable maldistribution of wealth, power, prestige, and immunity that is to be found everywhere. Thus a strongly implicit egalitarianism suffuses the thinking of the savants of the time, at least of those whose work still speaks resonantly to us. This may range from the openness to entrepreneurial innovation advocated by Smith to Rousseau's near-mystical celebration of the General Will and the unanimity of its votaries; but such distinctions seem more important, we submit, in hindsight. The key point is that it came to be seen that any system claiming to be based on natural justice must accommodate the concept that at some level all individuals are to be equally empowered by the fundamental political processes of the state. It hardly matters that at this level of generality such ideas are as ancestral to the apologies for free-market capitalism so dear to modern conservatives as to the garrison-state socialism of North Korea or Vietnam, and it hardly matters that the egalitarian view tended to be blind, now and again, to particular parts of the landscape.

It is fair to say, in short, that by the time of the French Revolution a certain suite of ideas had become regnant in European (and North American) political philosophy. The empiricism and rigor of the sciences were emulated in the analytic strategies of political thought; and this, in turn, was for the most part linked to an emancipatory project for the renovation or reconstitution of existing social systems. It is of course possible, and tempting, to speculate whether a similar system of scientific discourse might have arisen in an entirely different social context. Might it have been possible, for instance, in T'ang Dynasty China or under the Pax Romana? Or could science have matured only upon a substrate of subtly congenial social ideals and institutions, like those found in seventeenth-century Europe? Such speculation, though it continues actively and vigorously, is, in some sense, futile, for we are speaking of an event that is in essence unique and unrepeatable. Short of an utter collapse of our civilization on a global scale, the opportunity to reinvent science will not arise. So the association of Enlightenment ideas in the realm of politics with that era's celebration—indeed, near-deification—of science may be largely fortuitous. Nonetheless, at least to the extent that the political aspects come up against the authority of religion as well as the mythic power of other traditional rationalizations of the established order, science is a weapon to be wielded both specifically and emblematically. Laplace's famous explanation—"I did not find the hypothesis necessary"—of the absence of the Deity from his system of cosmology is both a succinct lesson in the explanatory parsimony of scientific thinking and a war cry of political and ideological defiance.

The disastrous failure of the French Revolution and the aftermath of that failure is, of course, perhaps the most ringing example of the triumph of inadvertence over intention in human history. It instilled in Western thinkers a full measure of skepticism concerning utopian systems and schemes for universal reform. Even before that, during the headiest moments of early republicanism, the canny Burke had already put his finger on the weaknesses of abstract philosophizing as a guide to the attainment of social perfection. Burke, however, is but one of a spectrum of thinkers who begin to show strong doubts about the deification of the merely rational. Far more emphatic and impassioned are the great figures of Romantic individualism, including Blake, Wordsworth, Coleridge, and, above all, Goethe. It is in literature and poetry that we first begin to encounter a reaction against Enlightenment values that reveals a specific distrust of science, as well as a strong reluctance to believe that mankind can be reformed along "scientific" lines.

This is a vexatious topic: to do it justice, one must be endlessly willing to draw distinctions. Blake is a very different animal, politically and philosophically as well as poetically, from the Olympian Goethe, and neither is very close in spirit to the reactionary Wordsworth settled into his endless counterrevolutionary old age. Yet, in point of attitude toward epistemological questions, and, quite explicitly, toward the authority of science, the poets are linked by a strong commonality of thought. Each distrusts the narrowly empirical and the strictly rational, each celebrates the vital importance of the intuitive, the irreproducible moment of insight and of direct access to truth in its unmediated essence. Each accuses science, especially in its schematic, mathematicized form, of blindness, or worse, stubborn refusal to see. Each fears a world in which scientific thought has become the sovereign mode, and recoils from the spiritual degradation and servility that, in his opinion, must inevitably come to characterize such a world. Blake makes his protest in the name of an ecstatic, antinomian, revolutionary vision that comforts neither Jacobins nor Royalists. Goethe speaks for an idiosyncratic classicism, neither fully pagan nor fully Christian, neither revolutionary nor reactionary, as singular as the great man himself. Wordsworth seems merely a self-satisfied old Tory. But beneath these divergent visions, we find an underlying distrust of straightforward, impersonal reasoning. The belief in direct, revelatory, intuitive truth to be had from communion with nature is the obverse of a deep epistemological skepticism about the kind of "systematic" truth that is the core of scientific knowledge. In this aspect, Romantic thought, even at its most revolutionary, is allied to the caustic, all-encompassing skepticism of that relentless reactionary Joseph de Maistre, whose most brilliant exercises in logic and empirical inference are expressly designed to demonstrate the unreliability and futility of logic and empirical inference.[2]

(We cannot resist the temptation to take note, in passing, of the fact that the Romantic exaltation of intuitive "Understanding" above merely cerebral "Reason" foreshadows the celebration of "holism" and "organicism" by contemporary critics of science, who are impatient with the disciplined analysis and methodological exactness of serious scientific work. Likewise, Maistre, in his counterrevolutionary ferocity, is the true spiritual ancestor of the "post-modern" skepticism so dear to the hearts of the academic left.)

Whatever its effect on the history of poetry and sensibility, however, the Romantic revulsion against the scientific worldview had virtually no effect on the development of science itself. The nineteenth century turned science into a profession. Its status as the preserve of gentlemen-amateurs and isolated virtuosi dependent on aristocratic patronage receded into history. The education of scientists was rapidly systematized, and the universities, especially in France and Germany, took on their now-familiar role as nurseries for aspiring scientists and sponsors of experimental, as well as speculative, work. The subdivisions of science came to be ever more clearly defined, and the intense specialization that marks the science of our own day took shape. At the same time, the link between theoretical science and direct technological innovation became concretized in the growth of institutions, both educational and commercial, that vastly expanded the scope of the engineering profession, while tying it ever more firmly to rigorous scientific foundations. The interval between the first systematic attempts to derive an adequate mathematical theory of electricity and magnetism—those of Gauss and Ampère, say—and the systematic construction first of telegraph networks, then of electrical systems to power whole cities is, by any standard, incredibly brief.

This fully symbolizes the degree to which Western culture, almost unthinkingly, entirely altered its own material underpinnings. To compare the European states, circa 1800 with, say, the Chinese or Ottoman empires is a historical and geopolitical exercise dealing with entities which, however greatly they differ, may be measured against each other in terms of economic, industrial, agricultural, navigational, and military capacity. By 1900, such a comparison is idle. The sudden disparity has little to do with the traditional ebb and flow of power, and everything to do with assimilation of the scientific enterprise into the heart of the Western social fabric. It seems to us doubtful that historians have yet come to grips with this development, in the sense of having found a language to tell us exactly what happened in the space of a few generations. We are still far too close to the scene, chronologically, to take the measure of this revolution; it is an upheaval yet in progress and its consequences cascade over us daily. We are too numb to grasp its magnitude: that privilege must fall to future historians.

We are nevertheless at liberty to make modest observations on the political concomitants of these transformations. In the first place, it is clear that millenarian hopes for the reconstruction of the social order along "scientific" lines hardly disappeared with the collapse of revolutionary idealism in France and the subsequent catastrophe of the Napoleonic Wars. The tradition of "social engineering" continued in the schemes of the Utilitarians, of Robert Owen and Charles Fourier, of the New England Transcendentalists, and of Auguste Comte.[3] Though of little historical consequence, these demonstrate how the habit of assuming that "society" is a tractable category, analytically and politically, had become ingrained in Western thought. They demonstrate almost equally well the rough equation of a more "scientific" social order *with a more egalitarian one*, and the opposition between a view of the world informed by science and one occluded by stagnant tradition. This observation takes on particular poignancy when we consider the curious intellectual trajectory of Karl Marx, an epochal thinker who eagerly admired science in the abstract, envied the inevitability of its logic, conscripted its prestige for his own polemical purposes, and still managed in the end to misunderstand it thoroughly.

In speaking of science and its social consequences in the nineteenth century, we cannot avoid the notion of "Progress" and its role in the generally optimistic view of historical process that held sway during that period. Con-temporary critics have told us repeatedly and with great sagacity how prob-lematical the idea of progress is. Progress for whom, in what direction, at what expense to which class? The progress of the upstart manufacturer in the English midlands or the New England mill town may well have affronted the seigniorial pride of the landed aristocrat or the Tidewater planter—hardly an outrage to one's democratic sentiments. On the other hand, the industrial-ist's prosperity was the millworker's hell. The technology that minted wealth for its owners forged chains for its servants. The superiority of the technolo-gized economic superstructure of Europe and the United States exacted a terrible tribute from millions of Chinese, Indians, Latin Americans, and Filipinos, who had no reason to praise the scientific virtuosity that showered them with shells and bullets.

In the final analysis, a real if grossly imperfect alignment persisted between the scientific outlook and the great emancipatory sentiments—abolitionism, women's rights, social reform, socialism itself—that drove the most idealistic souls of the era. To put it another way, the "science" that sustained the most ferociously antiegalitarian ideas—racist eugenics, "social Darwinism," and the like—has long since been effaced, while the claims put forth to bolster the egalitarian view have endured, on the whole, rather well. At any rate, if we are to judge a body of ideas by its worst enemies, it is simply absurd to impugn science as the tool of the most embittered reactionaries. Those forces, represented by Maistre and by Pius IX, the pope who denounced socialism, modernism, and the scientific outlook in a single breath, were convinced that their quarrel with science was a struggle to the death. Martin Heidegger was their recent offspring. To the extent that the liberatory and democratic ideals that roiled the nineteenth century and persist to our day with amplified force face the adamant resistance of dogmatic religions of one sort or another (hardly a dead issue in a world beset by a swarm of angry fundamentalisms), science, it would seem, has been and will be their strong-est and least dispensable ally.

Disillusion and Illusion

There are many reasonably well read people to whom the growing antagonism toward science on the part of a large number of left-wing intellectuals will come as something of a surprise. There is a tendency, mostly justified, as we have seen, to think of political "progessivism" as naturally linked to a struggle against obscurantism, superstition, and the dead weight of religious and social dogma. In this effort, the obvious ally and chief resource is scientific knowledge of the world and the systematic methodology that supports it, as these have developed over the past few centuries, chiefly in Western culture. Though the specific achievements of science are of some polemical importance for certain ongoing disputes, far more valuable and effective have been the modalities of critical and skeptical thought that have matured for the most part in a scientific context. The dissecting blade of scientific skepticism, with its insistence that theories are worthy of respect only to the extent that their assertions pass the twin tests of internal logical consistency and empirical verification, has been an invaluable weapon against intellectual authoritarianisms of all sorts, not least those that sustain social systems based on exploitation, domination, and absolutism. The notion that human liberation ought to be the chief project of the intellectual community is, it seems to us, coeval with the idea that superstition and credulity are among the most powerful foes of liberation, and that science, in particular, holds out the best hope for cutting through their fogs of error and confusion. Towering figures of political and ethical thought over the last three or four centuries make this point; one thinks, in this regard, of Galileo, Spinoza, Locke, Voltaire, Diderot, Lessing, Hume, Kant, Mill, Herzen, Turgenev, Russell, Einstein—the list could be extended endlessly. And, of course, one thinks of Marx, albeit with a sad irony that dwells on weaknesses in his mode of thinking, whose consequences and echoes will, to a great extent, comprise the focus of this argument.

Our era is singular, in that the commonplace wisdom cited in the last paragraph (wisdom we hold to be as valid as any generalization can be) has come under strident and increasingly scornful attack, not from reactionaries and traditionalists, who have always feared science, but from its natural heirs—the community of thinkers, theoreticians, and activists who challenge both the material injustices of the existing social system and the underlying assumptions and prejudices that perpetuate them. As Timothy Ferris observes in his appropriately skeptical review of a recent and popular antiscientific polemic, "The scientific community today, for all its faults, remains generally open and unsecretive, international and egalitarian: It is no accident that scientists are to be found at the forefront among those who call for global ecological responsibility, racial and sexual equality, better education, an end to hunger, a fair break for indigenous peoples, and other enlightened values."[4] Yet the alliance, so historically familiar that one is tempted to call it "natural," between the scientific worldview and the tradition of egalitarian social criticism, is not only under challenge but, from some points of view, may be said already to have dissolved. This has to be understood not as a hazy generality about the zeitgeist, but rather as an observation about a specific community, a particular, if rather limited, contemporary social formation: that of self-conscious left-wing political intellectuals and those who follow their work with attention and approval, and take a measure of inspiration from it.

We are particularly interested in the American left, although its pugnacity toward science is certainly echoed by left-wing intellectuals in Western Europe. Some of the key ideas, now common currency on American campuses—the "strong programme" in sociology of science associated with the Edinburgh school, the compendium of "postmodern" attitudes transcribed from Derrida, Foucault, Lyotard, Baudrillard et alii, to take a few obvious examples—are, in fact, imports. Nonetheless, the antiscientism of the American academic left has its own idiosyncratic resonances, if only because it is integral to a much broader array of challenges to received wisdom and settled ideas. For many left-wing thinkers, a radically skeptical attitude toward standard science is a means of burning one's bridges, of disavowing one's connection to a spectrum of liberal Enlightenment values, moral as well as epistemological. It is a symptom, therefore, of profound distress at the inability of that value system to deliver on its promises. In the present situation, the orthodoxies of liberal humanism seem to have curdled, and the resplendent intellectual achievements that symbolize the worth of liberal humanist attitudes seem ripe for dismissal. The defiant bravado that marks the various critiques of science (ill advised and vainglorious though we think it to be) is an index of the pained confusion of the left in the face of a world that seems impervious to its insights, however brilliantly thought out or passionately expressed.

For many left-intellectuals, social justice and economic equity seem ever more elusive as practical possibilities. American society and the global capitalism of which the United States is still the epicenter go their own way without taking much notice of left-wing thought. The problem of race in this country seems to be more intractable than ever. The changing demography of the American population seems to promise not an amiable and beneficent polyculturalism, but rather an increasingly venomous tribalism and nativism. Feminists see themselves as driven into a defensive circle, and the agitation for equitable treatment of homosexuals seems often to be answered by paranoia and violence. The hope, which was never quite absent from the heart of even the most disillusioned leftist, that "actually existing socialism" in the Soviet Union and Eastern Europe might finally be able to escape the horrors of its Stalinist past and take on the task of building a worthy alternative to capitalism, is irredeemably dead. There may be a resurgence of political liberalism in this country, but, if so, it will be, at best, a pallid and compromised liberalism, unlikely to accommodate very much in the way of redemp-tive social design. The contemporary, popular definition of "liberalism" is a political tendency to leave things pretty much alone, except that they are to be funded, whenever possible, and monitored, by agents of a wise and benefi-cent government.

Meanwhile, the historical constituency of the American left is frag-mented. The traditional moral language of the left, deriving as it does from Enlightenment humanism, seems to have lost its power to exhort and unite. It is hard, for example, to imagine a contemporary black militant employing the rhetoric of Paul Robeson or Martin Luther King; Malcolm X seems to be the only relevant historical figure. Feminism has long since wandered into its own discursive universe. The new immigrant groups from Asia and Latin America have little familiarity with the themes of working-class emancipa-tion that inspired Irishmen, Germans, Swedes, Italians, and Polish Jews in the factories, sweatshops, mines, and rail-yards of America a century ago.

We ought not to wonder, then, that so many academic leftists (as opposed merely to Democrats who win elections), finding themselves in a dispiriting historical corner, are in a sullen mood, a mood in which it seems that the most immediate solace comes from devising reasons for discounting and minimizing the proudest accomplishments of the smug society that surrounds them. The history of Western artistic and intellectual achievement no longer provides hope or inspiration—on the contrary, it taunts and irritates. As the wholly owned subsidiary of a despised culture, it becomes the target for contempt and disparagement. The philosophical concomitant of this atti-tude is, unsurprisingly, a defiant relativism.

True enough, these instincts find expression in what often seems like a positive program. The dethronement of the literary and artistic "canon," for instance, is packaged carefully and announced to all who hear and read as a movement to empower the unempowered by letting us all hear the voices of those heretofore silenced. New modes of doing sociology and anthropology are proposed as ways of rescuing historically subordinate peoples from the ignominious position of "objects of study," and endowing them with agency and meaningful historical will. This is by no means a hypocritical or disin-genuous pose. These arguments do have moral force. They have to be reck-oned with (although not to the exclusion of countervailing ideas) by anyone concerned with equity and the redress of historical injustice. Nevertheless, the aroma of sour grapes is in the air. The urge to redeem slides easily into an eagerness to debunk for the sake of debunking. New candidates for veneration—writers, artists, musicians, philosophers, historical figures, non-

Western "ways of knowing"—are put forward not for what they are but for what they are *not*—white, European, male.

It is impossible to understand fully the academic left's attack on science without taking into account how much resentment is embodied in it. Science is, if anything, a more natural target for the frustrated spite of the left than literature or art or other aspect of high culture. Shakespeare, Beethoven, and Rembrandt may adorn the theaters, concert halls, and museums of the rich, but they are long dead; and, in any case, there is a venerable tradition of regarding artists per se as rebels, malcontents, and social critics. Science, on the other hand, is anything but antique. It thrives—or, as its critics would have it, fulminates—in the heart of the contemporary world. What is more, it is an indispensable prop to the politics and commerce of that world. It builds the bombs for the Pentagon and fiber-optics networks for the stock exchanges of the world. It computes the macroeconomic projections of the neoclassical economists and the demographic projections of cynical political operatives. It creates an enormous environmental mess and then charges us an arm and a leg to clean it up! It has all of us by the throat.

Resentment is a strong force in human affairs; philosophical caution is deplorably weak. The left's resentment of science is no sillier than that of, say, religious fundamentalists. Typically, it is expressed with incomparably greater cleverness and verbal agility. Nonetheless, resentment is not a trust-worthy ally in any intellectual endeavor. In the present case it has betrayed left-wing intellectuals into futility.

Neither of us is a professional historian; yet we have undertaken a study that has important historical dimensions. We cannot ignore them or dismiss them with the currently fashionable glibness. The left's flirtation with irra-tionalism, its reactionary rejection of the scientific worldview, is deplorable and contradicts its own deepest traditions. It is a kind of self-defeating apos-tasy. But it is not the result of a sudden whim or of spontaneous mass hysteria. It has a history. We owe the reader some sense of our understanding of that history, non-professional as it may be, before proceeding to the details of our critique.

History of the American Left, Briefly Considered

Socialist radicalism in this country derives, in the main, from the labor struggles of the second half of the nineteenth century in such industries as steel, textiles, mining, and railroading, and from the efforts of poor farmers to free themselves from the dead hand of bankers and middlemen.[5] The emergence of explicitly socialist parties in the 1880s and 1890s was facilitated by the accession of immigrant workers, already steeped in European radical traditions, to the ongoing efforts of native-born Americans. The Socialist party of the early 1900s was the main vehicle for the dissemination of radical ideas and the main platform for electoral activity on behalf of those ideas. Its period of greatest growth followed hard upon the failure of the less ideologically explicit, antiplutocratic populism championed by William Jennings Bryan. As is seemingly inevitable in left-wing politics, splits and factions produced other socialist and anarchist organizations of varying importance; but, under the leadership of Eugene Debs, the Socialist party commanded by far the largest following. Less formally, indigenous radicalism owed much to older movements, notably the Abolitionist and Suffragist struggles. Looking back from the point of view of contemporary militancy, one must also take note of the conservation movement, inspired by such writers as John Muir and Gifford Pinchot, although at the time of its origin, that movement largely drew its support from the genteel classes and had little, if any, connection with the discontents of exploited urban workers or impoverished midwestern farmers.

The popular appeal of American socialism is usually considered to have peaked in 1912, with the most successful of Debs's presidential campaigns. However, the coming of World War I proved as disastrous for the American Socialist party as it had for its sister parties in France and Germany. The principled opposition of the party to American participation in the European bloodbath, which it rightly viewed as a family quarrel among capitalists, a quarrel whose costs were largely to be reckoned in the death and dismemberment of workingmen, made it vulnerable to accusations of disloyalty and lack of patriotism, the more so because so many of its followers came from the foreign-born population. The popularity of the party sank rapidly; Debs was imprisoned on absurd sedition charges; and many of the faithful were imprisoned by dint of similar legal outrages, while others were simply deported as aliens whose radicalism made them unfit for residence in America.[6]

Equally fateful was the October Revolution in the one-time Russian Empire. Quite naturally, it inspired hope among American radicals for a worldwide revolution leading to a bright socialist future. At the same time, however, it precipitated a rancorous split within the Socialist party, out of which its eventual bitter antagonist, the Communist Party of the United States of America, was born.[7] Both wings of the divided socialist movement struggled subsequently, through the twenties, with little notable success. Furthermore, by the end of the decade the American Communist movement had become, for all practical purposes, a shameless satellite of the Stalinist regime, whose grip was beginning to throttle whatever genuine revolutionary idealism remained in the Soviet Union.

With the advent of the Great Depression, however, both branches recovered a measure of their former influence among the millions of workers and farmers displaced and despoiled by the collapsing economy. The Communist party was especially successful: at its height its power and influence rivaled that of Debs's prewar Socialist party. This transitory success is probably best explained by the widespread fear of international fascism and the sense that the Communist party, with its discipline, strength of purpose, and alliance with the "actually existing socialism" of the Soviet Union, was the most reliable bulwark against Mussolini, Hitler, and Franco, and the most implacable agent of genuinely fundamental change. A view leavened with appropriate skepticism would also suggest that the CP's often-cynical political opportunism contributed to its popularity for a time, even more than the appeal of its supposedly immutable principles.

This success was, of course, short-lived. The Moscow purge trials and the Hitler-Stalin Pact disabused many of the party's sympathizers, especially among the intelligentsia; and the rise of Trotskyism provided many of them the means of defying the party while remaining loyal in their own minds to the single-minded revolutionism of Marx and Lenin.[8] As for the other branch of the original split, Norman Thomas's Socialist party, the reprieve brought about by the Depression revived its fortunes only moderately, and increasingly it became an aggregation of intellectuals equally disaffected by capitalism and Stalinism, rather than an active electoral organization or a militant corps of grass-roots organizers.

World War II brought a truce of sorts between the Communist party and its customary foes in the government; the interests of Stalin, Churchill, and Roosevelt had converged momentarily, and the acquiescent American party subordinated both its ideology and its organizational tactics to the needs of the war effort. However, the ink was hardly dry on the surrender instruments of the Axis powers before the cold war had broken out, and the postwar repression of the Communists, echoing the antiradicalism of the early 1920s, effectively quashed the party as a meaningful political force. A curious aspect of this period is the degree to which non-Communist radicals joined forces with traditional anti-Communist conservatives to repress the CP and expunge its influence, while continuing to think of themselves as loyal and principled socialists. In its early days, for instance, the CIA recruited a substantial number of such people. This is explained not only by the enormities of Stalinism in the Soviet Union and Eastern Europe, but in equal degree by the distrust and loathing with which many radicals and ex-radicals had come to regard the American party by reason of its duplicity and manipulativeness. Many of those appalled by McCarthyism were chiefly outraged by the witch-hunters' refusal to allow ex-Communists to distance themselves from the party with honor and dignity, rather than by the misfortunes of the hard-core remnant of the party itself.[9] In any case, by the end of the 1950s American Communism per se was little more than a comical relic, sustained chiefly by the contributions in cash and nominal membership of Hoover's FBI, which felt a desperate need to keep its ritual enemy alive, or, if not alive, at least in the semblance of a minimally animated corpse.

On the other hand, anti-Communist socialism of the period fared little better. The fifties were the local high-water mark of American postwar ascendancy, economic as well as in military strength: the material condition of American workers and farmers, at least among the white population, improved rapidly. The rapid expansion of higher education, sustained by such measures as the GI Bill, opened in a remarkable way the upper-middle-class professions to a large number of men and women of modest social origins. They attained a prosperity undreamed of by their parents and unmatched elsewhere in the world. Untroubled by foreign competition, and bolstered by an industrial and agricultural economy of unrivaled strength, the American social and political system felt itself poised on the brink of an era in which the most grievous sources of conflict and discontent were bound to disappear, leaving no room for serious radical thought, let alone for active radical movements.

The 1960s were rapidly to shatter the delusion. The appalling inequities of the racial situation in America, sustained since Reconstruction by structures of legal and informal segregation, had bred resentments and rage that could no longer be contained; and the struggle to enforce the legal end of school segregation quickly burgeoned into a militant and massive civil rights movement, whose demands were wide-ranging and which in the context of ossified racial attitudes effectively constituted a call for radical social change.[10] Uneasiness over the cold war and its attendant arms race inspired the peace and disarmament movement to recover its nerve after more than a decade of McCarthyite intimidation. But it was the Vietnam War, with its clear record of governmental blundering and deceit, its lack of justification either by history or as genuine *realpolitik*, and its disproportion of means to ends, that truly revivified the American left and gave it the sense that a mass constituency receptive to its views was about to coalesce.

This phenomenon was especially strong on college campuses, particularly at elite schools where a tradition of intellectual independence had always been encouraged. Indeed, apart from the university, it is hard to think of social loci wherein the newly revived leftist dream of a mass movement ever moved beyond wishful thinking or outright self-delusion. Even in the black communities, where support for the civil rights movement, as well as far more intransigent forms of militancy, was ubiquitous, the endorsement of a specifically *socialist* vision of social and economic change was rare and at best equivocal. Only among students (and a substantial number of sympathetic faculty) did a more or less coherent radicalism exist, a movement that combined concern about theoretical issues with a genuine awareness of its own historical roots. Sectlets that had limped along for decades suddenly discovered a new generation of recruits. Even more important, organizations that had come into existence as mildly social democratic lobbying groups found themselves transformed, after a few short years, into foci of the most apocalyptic and intransigent radicalism. The most obvious example is Students for a Democratic Society, whose trajectory from cautious reformism to a fullblown Maoism was bewilderingly swift.

As the war continued in Asia and the civil rights struggle raged amidst rhetorical and real violence, the student left gathered strength and sympathy; and it acquired new causes. Wide-ranging discontent with contemporary capitalism and its dislocations was in the air. Environmentalism, as a radical and transformative view of the world, began to stir. The tactics and rhetoric of the black civil rights activists began to infuse the thinking of other minorities—American Indians and Latinos. By the end of the decade, feminism, in its modern incarnation, had taken shape and had become yet another article of faith for campus radicals. The concerns of lesbians and homosexuals, long trivialized and regarded as "apolitical," were quickly made part of the overall radical agenda. In terms of constituency, the left seemed to be constructing a wide base.

Moreover, student activists had apparently genuine grounds for thinking of themselves as participants in, or even leaders of, a serious political force. The protest at the 1968 Democratic Convention, together with its repression by clubs and tear gas, seemed to them (and to their parents) an event of profoundest historical significance, especially since it resonated so accurately with the Events of 1968 in France, the ongoing student militancy in Germany and Japan, and the Chinese Cultural Revolution—widely (and absurdly) perceived at the time, among leftists of all stripes, as authentically democratic in spirit and practice. The "counterculture" was at its height, with its twin promises of political and cultural liberation. A genuine and infectious euphoria gripped the most committed of university radicals; their abstractions seemed finally to have become incarnate as authentic social possibilities.

Within an astonishingly few years all those hopes were substantially dead.

A dozen reasons can now be given for the disappearance of a "mass" left, even within the hothouse venue of higher education. The left's perennial predilection for factionalism and internal bickering certainly had something to do with it, as did the overeager migration of the most ardent "theoreticians" within the movement to the far shores of doctrinal extremism. The winding-down of the Vietnam War relieved millions of young men (and, again, their parents) of the besetting anxiety—of being drafted—that had motivated them to consider themselves de facto supporters of the antiwar movement. The failure of the civil rights movement to meet unrealistic hopes for an immediate abolition of racism and its economic consequences led to an increasingly emphatic black militancy, which embraced the most violent rhetoric and—occasionally—distressingly brutal and violent tactics as well. For many white students, fear replaced visions of brotherhood. The feminist movement embarked on its own separatist path, obsessed by its special concerns and its own brand of hermetic theory. Finally, the foreign models that had by their example energized the American left throughout the sixties swiftly lost their power to inspire during the seventies. Events in China, Cuba, Cambodia, Vietnam itself revealed painfully the moral shortcomings —more bluntly, the moral *crimes*—of Third World revolutionism. American radicalism became an ideological orphan.

The dream of a unified, militant left with a widespread constituency belongs once more to the realm of wistful speculation. The civil rights movement that once stirred the conscience of the nation and seemed the rightful heir to everything noble in American tradition has devolved into a morass of bitter resentments, susceptible to tribalistic fantasies and demagoguery, but unable to formulate coherent goals or effective strategies. True, the women's movement retains wide support, if one's criterion is support for such key doctrines as reproductive rights and equal status in the workplace. But there is a sharp gradient separating mainstream feminism of this sort from the acute and apocalyptic oppositionism of "academic" feminism. (There is no realistic sense in which the former can be called "radical"; such categorizations are out of place everywhere but in the fantasies of the radical right.) The Marxist tradition has deliquesced into a mere oppositional posture decorated with a traditional lexicon but severed, apparently forever, from the struggles of an organized or organizable working class. The left, in sum, is at the moment the surviving squad of theoreticians of a nonexistent mass movement.

Nonetheless, the radical style of the sixties left traces that persist. First, there is the enduring relation between left-intellectuals and American universities. The campus constitutes the *only* environment in which recent radicalism became naturalized. Even as leftist rhetoric denounced higher education as the breeding ground for unquestioning servants of the bourgeoisie, leftist intellectuals, almost inadvertently, were forming a network of personal and professional ties with the institutions themselves. The scholarly community was the inevitable refuge to which activism retreated as its concrete political possibilities melted away.

Secondly, there is an intellectual heritage. Most of the thinking, as well as the rhetoric, of the sixties left was built around the theme that liberation for the oppressed can only be won when the oppressed group acts as the autonomous agent of its own revolutionary process. This notion of the "special competence" of the oppressed was deeply ingrained and has become unchallengeable within leftist circles. Only blacks, it was held, could define the terms of the black liberation struggle; all ideas, as well as all decisions, had to come from black revolutionaries themselves. Whites could function only as agents of tactical support. The same assumption was extended to other peoples "in struggle"—Native Americans, Chicanos, and so forth. Of course, it was always understood that American radicals had no right to criticize the tactics and strategies of Vietnamese revolutionaries; again, they were relegated to the function of uncritical support. The corresponding maxim was applied to radical feminism as that movement took shape—men, however sympathetic, could be spear carriers but never theorists or analysts, let alone leaders.

These attitudes, recurring from context to context, have a theoretical counterpart, a doctrine declaring that a group traditionally "privileged" has no right to define reality for others. It goes further; *the very state of being oppressed is somehow supposed to confer a greater clarity of vision*, a more authentic view of the world, than the bourgeois trappings of economic, racial, and sexual hegemony.

Finally, and quite expectedly, there is a lingering distrust of science and technology. This obviously derives from the long tradition of fear and loathing toward the nuclear arsenals of the world and the technocrats who created them. It was greatly intensified by the brutal spectacle of the war in Indochina, where all the technical ingenuity of the most scientifically advanced culture in the world seemed to have been conscripted to inflict butchery on a peasant culture. Finally, the misgivings of the environmental movement toward technology as such became common currency within the left, thereby widening still further the rift between contemporary radicalism and the Enlightenment tradition of science as the ultimate product of human wisdom and the staunch ally of liberation. All of these factors come into play, as we believe, in generating the peculiar amalgam of ignorance and hostility that glides beneath the surface of virtually all of the "critiques" of science that leftist theorizing has brought forth.

The Face of the Enemy

As we have proposed to use the term, *academic left* refers to a stratum of the residual intelligentsia surviving the recession of its demotic base, a stratum that must now, for the most part, content itself with inward meditations and hopes for the eventual revival of mass participatory politics on the left. However, the image of this body of intellectuals as a beleaguered relic is bathetic. In terms of their relations with this country's formal institutions of higher education, particularly those at the elite level, left-wing thinkers have never enjoyed anything remotely close to the current hospitality. Prestige-laden departments in the humanities and the social sciences are thickly populated—in some by now well-known cases we might say, without op-probrium, "dominated"—by radical thinkers. Despite all protestations to the contrary, entire programs—women's studies, African-American (or Latino or Native American) studies, cultural studies—demand, de facto, at least a rough allegiance to a leftist perspective as a qualification for membership in the faculty. A considerable number of high-flying academic stars are leftist celebrities who hop from one first-rank institution to the next at their own pleasure, and who must be given substantial perquisites (as these things are measured in academia) in order to put down institutional roots. Scholarly associations are often dominated by these same stars,[11] with the substantial assent of a rank and file in whom similar political sympathies run strong. Administrators who are also prominent left-wing figures are no longer anoma-lous. Even more significantly, university officials whose own politics may, in reality, be considerably more bland, have, in many places, come to treat the local campus left as an important and stable segment of the academic commu-nity, whose views must be taken into account and whose misgivings must continually be placated. Often, when administrations take official positions on social issues—particularly those involving race, ethnicity, and gender questions—the tone, and the jargon as well, is indistinguishable from that of the militant left.

These days, academic presses pour out dozens upon dozens of volumes grounded in left-wing theory. Proportionately numerous are learned journals, many of them brand new and as slick as MTV, whose purpose is avowedly political and unapologetically leftist. Universities by the score are delighted to host conferences and symposia whose podiums resound with rhetoric whose basic sentiments and passionate delivery would delight the soul of Emma Goldman (though she might find the terminology a bit off-putting).

This fact—this naturalization of the left as a well-dug-in sector of the university community—presents us with a considerable puzzle in view of the isolation and neutering of significant left-wing sentiment in the world of "real" politics. There is no strong—or even anemic—left-wing constituency out there standing godfather to the academic careers of its theorists. Of course, there is the fact that a surprising number of foundations, including some of the biggest and most munificent, are strong patrons of the left intelligentsia, and this affords it much academic leverage. But then again, the left-wing sympathies of the foundations themselves are a part of what has to be explained. In any case, grantsmanship is a minor factor.

The best explanation that we can come up with, ad hoc, is that over the last twenty-five years the entire process of recruitment into academic careers, especially outside the exact sciences, has been altered in a way that lures people with left-wing sympathies and hopes for radical social change into scholarly careers, while simultaneously bright young students of conservative bent are less and less enchanted at the prospect of joining the professoriate. This is not just a matter of "affirmative action" or of special enclaves for women's studies, black studies, and the like. It is a much more diffuse phenomenon, largely inadvertent and unplanned. The glamorization of high-powered careers in business, finance, and corporate law has something to do with it. Absent an initial emotional commitment to a radical political vision, a bright young person is likely to be far more strongly tempted by the prospect of having his or her own stretch limousine and Lear jet than by even the cushiest faculty appointment, especially in a mythic atmosphere in which twenty-six-year-old self-made billionaires abound. Moreover, academic recruitment is a process that significantly involves positive feedback; the greater the density of campus leftists, the more quickly that density grows. The reputation of the campus as the place to be for radical action seeded higher education with a substantial population of militants during the sixties; and this seems to have started the ball of exponential growth rolling.

Finally, the process has had the crucial goodwill of a kind of academic "silent majority," the great body of professors who, while they may distance themselves from doctrinaire ideological formulations and exotic new social theories, somehow continue to believe vaguely that the left, broadly construed, remains (after all these decades) "the party of humanity," the locus of right thinking; and that it deserves to be nurtured and encouraged even if it goes overboard from time to time in the vehemence of its views.

At this point we would seem to be well-launched into the standard right-wing tract decrying the infestation of the university by monomaniacal leftists, feminists, and black nationalists, and calling for the return to the tweedy good old days when the devotional poetry of John Donne and T. S. Eliot was solemnly celebrated in English departments everywhere, when acceptance of America's responsibilities as leader of the Free World was an article of faith in all courses on political and social matters, when culture meant Bach, Shakespeare, and Henry James and was completely disjoint from the iconography of pop musicians who wear their underwear on the outside, or compose chants in honor of killing white cops. Under that presumption, we ought now to start demanding that our campuses whip themselves into shape and that all these troublesome agitators should be sent into outer darkness, taking their ideological baggage with them. At the very least, we should be calling for the "depoliticization" of the classroom and for an end to the favoritism now so lavishly bestowed on left-wing doctrine and its proponents.

This, however, is far from our intention. There is a counterargument that has merit and is worthy of attention. We are living in a society where homogenization of political and ideological opinion is increasingly hard to resist. The number of newspapers has shrunk to a tiny fraction of what it was a hundred years ago in a much less populous country; indeed, newspapers no longer seem independent publications but rather compendia of news stories, columns, and editorials produced by national syndicates and merely bundled for local consumption. Broadcast news, similarly, is the mechanical dispersion of centrally produced material, and television journalism, whether from the traditional networks or the new cable services, is largely a matter of indistinguishable talking heads, all doling out the same narrowly conceived accounts of the same spectrum of *stories*. As a mode of discourse, contemporary journalese is as full of formulae and epithets as the most debased tradition of oral folk-poetry. "In depth" public issues programming offering "debates" between "liberals" and "conservatives" often provide a right-of-center, nominally Democratic neoconservative as representative of the former, and a neofascist as spokesman for the latter. The range of discussion of serious political and social issues in the mass media is, in short, so severely attenuated, so formulaic, and so castrated by the absence of an honest critical vocabulary that to call it superficial and inadequate is to praise it altogether too highly.[12] There are probably some honorable exceptions, even in the *mass* media, but their ratings must be insignificant.

Honest and undogmatic intellectuals, whatever the cast of their views, ought to ask themselves the following question: Shouldn't an alternative political culture, one with a long and often (if not always) honorable history, a great deal of intellectual energy and theoretical acuity, and a habit of asking discomfiting but morally urgent questions, have some sort of venue, some institutional homeland within this vast and still wealthy country? The indignant conservative who denounces the supposed ideological monoculture of the "radical" universities has a point; but it is a modest one. Why should the doctrinal narrowness of black or women's studies departments be more objectionable than that of some schools of business administration, quite a few military science departments, or even athletic departments? No fire-breathing feminist zealot has ever had the power over the lives and minds of her charges that is exercised routinely by the football or basketball coach at a school with a major "program," that is, one aspiring to be a significant NFL or NBA farm team. So far as the average student is concerned, degrees are pursued in marketing or chemical engineering or pre-med much as they have always been, with little input from the cryptic rituals of the postmodern, cutting-edge critical theorists. For such a student, an encounter in an expository writing course with a graduate teaching assistant who is, shall we say, a little too hyped on Foucault, Lyotard, and a gaggle of post-everything feminists, will probably do no lasting harm and might, possibly, do some good.

All this in no way implies that there is something out of bounds in principled opposition to one's left-wing colleagues over issues concerning curricular content, speech codes, affirmative action, or "studies" departments of one sort or another. Still less does it entail deference to whatever trendy doctrine or windy generalization has taken the inconstant fancy of the left. Such theorizing is obliged to take its chances in the rough-and-tumble of debate. It is a test of the maturity of the academic left whether it can deal with such criticism without imputing dire political motivations to the critics.

We thus come round to our own announced intention—to analyze and refute the critiques of science—its methods, assumptions, conclusions, and social aspects—that have arisen among left-wing scholars, or, more precisely, that specialized subset we have styled "the academic left." We concede yet again, at risk of trying the patience of the reader, that the term is not felicitous.[13] Perhaps we should have used something different, "postmodern left," say; but this seems a trifle too confining. "Post-Marxist left" or "hyper-theoretical left" might have done, but again, these smack of jargon and imply a higher degree of uniformity than is warranted. Neologisms should not be multiplied beyond necessity and will not be, at least by us.

Still, there are distinctions and characterizations that need to be made. Our notion of what constitutes the academic left must, to the extent possible, be clarified. We are not a credentials committee. Nor is this book in any sense an update of the *Malleus Maleficarum*; we shan't give our readers detailed instructions for finding the witch's mark. We do, however, need a basic intellectual and attitudinal portrait, one that sharpens the contrast between the academic left as such and those scholars holding left-wing opinions whose research is basically apolitical or who pursue a political agenda largely uncolored by those doctrinal singularities that give the academic left its particular character.

It seems to us that the central tenet of the various schools of thought that make up the academic left is one that may be labeled "perspectivist." The basic thrust is that various bodies of ideas that have been favored and championed by Western culture over the centuries must be stripped of their claims to universality and timeless, uncontextual validity. They are at best the expression of local "truths" or "structures" that make sense only within a certain context of social experience and a certain political symbology. On the other hand, they may be justificatory myths meant to uphold authority and hierarchy. In either event, they are always deeply marked by the power relations that govern the societies in which they arose.

By the same token, perspectivism is highly sympathetic to the claim that the heretofore disempowered have the right to have their own "narratives," their own particular accounts of the world, taken *as seriously as* those of the standard culture, notwithstanding differences and outright contradictions. The intellectual apparatus of the post-Enlightenment West, it is held, affords no special leverage for deciding among competing versions of the story of the world. Such methodologies have been deferred to in the past, but that is because they have been arbitrarily "privileged" by the historical ascendancy of Euro-American capitalism, a merely contingent circumstance. They occupy no firmer epistemological ground than the accounts produced by women, descendants of black slaves, Third World revolutionaries, or even a reified and personalized Nature. The latter thus become immune from criticism by the reigning Western paradigm—and from white European males, dead or alive.

The dethronement of Western modes of knowledge and their claims to objectivity is said to be justified on a number of grounds. To some, it is the inherent instability and cloudiness of language that does the job. Others appeal to fairly traditional Marxist notions of class consciousness. Feminists champion "women's ways of knowing," while Afrocentrists have their own version of the blood-and-soil myth. The important point, however, is that each faction thinks the job is complete and that Western paradigms have been effectively demolished.

Among academics, such attitudes are nowadays extremely common. They are conjoined, however, with other habits of thought characteristic of intellectuals as a class. There is, for instance, an abiding cabalistic faith that excursions into *theory*, if pursued at great enough length with sufficient intensity, will tease forth all the deepest truths of human experience. This adds considerably to the impression, common outside of academic-left circles, that the "critical theory" in which academic leftists take such delight is a swamp of jargon, name dropping, logic chopping, and massive attempts to obliterate the obvious. The irony is that this faith in the omnicompetence of theory runs particularly strong in those who claim to abhor "totalizing" theories. "Both [Derrida and Foucault] have, in different ways, actually stimulated a return to a form of scholasticism, to those abstract and totalizing methods of the traditional Western humanist the new theory claimed to reject."[14]

The evidence that this point of view thrives in the academic world is easy to find. We take as illustrative a series of announcements that recently appeared in *Lingua Franca*, a journal for au courant academic humanists that treats feminism, postmodernism, critical theory and the like with a piquant mixture of respect and irreverence. In a recent issue we find assembled a number of calls for manuscripts for new series of scholarly books. The editors of "Pedagogy and Cultural Practice" (University of Minnesota Press) tell us that

> the series analyzes the diverse democratic and ideological struggles of people across a wide range of economic, social and political spheres and provides an opportunity for cultural workers from many fields to submit manuscripts that link the pedagogical and the political around new forms of cultural practice.

The organizers of "Theory out of Bounds" (also University of Minnesota Press!) announce:

> The works we seek join cultural analysis with tactics of cultural resistance as they enact their own critical ground and create new spaces of cultural invention.
>
> The series will deal with such issues as the affective constitution of the body, the politics of cultural appropriation, the social production of subjectivities, *feminism and the philosophy of science*, and the formation of communities outside of identity. (Emphasis added.)

"Ideologies of Desire" (from Oxford University Press) offers a new forum for "cultural studies" and declares:

> These new approaches have also traced the inscription of sexual meanings in widely scattered fields of cultural production, while detecting the inscription of diverse cultural meanings in the practices and discourses of

> sex. . . . The project ultimately has an oppositional design: its purpose
> is to map more precisely the available avenues of cultural resistance to
> contemporary institutional and discursive practices of sex.[15]

Finally, "Re-Reading the Canon" (Penn State Press) merely asks, in somewhat less trendy language, for "feminist reinterpretations" of important philosophers—Plato, Aristotle, Locke, and Wittgenstein, as well as academic leftist icons such as Marx, de Beauvoir, Foucault, and Derrida.

Clearly, the academic left thrives as a burgeoning industry within the scholarly community. The quotations above make clear how easy it is to spot; any writer who uses "discourses," "cultural," "practice," and "inscription" in the same paragraph is a member ipso facto in good standing. The extent to which the perspectivist attitude, in one form or another, informs its thinking is equally obvious. Each of the cited examples obviously enthrones some version of the perspectivist approach.

In sum, we were finally tempted to supplant the term *academic left* by *perspectivist left*, and might have done so but for our aversion to conducting debates as though that were a matter of affixing labels. *Perspectivism on the left is the true legacy of the activism of the 1960s and early 1970s*, a time when it was assumed that the oppressed are endowed with uniquely privileged insights, and that the intellectual, as well as moral authority of victims is beyond challenge. Like many philosophical stances, perspectivism points to some genuine issues and may lead to some valuable observations, as long as it is not run into the ground. But the overenthusiasm of the academic left consists precisely in its eagerness for and growing success in running things into the ground. So far as we are concerned, perspectivism, in its soberest and most prudent form, has interesting things to say about the history of science, the shape of modern science as a social institution, the rhetoric of scientific debates. When it comes to the core of scientific substance, however, and the deep methodological and epistemological questions—above all, the incredibly difficult *ontological* questions—that arise in scientific contexts, perspectivism can make at best a trivial contribution. The attempts to read scientific knowledge as the mere transcription of Western male capitalist social perspectives, or as the deformed handicraft of the prisonhouse of language, are hopelessly naive and reductionistic. They take no account of the specific logic of the sciences and they are far too coarse to deal with the conceptual texture of any category of *important* scientific thought.

We are unkind enough to wish to compare the academic left's recent attempt to advance perspectivist accounts of science to the "cargo cults" that flourished on some Pacific islands in the wake of World War II. During the war, a number of technologically primitive tribal cultures acquired new neighbors in the form of the military bases of the warring powers. What chiefly impressed the indigenous peoples were the airstrips where giant machines would periodically land to disgorge vast quantities of goods, some of which found their way into native hands by barter, as gifts, or simply by being left over when the armies departed. After the war, sects grew up around the idea that the airplanes, with their loads of precious goods, could be induced to return by magical means. In some places, the tribesmen built their own "airplanes," with the idea that ritual might transform them into the real thing. In our view, the model of "science" constructed by perspectivist theorists is a lot like the wicker-and-mud mock-up of a C-47 built by the cargo cultists. It bears only a vague and superficial resemblance to the real thing, and its internal logic is laughably different. Still, those who built it hope, with the aid of their theoretical magic rituals, to gain control over the real thing.

The remainder of this book deals with a body of research and scholarship, examining its assertions, its weaknesses, its influence, real and potential, on social activism of one sort or another, and the underlying motives, acknowledged or not, that inspire it. For the most part, we allege, this work is deeply unsatisfactory. In scholarly quality, it ranges from seriously flawed to hopelessly flawed. It is infested with tendentious arguments, special pleading of one sort or another, and the rhetoric of moralistic one-upmanship. Yet it has been well received by a substantial part of the academy, where it is promulgated as a breakthrough for current social theory. Textbooks and courses are built around it and students and scholars with little background in the sciences turn to it first when they want to understand something about science.

The left, in this country as elsewhere, has a history of putting important questions on the table. It is not for this book to judge the moral worth of its prescriptions for the constitution of society or the political practicality of putting them into effect. Those arguments will resound in one way or another for the foreseeable future. As for the unforeseeable future, it is possible that socialists may prove, after all, to have been prophetic. We are obliged to observe, however, that leftists have a long history of weaving philosophical phantoms into fantasies of universal redemption. We are convinced that the academic left's recent attempts to theorize "science and society" are further instances of the same thing.

CHAPTER THREE

The Cultural Construction of
Cultural Constructivism

The point is that neither logic nor mathematics escapes the contamination of the social.
STANLEY ARONOWITZ, *SCIENCE AS POWER*

So long as authority inspires awe, confusion and absurdity enhance conservative tendencies in society. Firstly, because clear logical thinking leads to a cumulation of knowledge (of which the progress of the natural sciences provides the best example) and the advance of knowledge sooner or later undermines the traditional order. Confused thinking, on the other hand, leads nowhere in particular and can be indulged indefinitely without producing any impact upon the world.
STANISLAV ANDRESKI, *SOCIAL SCIENCES AS SORCERY*

Natural scientists—at least those with a sense of fair play—are usually diffident in confronting the disciplines that study science from a social and historical point of view. They do not feel that their particular expertise in some area of science automatically endows them with insight into the human phenomenology of scientific practice, or that their familiarity with the recent results and the liveliest questions of their specialty qualifies them to pronounce on its evolution as that relates to the course of human development. Apart from the most arrogant, they concede that the psychological quirks and modes of personal interaction characteristic of working scientists are not entitled to special immunity from the scrutiny of social science. If bricklayers or insurance salesmen are to be the objects of vocational studies by academics, there is no reason why mathematicians or molecular biologists shouldn't sit still for the same treatment.

Accustomed to regarding specialists in a scientific subject remote from their own with a certain courtesy or even deference, natural scientists are usually quite willing—perhaps even too willing—to adopt a similar attitude toward apparently competent scholars whose chief interests lie outside of science even as they try to come to grips with the relation between science and their own fields. Above all, natural scientists are reluctant to take a haughty and dismissive attitude toward the hypotheses and theories of outsiders merely because they seem at first paradoxical or are expressed in a recondite language. They are aware that some matters of the greatest professional concern to them may strike an outsider as abstruse, bewildering, perhaps even nonsensical. Consequently, their initial inclination is to credit the sociologist or historian of science with having reliable intellectual tools and a sense of responsibility in applying them, however jarring the language of the unfamiliar discipline.

This, at least, has been the traditional attitude. We must report, however, that it is changing to one of skepticism and even revulsion in the face of what scientists—at least those few who have so far taken a serious interest in the question—have come to see as a growing tendency among a particular breed of historians and sociologists of science to spin perverse theories. These seem often to escape mere inaccuracy and rush hell-for-leather toward unalloyed twaddle. Such words may strike the reader as splenetic; but they seem to us justified in view of certain recent developments in the social-scientific analysis of the natural sciences, specifically those that can be lumped together under the heading "cultural constructivism."

Cultural Constructivism, Weak Form

We are all, in a commonsensical way, cultural constructivists in our view of science. Science is something that human culture has, indeed, "constructed," after seventy thousand years or so of false starts and dead ends. Our thankfulness for it knows no bounds. Pieties aside, however, we can accept many of the views that historians and sociologists of science promulgate by way of asserting that science is, in some sense, a cultural construct. It would be idle to pretend that the projects taken on by science, the questions that it asks at any given period, do not reflect the interests, beliefs, and even the prejudices of the ambient culture. Clearly, certain kinds of research get the strongest encouragement—funding, recognition, celebrity, and so forth—in response to the recognized needs of society. To take an obvious current example, research in high-temperature superconductivity is avidly pursued by increasing numbers of physicists and chemists, not only because the subject is fascinating, beautiful, and difficult, but because the potential utility of high-temperature superconductors is enormous. They might very well have technological and economic reverberations comparable to those of semiconductors. And, at least for a time, a few years ago, when new superconductor materials were appearing thick and fast, the work got almost as prurient attention from the media as did cold fusion. To say that in this sense science is culturally constructed is tautological.

Naturally, some social theorists would extend the analysis to suggest that the topics scientists focus upon are determined by socially derived attitudes, aspirations, and biases less forthrightly instrumental, and that there are negative aspects as well—certain areas of potential research are avoided in obedience to assumptions that are rarely articulated in undisguised form. While this is not incontestable, it has some plausibility, and we would not deny it out of hand.

An even stronger assertion is that in scientific debate and in the process by which a preference for one paradigm over another emerges, attitudes of mind come into play that are in some measure dictated by social, political, ideological, and religious preconceptions. For example, Stephen J. Gould[1] has recently argued that Darwin's view of sexual selection as an important evolutionary mechanism was slow to win acceptance because it offended the prejudice, obviously tied to prevalent ideology, that females are by nature passive and lack sufficient volition to make the crucial choices that Darwin's—extended—model required. We accept such ideas as reasonable in principle, even though they are much oversold these days. We caution, however, that the areas of science in which such direct intrusion of ideology becomes possible are few. Our reading of the history of science suggests, moreover, that theories leaning *heavily* on such props tend to be fragile and ephemeral, and that part of the increasing power of scientific methodology derives from always-increasing awareness of the danger that reasoning can be corrupted in this way if one is not careful. Nevertheless, we are obliged to listen with interest to historical and sociological accounts of the effect. Thus we accede in principle to what might be called the "weak" version of cultural constructivism.

Such a point of view can produce illuminating research. Like any other point of view, it can also be driven into the ground and employed in doctrinaire fashion to substitute for evidence. Good work and bad can be done in its name. A further danger, frequently in evidence in the writings we consider below, is that analyses and case histories counting as reasonable instances of weak cultural construction are slyly adduced as justifying a far more radical and dubious theory, a version of philosophical relativism and conventionalism that merits the name "strong cultural constructivism." This is another part of the theoretical woods entirely—although many historians, sociologists, and even philosophers of science are insufficiently vigilant in maintaining the distinction.

Cultural Constructivism, Strong Form—Science as Convention

In strong form, cultural constructivism (sometimes, another phrase such as "social constructionism" may be used, depending on the terminological preferences of the expositor) holds to the following epistemological position: science is a highly elaborated set of conventions brought forth by one particular culture (our own) in the circumstances of one particular historical period; thus it is not, as the standard view would have it, a body of knowledge and testable conjecture concerning the "real" world. It is a *discourse*, devised by and for one specialized "interpretive community," under terms created by the complex net of social circumstance, political opinion, economic incentive, and ideological climate that constitutes the ineluctable human environment of the scientist. Thus, orthodox science is but one discursive community among the many that now exist and that have existed historically. Consequently its truth claims are irreducibly self-referential, in that they can be upheld only by appeal to the standards that define the "scientific community" and distinguish it from other social formations.

It must follow, then, that science deludes itself when it asserts a particular privileged position in respect to its ability to "know" reality. Science is "practice" rather than knowledge; and practice involves convention and arbitrariness. Questions can be asked only when they conform to the modalities of existing discursive habits; likewise answers can be formulated and recognized only to the extent that they are accommodated within that template. The verification of these supposed answers is perforce a discursive—in the broad sense, a linguistic—event, in that it involves dialectical manipulation of accepted semiological conventions. Even—no: especially!—the practices that most particularly embody the sacred "objectivity" of science—experiment and observation—are inescapably *textual* practices, meaningless outside the community that endows them with meaning.

The attentive reader will have noted that this point of view rigorously applied leaves no ground whatsoever for distinguishing reliable knowledge from superstition. Indeed, there are various contexts in which that would seem to be exactly the point of the exercise. Given the long history of progressive Western thought in which science has been linked, by and large, with the efforts of human liberation, it will seem surprising if not positively bewildering that this complex of ideas has for the most part been developed and embraced by self-identified left-wing intellectuals.

True enough, critiques of science employing a similar logic appear from time to time in defense of theistic views of the universe that place scientific materialism in its more accustomed role as the enemy of religion and revealed truth.[2] Nowadays, however, the left seems far more eager to sponsor such views. Its motivation, its polemical point, seems to be as follows: Scientific questions are decided and scientific controversies resolved in accord with the ideology that controls the society wherein the science is done. Social and political interests dictate scientific "answers." Thus, science is not a body of knowledge; it is, rather, a parable, an allegory, that inscribes a set of social norms and encodes, however subtly, a mythic structure justifying the dominance of one class, one race, one gender over another. This, at any rate, is the message that permeates the culture of the academic left, setting the terms of its view of science. It is the motto on the banner flown by cultural constructivism when it functions as a political force.

A typical example of the discourse of the cultural constructivists, certain to startle a scientifically literate person who has never encountered the genre, can be found in *The Science of Pleasure: Cosmos and Psyche in the Bourgeois World View*, by sociologist Harvie Ferguson. He summarizes a key development of twentieth-century physics as follows:

> The inner collapse of the bourgeois ego signalled an end to the fixity and systematic structure of the bourgeois cosmos. One privileged point of observation was replaced by a complex interaction of viewpoints.
>
> The new relativistic viewpoint was not itself a product of scientific "advances" but was part, rather, of a general cultural and social transformation which expressed itself in a variety of "modern" movements. It was no longer conceivable that nature could be reconstructed as a logical whole. The incompleteness, indeterminacy, and arbitrariness of the subject now reappeared in the natural world. Nature, that is, like personal existence, makes itself known only in fragmented images.[3]

We assure the reader that Ferguson is referring unambiguously to Einstein's relativity theory, not to some broader and murkier notion of "relativity"! He means literally, and reaffirms throughout the book, that developments in physics are not only conditioned, but *dictated* by the evolution of something called "bourgeois consciousness," whose course is in turn determined, in proper Marxist fashion, by "commodity relations." People moderately expert in modern physics and minimally familiar with its history will not find such pronouncements plausible in the slightest, nor will they concede very much to the supporting arguments. The latter merely decorate a compressed version of the standard history of modern physics with bizarre assertions to the effect that throughout it all, the desperate bourgeois ego was frantically supervising developments, a sort of crazed dramaturge. Such propositions have all the explanatory power of the Tooth-Fairy Hypothesis. Still, hundreds of left-wing social theorists dote on them.

Quite naturally, cultural constructivism—in its strong form—is one of the starting points and chief ideological mainstays of the feminist critique of science. Likewise, it fuses with the spectrum of doctrine and attitude that comprises the so-called postmodern intellectual stance, when that viewpoint attempts to give its own account of the sciences. We shall, however, address both the feminist and postmodernist analyses of science in separate chapters, recognizing that this involves quite arbitrary distinctions. Here we consider strong cultural constructivism as it is practiced by historians, sociologists, and other students of natural science as a social phenomenon. Most of these are committed to a leftist political position; they regard their study of science as part of an overall program of radical analysis and demystification of bourgeois sacred cows.

We shall not attempt immediately an exploration of why this should be so; but some points are clear enough. First of all, in the face of an increasingly monolithic social and political structure, whose capacity for self-perpetuation and extension seems endless, it is difficult for radical intellectuals to accord an exceptional status to science, leaving it exempt from what they regard as the omnivorous tendencies of capitalism. They are highly unwilling to view science as an activity of the autonomous and unfettered intellect. It is easy to see their point. Science is, after all, well integrated into the technological, industrial, and military machinery of the capitalist system; in turn it relies on that system for the material basis of its continuing progress, at least in those fields where a substantial investment of money is necessary for fruitful research. For working scientists in the belly of the beast, of course, the situation seems far more subtle than that. In fact, from a variety of perspectives, scientists and intellectuals in general might honestly (and correctly) view the present culture as a historical paragon, to the degree that it fosters and encourages autonomy of thought and freedom of ideas. On the other hand, the social critic who identifies with a long tradition of militant intransigence, and for whom positive social change invariably requires discontinuity, remains unmoved by such considerations. This critic views the scientist's claims to independence as part of the constructed ideology that imprisons and in the end directs him. To the analyst of cultural constructivist bent, matters of scientific truth are "always and everywhere matters of social authority."

Furthermore, to the extent that conventional science can be deposed from its position as a uniquely accurate way of finding out about the world, contending perspectives, especially those arising from the demotic substructure of society or from oppositional movements, rise at once to a higher epistemological dignity. Belief systems that, on the scientific view, are little more than superstition are at least provisionally validated by the cultural constructivist hypothesis, while the results arising from scientific investigation, if ever they appear irritating or unwelcome, become "contestable."

This is a book about politics and its curious offspring, not about epistemology or the philosophy of science; we cannot therefore refute, in abstracto, the constructivist view either in the strong form outlined above or in some of its more qualified but still erroneous versions. Nor are we obligated to do so: serious philosophers of science have been at it for decades. Nevertheless we record now, as scientists of long experience who have not been indifferent to philosophical questions, our emphatic rejection of that view. In our opposition, we undoubtedly have the concurrence of the majority of practicing scientists over a broad range of disciplines. Still, a sense of honor compels us to sketch at least one common argument against the constructivist view.

Consider how the theory itself is built up and defended. There is an obvious appeal to rather conventional veridical standards. A model of a phenomenon is proposed and given coherent logical form. Evidence for that model is adduced with every indication that it is evidence of a specifically factual kind.

This putative evidence is made to articulate with the presumptive model by means of arguments whose canons of logic and relevance are entirely unexceptional. Inferences from the model—specifically, those inferences we have summarized as constituting the core of cultural-constructivist doctrine—are likewise arrived at by the presumed application of ordinary logic, that is, deduction. Thus, the cultural-constructivist case is brought into being by an intellectual process that implicitly accepts the same methodological paradigm as the empirical sciences it presumes to analyze!

This is not to suggest that the paradigm is particularly well served in this instance. In fact, the logic of cultural constructivists seems to us sloppy and full of holes in the matters under discussion, their evidence dubious, and their case corrupted by special pleading and covert appeals to emotion and the prejudices of a certain audience. These objections aside, however, we note that the very form of their argument makes the cultural constructivists self-subverting. They appeal to the same canons of judgment that their argument seeks to condemn.

There is, of course, a smug habitual rejoinder to the foregoing. We are challenged to consider that if in fact the empiricist logic that undergirds the sciences can be demolished by an application of the very same logic, then there must be something wrong with empiricism in the first place. It, rather than cultural constructivism, must be the locus of self-contradiction. This is glib, but entirely unconvincing. It assumes that the arguments put forth by the cultural constructivists are airtight from both the formal and evidentiary point of view—a point, as we have said, that hardly needs to be conceded.

We invite the reader to judge where comparisons must ultimately be made. The cultural-constructivist thesis seems at first to be arguing against epistemological justifications of science, against attempts at finding foundations. Yet, inasmuch as the specific content of the thesis challenges the reliability of scientific conclusions—this is what it asserts in the final analysis, and not merely the inadequacy of foundational arguments—and inasmuch as it does so, roughly speaking, on the basis of the same argumentative paradigm as scientists use in practice, the logic, evidence, and pertinacity of the thesis must be weighed against that of specific scientific arguments.

In other words, in order to claim that they have made their case, cultural constructivists must demonstrate that their arguments for unreliability outweigh those of conventional scientific papers for reliability *in the realm of phenomena addressed by the latter.* They must show that their arguments are stronger than those put forth by Professor X in his paper on the role of transforming growth factor beta in the morphogenesis of the optic tectum, while simultaneously outweighing those of Dr. Y in his monograph on the classification of compact Lie group actions on real projective varieties! If they are to demonstrate that their arguments *contra* science are anything but sheer bluff, then clearly they must play on the scientists' court. At this point, we think a simple *res ipsa loquitur* is in order. There surely are scientific papers that are inaccurate for reasons more or less implicit in the cultural constructivist hypothesis, but these are on the whole rare and exceptional; they cannot be used to prove a hypothesis of such stupefying generality. To put the matter brutally, science *works.*[4]

We do not present this as a complete critique. Many other lines of analysis are available.[5] But there is no need. The state of affairs is best summarized, probably, by the philosopher Paul Feyerabend, one of the thinkers directly responsible for initiating the chain of ideas leading to the cultural constructivist view of science (and, next to Thomas Kuhn, the most often cited), who now expresses deep reservations about the outcomes of this line of thought. "How can an enterprise [science] depend on culture in so many ways, and yet produce such solid results?" he asks. "Most answers to this question are either incomplete or incoherent. Physicists take the fact for granted. Movements that view quantum mechanics as a turning-point in thought—and that include fly-by-night mystics, prophets of a New Age, and relativists of all sorts—get aroused by the cultural component and forget predictions and technology."[6]

Whom does Feyerabend have in mind? Many, certainly, among those thinkers on the left end of the political spectrum who have developed a cultural constructivist (or relativist, or contextualist, or perspectivist) critique of science with a view to extending their general indictment of Western capitalist social structure. We turn to a series of examples, work well known in the community that views the cultural constructivist program as both path-breaking and fully credible. Our exposition must be brief, yet we hope that our objections rise to a higher level than mere *amour propre*.

Science as Power

A paradigmatic example of the constructivist program may be found in the recent work of Stanley Aronowitz, a sociological theorist whose wide interests reflect many of the concerns, trends, and attitudes prevalent on the academic left. Aronowitz has long been active in the causes embraced by the left; indeed, he has the distinction, almost unique among university-based theoreticians, of having worked in the trenches as a union organizer and shop-floor politician. He is a leading figure in the Democratic Socialists of America,[7] an editor of such theoretical journals as *Socialist Review* and *Social Text*, and a ubiquitous presence at leftist symposia. His interest in science is relatively new but characteristically sweeping and ambitious, despite the fact that he has little formal training or technical facility in any branch of it. He is a professed admirer of Feyerabend, which makes it all the more ironic that Feyerabend's strictures fit him so well.

Aronowitz's major work on science is a turgid and opaque tract entitled *Science as Power*. It constitutes a major attempt to justify the cultural (or social) constructivist viewpoint and is clearly motivated by the belief that since science and technology are key elements in the substructure of modern capitalism, it is one of the duties of the oppositional social critic to demystify science and topple it from its position of reliability and objectivity. The major premise from which this work of demystification proceeds is that science is "situated" knowledge, conditioned by the historical circumstances that engender it and reflective of the ideological patterns of dominance and authority that prevail in the society.

Ambition, however, is one thing and achievement quite another. Aronowitz's book is notably clumsy in its approach to argument. Its chief method seems to be to invoke from the philosophy of science as many names as possible, in as small a space as possible, and to present their views, as paraphrased by Aronowitz himself, briefly and cryptically, cementing the whole business together, finally, with a wash of the author's pontifications. Very few specific positions are analyzed at great enough length to make them coherent; names and phrases are simply run in and out of the text as props for Aronowitz's views. This is done in a context emphatically illustrative of the pertinence of Feyerabend's remarks. *Science as Power* certainly does get highly aroused by the "cultural context" of science; it is intent on taking the development of quantum mechanics as a solemn turning point for Western science; it is unblushingly relativistic as to underlying philosophical doctrine; and it is, when all is said and done, incoherent.

We begin by considering the treatment of quantum mechanics and its philosophical implications. Aronowitz devotes an entire chapter, "History and Philosophy of Modern Physics," to this issue. Unfortunately, this treatment is greatly marred by the author's evident (at least, this seems the obvious inference from the text) ignorance of the particulars of physics. Despite the chapter's self-confident title, the reader who knows little of twentieth-century physics going in will remain, in point of specific knowledge, equally (or perhaps more) ignorant on his way out.[8] Here and elsewhere in the book what we get, instead of a pertinent history of contemporary (that is, seventy-year-old) physics, is a series of solemn and fetishized invocations of the uncertainty principle and, more generally, the quantum-mechanical challenge to classical determinism.

Now, the uncertainty principle is undoubtedly one of the cornerstones of quantum mechanics and one of the most philosophically provocative developments in the history of science. Under Aronowitz's description, however, it seems rather to refer to a kind of epistemological and spiritual malaise, plaguing the minds and souls of contemporary physicists. The argument, roughly but accurately paraphrased (and all too familiar from New Age tracts, among other things), is that since physics has discovered the uncertainty principle, it can no longer provide reliable information about the physical world, has lost its claim to objectivity, and is now embedded in the unstable hermeneutics of subject-object relations. This, alas, demonstrates depressingly well the connotative power of words when they are allowed to drift apart from their contextual meaning. If Heisenberg and company had chosen a less evocative term, an awful lot of nonsense of this sort might never have seen the light of day. Philosophical and pseudophilosophical posturing has dreadfully befuddled discussion of the issue addressed to nonspecialists.

Once obscurantism has been stripped away, we recognize that the uncertainty principle is a tenet of physics, a predictive law about the behavior of concrete phenomena that can be tested and confirmed like other physical principles. It is not some brooding metaphysical dictum about the Knower versus the Known, but rather a straightforward statement, mathematically quite simple, concerning the way in which the statistical outcomes of repeated observations of various phenomena must be interrelated. And, indeed, it has been triumphantly confirmed. It has been verified as fully and irrefutably as is possible for an empirical proposition. In other words, when viewed as a law of physics, the uncertainty principle is a very certain item indeed. *It is an objective truth about the world.* (If that were not so, there would never have been so much fuss about it!)

Aronowitz's incoherent account completely occludes that simple fact. He insists on adverting only to the most mystical views of the matter (those of Heisenberg *qua* philosopher-oracle, for instance) and ignores the particulars of the lively debate among physicists attempting to clarify what the predictive success of quantum mechanics really tells us about the physical universe. He naively echoes, for example, the view that the causal and deterministic view of things implicit in classical physics has been irrevocably banished. This is simply wrong.[9] He propounds, moreover, the undocumented and egregiously unlikely notion that the source of these developments lies in a general malaise that afflicted European culture in the wake of World War I. On Aronowitz's account, the pioneers of quantum mechanics were merely clever artificers obedient to the society's peremptory demand for an abolition of determinism and causality. Genuine familiarity with the history and content of the work of Heisenberg, Schrödinger, von Neumann, de Broglie, and others makes such a proposal hallucinatory.

Aronowitz's treatment, in short, gives no indication that he really understands the underlying physics and mathematics of the situation. He seems to be expostulating on the basis of dilute paraphrases or worse, vulgarizations of paraphrases. It undoubtedly seems snobbish to say so, but this field of speculation is notoriously unkind to amateurs. Feyerabend has in fact understated the case—getting aroused about the cultural component of physics has not only led Aronowitz to forget about technology and prediction, but has induced him to ignore the physics as well.

At a later point in the book we find the dictum cited as an epigraph above: "The point is that neither logic nor mathematics escapes the 'contamination' of the social."[10] The argument leading to this emphatic conclusion precedes it and gives us an excellent opportunity to examine the coherence of Aronowitz's own logic, approvingly borrowed from a certain David Bloor. Bloor, a sociologist identified with the Edinburgh School of unyielding cultural constructivists, points out that by report of the anthropologists, certain tribal peoples reason thus: Everyone, without exception, is a witch; on the other hand I know lots of people who aren't witches. Bloor and Aronowitz deny that this is a logical error: it is merely the logic of another culture! Aronowitz goes on to insist that our own culture has its own weird logic, largely un-acknowledged. Here is what he supposes to be a telling example: We define murderers as people who deliberately kill people; bomber pilots deliberately kill people; yet we deny that bomber pilots are murderers. Thus our own culturally generated logic is skewed and is not the disinterested thing claimed by defenders of the ideational autonomy of mathematics!

The confusions evident in this passage are comical. Observe, first of all, that in this culture we *don't* define murderers simply as people who deliber-ately kill people. A woman who shoots and kills a potential rapist is not a murderer in this culture, nor is a shopkeeper who dispatches a holdup man with the .38 he keeps under his cash register. A psychotic who kills a nun under the belief that she is a five-foot penguin is not a murderer either. Nor is a member of the military who, under official orders, kills large numbers of people, including innocent civilians. Indeed, in the last example, the person becomes a criminal if he *refuses* to kill other people. From an ethical or religious viewpoint, it is quite possible to quarrel with this exemption from the category "murderer"; but that dispute is about *ethics*: formal logic has nothing to do with it!

On the other hand, examples of the error made by the cited tribesmen can easily be found in this or any other culture. An anthropologist from the Death Planet, in disguise as one of us, recruiting informants from the shoppers at a typical suburban mall, will surely find someone who says something like the following: All politicians are crooks, without exception; President Clinton is certainly a politician; but President Clinton is not a crook since he's the commander-in-chief. The point of course is not that logic (and therefore mathematics) is culturally contaminated, but that cogent, self-consistent, logically coherent thinking is not ubiquitous. Many people shrink from it, in fact. It is an art and a skill that must be mastered; and it requires patience, diligence, humility, and intellectual energy. That it is not universally prac-ticed is a fact that does nothing to uphold Aronowitz's thesis.

There is a further point to be made about Aronowitz's argument, however—rhetorical rather than logical. The example about the bomber pilot is advanced precisely on account of its capacity for moral intimidation. The lurking suggestion is that if the reader fails to accept this argument, then he is implicitly condoning the bombing of innocent, Third World civilians by imperialist U.S. pilots. In left wing circles, such forensic flim-flam has considerable weight, despite its being worthless in logical terms. This sort of trick alone accounts for much of the popularity of cultural constructivism on the academic left.

Apart from his book, Aronowitz continues to speak out on questions regarding science. His newest statements do not reflect a chastened or cautious mood. He remains a true believer in the strongest form of the cultural constructivist dogma. A recent lecture,[11] for example, was studded with bizarre pronouncements unsupported by evidence or plausible argumentation, but delivered with that combination of blithe self-assurance and moralistic bullying that audiences with strong political sentiments but weak backgrounds in science find so intimidating—or exhilarating. To take but one example from among several dozen—and we stress that it is entirely typical in its pugnacious illogic—we hear Aronowitz denounce the idea that Einstein conceived special relativity by sitting around the cafes of Bern and meditating upon the Michaelson-Morley experiment. It is an idle legend, Aronowitz asserts, a mystification designed to conceal how the social, political, and economic currents of late-nineteenth-century European capitalism and imperialism brought to birth the theory now famously associated with Einstein. Similarly, he tells us how the development of celestial mechanics by upper-class gentlemen-scientists *such as Newton* cannot be understood apart from an analysis of seventeenth-century mercantilism. Aronowitz is probably right on a couple of points—but for reasons that do him no credit whatsoever. As it happens, it is most likely that Einstein did not know of the Michaelson-Morley experiment at the time he was working on special relativity.[12] Aronowitz is childishly wrong on other points (the class origins of Newton, for instance). But these are minor matters. What is staggeringly silly is the essence of his thesis.

The history of calculus, mechanics, and relativity and the biographical details of Einstein's and Newton's lives have been studied and documented at sufficient length for us to know very well that the "Bern cafe" myth is true in its essentials, as is the story of the solitary Newton's incredible burst of creativity during the Plague Year, 1665, when he invented modern mathematics and modern physics simultaneously. This is apparently not good enough for Aronowitz; for him, it is vital that both these achievements be seen somehow to encode, in a quite literal sense, ideological and social perspectives. He never makes it clear how this is supposed to have worked. Neither his book nor his lectures set forth a coherent account of the putative transcription process.

What is needed to make a theory of this sort even weakly plausible is a demonstration of specific intellectual correspondences between the details of

the physics and the hypothetical complex of "social" or "economic" attitudes. As well, some kind of argument would have to emerge that these
correlations were at least comparable in importance to the internal logic of
physics and mathematics as it influenced Einstein's thought or Newton's. No
such case is to be found in Aronowitz's work, if only because the physics is terra
incognita to him. What takes its place is the arrogance of the dogmatist. This is
a common failing in cultural constructivist histories of scientific achievement.

The notion that such work as that of Newton and Einstein was "needed" by
the technological infrastructures of their respective societies is plain nonsense. Seventeenth-century merchants and navigators needed innovation in
the form of an accurate chronometer, not an explanation of Keplerian ellipses
in terms of the inverse square law. Turn-of-the-century industrialists were not
sending out desperate requests for a more subtle understanding of the invariance of physical law under change of inertial frame. To offer this sort of
"explanation" as an account of profound intellectual developments is to show
unlimited contempt for the very notion of explanation, as well as a boundless
ignorance of the phenomena one is trying to explain. Aronowitz's thesis is no
more than an unsupported dictum that declares, in effect, that by some
mystifying process, the Zeitgeist Fairy of 1665 contrived to tickle Newton's
brain cells with her magic wand, while her counterpart of 1905 did the same
for Einstein! This is not intellectual history, sociology, philosophy, or anything else worth a scholar's serious attention. We leave the last word on
Aronowitz's work to Michael Sprinker, a sophisticated Marxist who still
hopes to retain a place for Marxist notions in discussions of the philosophy of
science. He has the good sense to understand that this involves putting
oneself at a considerable distance from the naive and uninformed constructivism evinced in *Science as Power*. His caustic review of that book goes to the
heart of the matter. "If we are to doubt the findings of the empirical sciences,
we need to be given better reasons than that they have arisen from and been a
necessary adjunct to capitalist social relations."[13]

Explanation

In general, even when cultural constructivists make a more serious effort to
put forth an account of what is supposed to go on during this process of
cultural construction there is a strong flavoring of circularity. It is assumed *ab
initio* that cultural construction has taken place. Thereupon, the historical
and scientific record is subjected to a strained and arbitrary reading that
decodes it with the help of a great deal of interpretive contortion and her-
meneutic hootchy-koo. At last an account is produced that "explains" how

the culture has constructed the theory. This is then put forth as a confirmation of the cultural constructivist hypothesis. In form and soundness, this
procedure closely resembles the methodology through which the tenets of
psychoanalysis are "confirmed" by the interpretive prowess of the psychoanalyst.[14] In both instances, circularity and special pleading rule the day, and
little worthy evidence emerges.

In saying this, we are *not* trying to deny that social interests and nonscientific belief systems often enter into the very human business of doing creative
science, sometimes to catalyze the process, more often to retard or deflect it.
The work of Stephen J. Gould[15] (who must be recognized as holding strong
leftist views) is replete with incisive essays on examples of this, presented in
minute detail. But Gould's well-informed work is by no means comparable to
the cultural constructivist program. Gould knows perfectly well that in the
long run logic, empirical evidence, and explanatory parsimony are the masters (with apology to our feminist friends for the metaphor) in the house of
science. In this he echoes Thomas Kuhn,[16] whose work has so often been
vulgarized and distorted by the cultural constructivist school.[17]

Cultural constructivism, at least in the full-blooded version of ideologues
like Aronowitz, is a relentlessly mechanistic and reductionistic way of thinking about things. It flattens human differences, denies the substantive reality
of human idiosyncrasy, and dismisses the ability of the intellect to make
transcendent imaginative leaps, in a way that O'Brien, *1984*'s master manipu-
lator of consciousness, would cheerfully approve. According to the constructivist canon, all are puppets of the temper of an age, and science is just
another inadvertent ratification of its ideological premises. Only the cultural
constructivists themselves (of course) are licensed to escape the intellectual
tyranny of this invisible hand. For their part, mathematicians, physicists,
chemists, and biologists must all succumb.

Aronowitz represents cultural constructivism with all its philosophical and
political cards on the table, so to speak. His program is maximalist in both
respects and forthrightly asserts its prescriptive ambitions even as it makes its
sweeping descriptive judgments. Other theorists and publicists of the constructivist school are more circumspect in their claims and cagier in their
tactics. They are content, for the time being, to conduct an irregular guerrilla
war on behalf of their doctrine, while Aronowitz insists on undisguised frontal assault. Typically, in the face of all-out challenges from scientists and
philosophers armed to do intellectual battle, they edge away from the strong
version of the constructivist claim and retreat to the proper territory of
sociology or history. In the presence of a different audience, one primed to

hear science contextualized, relativized, and revealed as the deformed off-spring of capitalist hegemony, the constructivist claws come out once more.

Science as Power Struggle

In this respect we cannot avoid citing the work of Bruno Latour, a sociologist, anthropologist, and social philosopher whose work on science as social prac-tice has been as much of an inspiration to the constructivist camp as that of Thomas Kuhn. In contrast to Kuhn, however, this does not reflect any inad-vertence on Latour's part. He clearly relishes his role as self-appointed heretic and gadfly. His reputation and the substance of his claims rest on his record as an "anthropologist" of science, who does fieldwork at research facilities rather than among the denizens of New Guinea. He is not loath to let it be known that he has brought back amazing tales from his sojourn among the trog-lodytes. He claims, with no particular modesty, to be the first modern thinker to discover what scientists *actually do*, as opposed to what they *say* they do or think they do. His tools are those of the microsociologist; in his primary research he concentrates on small groups and personal interactions in which quirks, prejudices, and local hierarchies obviously play a role. Consequently, unlike theorists such as Aronowitz, who derive from a Marxist tradition that reifies such grand abstractions as "relations of production" and who impute far-reaching powers to them, Latour works in the interpersonalist tradition of Erving Goffman and his disciples, as well as that of field anthropology in the classical mode.

What provokes and titillates in his work is that he places full-fledged members of the scientific and technological elites in the "object-of-study" position usually reserved for inhabitants of the Trobriand Islands or the head-waters of the Orinoco. For Latour, the Heart of Darkness is the solid state physics laboratory. Notwithstanding the specificity and locality of his direct investigations, Latour is eager to emerge with far-reaching generalizations and epistemological laws. These are embedded in an expository style as unconventional as the theses it propounds. His major work, *Science in Action*, is studded with aphorisms, diagrams, cartoons, and doodles, and is charac-terized by a mercurial, gnomic wit; but his purpose is seriously iconoclastic. Here, for instance, is his "Third Rule of Method": "Since the settlement of a controversy is the Cause of Nature's representation, not the consequence, we can never use the outcome—Nature—to explain how and why a controversy has been settled."[18] This would seem to be an instance of unbending relativ-ism and antirealism. What it seems to say is that nature is purely a social convention, and that scientific controversies are settled by a dialogic process within a scientific community resulting in a general agreement about the details of that convention. Thus, to read this as it applies to a concrete situation, we must believe that William Harvey's view of the circulation of the blood prevailed over that of his critics not because blood flows from the heart through the arteries and returns to the heart through the veins, but because Harvey was able to construct a "representation" and wheedle a place for it among the accepted conventions of the savants! In other words, it is not to be admitted that nature might provide a template in conformity to which these "representations" are tightly molded.

A homely example will serve to clarify this point. Imagine that a few of us are cooped up in a windowless office, wondering whether or not it's raining. Opinions vary. We decide to settle the issue by stepping outside, where we note that the streets are beginning to fill up with puddles, that cars are kicking up rooster-tails of spray, that thunder and lightning fill the air, and, most significantly, that we are being pelted incessantly by drops of water falling from the sky. We retreat into the office and say to each other, "Wow, it's really coming down!" We all now agree that it's raining. Insofar as we are disciples of Latour, we can never explain our agreement on this point by the simple fact that it *is* raining. Rain, remember, is the outcome of our "settlement," not its cause! Baldly put, this seems ridiculous. Nevertheless, if we accept the validity of Latour's putative insight, we are ineluctably obliged to accept this analysis of a rainy day.

It is clear that a rather light-footed style is needed to get away with such stuff, which drives more earnest and responsible philosophers of science into paroxysms of disgust when confronted with it. (Scientists themselves, less oppressed by a professional obligation to grapple with every piece of gaudy nonsense that comes down the highway, simply go about their business.) The idea that Latour's reports on the activities of scientists *are* to be accorded factual status, while scientists' reports on nature are *not*, involves a metaphysical conceit (in both senses of the word) of astounding proportions. True, no one can object to the observation that the world of science is a human world, and that "laboratory politics" plays a significant role in it from time to time—although this is hardly a new insight. But Latour's sanctimonious insistence that such politics accounts for science as such and is the real story behind the emergence of scientific theories is in itself a signal instance of politics dictating theory.

Latour's picture of science is bleak and ominous: a war of all against all! Science is presented as a savage brawl in which, from day to day, the dominant chieftain is he who assembles, by dint of wealth, prestige, and warrior cunning, the biggest and nastiest gang of henchmen (i.e., a "network," in Latour's parlance). We must remind ourselves—with a pinch if necessary—that this process is alleged to account for the emergence of celestial mechanics, Maxwell's equations, the periodic table of the elements, plate tectonics, the genetic code, algebraic topology, quantum mechanics, massive parallel processing, and a million other insights and advances, modest as well as exalted. Empirical verification is dismissed as a species of bluster, or as a kind of collective hallucination of the power-crazed.

From the example above, it is easy to see why Latour has frequently been classified as an unreconstructed constructionist. Nonetheless, to say this is to miss an important aspect of his intellectual cunning and his seductive charm. Latour is always ready to recast and, in effect, retract what he has previously said. In other contexts he will, with an apparently straight face, admit that there is a natural universe "out there" and that scientific theories are shaped by it in important ways. Simultaneously, he will censure rigorously the dogmatics of strict cultural constructivism. Just as he pictures (literally) the mind-set of science as a Janus-faced dualist, he too is constantly springing from one side of a dichotomy to the other.

An interesting instance comes directly to hand in connection with the very example (used by Stanley Aronowitz) we evaluated above, that of the supposed parity between the tribal people who can't reason syllogistically from the proposition that "everyone is a witch" and Westerners who exhibit the same deficiency with respect to the proposition that "all who kill are murderers."[19] Latour criticizes the analysis on which Aronowitz relies, and he does so by dint of arguments that overlap, in most ways, the ones we advanced above.[20] He acquits the Westerners of illogicality on the very same grounds we cite. In fact, he goes further and acquits the "primitives" as well, asserting that it is the anthropologist's ignorance, rather than that of his subjects, that engenders the example! On this view, the ethnographic researcher has been guilty of *insufficient familiarity* with the nuances of the culture, of the implicit subtleties and unstated exceptions that his informants tacitly invoke when they discuss certain cultural categories. Thus, to some extent at least, Latour seems to be arguing that canons of logic really exist, and—though one might regard this as rather hopeful—that all sorts of people are reasonably good at adhering to them. Understandably, one might tend to view this position, of itself, as an argument contra relativism. But then, putting a characteristically paradoxical twist on the matter, Latour pushes on to the conclusion that people are *hardly ever* irrational! In particular, he argues that refusal to accept scientific argument or evidence is virtually never grounds for impeaching a position or those who hold it as "irrational." Thus his so-called Fifth Princi-

ple: "Irrationality is always an accusation made by someone building a network over someone else who stands in the way; thus, there is no Great Divide
between minds, but only shorter and longer networks; harder facts are not the
rule but the exception, since they are needed only in a very few cases to
displace others on a large scale out of their usual ways."[21] The net result, of
course, is that the most indulgent relativism is now back with a vengeance,
and adherents of faith healers, palm readers, cancer quacks, and "creation
scientists" may now go their way—courtesy of Latour and the Harvard University Press—in the full assurance that they are every bit as rational as their
scientific critics.

This example illustrates an important aspect of current intellectual life,
especially among the trendier doctrinal movements to which the academic
left has proved susceptible. Self-consistency is no longer considered to be
much of a virtue; and logical coherence, in the version that working scientists
are obliged by their peers to honor, is viewed as a chimera. One must understand that a large part of the reason for Latour's success and celebrity is
rhetorical. He provokes and challenges with his insistence on paradox and
contrarian whimsy. His reader is constantly reminded that to reject Latour's
maxims is to mark oneself as hopelessly stodgy, humorless, and tradition
bound. It is no accident that his style stands in such contrast to the single-
minded, rather ponderous linearity of the papers and monographs of the
scientists he studies. He is, despite his proclaimed fascination with science
and technology, a Panurgian imp, come to catch all those solemn scientists
with their pants down, a project that delights his largely antiscientific audience.

Questions must be raised, however, not only about the depth and accuracy
of Latour's claimed insights, but about the soundness of his own observational
technique. This is best illustrated, in our view, by lectures, dealing with his
work, sponsored by the French government, on a social analysis of the "Aramis" project. "Aramis" was the glamorous code name for an ill-fated attempt
to build a high-tech public transportation system for Paris. The basic idea
involved the construction of an elaborate network of trackage, full of switches
and cutoffs. These were to accommodate not trains but a fleet of six-passenger
self-propelled tram cars, each controlled by computer. The idea was that the
traveler, upon entering a station, was to signal his presence and his destina-
tion to the central computer. Thereupon, a nearby car with a compatible
itinerary would come by to pick him up and carry him to his target station,
picking up and dropping off other patrons along the way as convenience and
capacity dictated. Thus, Aramis was a kind of automated van service.

Attractive as the idea might be on paper, Aramis was from the start a troubled and problem-plagued project. Yet it continued under high-level funding for about a dozen years before being supplanted by a much more conventional urban railway system. Why was it so durable? For much the same reasons that any boondoggle hangs on long past the point at which its wastefulness becomes clear. It provided jobs for technical and industrial workers, research funds for high-tech consortia, power for bureaucrats, and photo-ops for politicians eager to show their commitment to a high-tech future. One would think that a highly politicized sociologist like Latour would positively salivate to sample the delights of this particular pork-barrel, and to tell us where all the money went. But he barely touches the issue; certainly not enough to embarrass anyone. By the usual muckraking standards, this is a pretty poor performance.

Why did the project fail? To us there seem to be two obvious reasons (there may be many others, of course). The first is technical. In a proposed system like Aramis, the chief head-ache will inevitably be software. Real-time algorithms must be devised for running such a system efficiently, which means minimizing station-to-station travel time for each passenger, maximizing utilization of each car, and avoiding the sort of instabilities that cause cars to bunch up in one region, leaving others bereft of service. This is a formidable undertaking! It involves all the notorious difficulties of the "traveling salesman problem," the paradigmatic holy grail of combinatorics and operations research. Compared to this, the "hardware" problems of building trackage and computer-controlled cars are trivial. How does Latour deal with this (or any other relevant *technical* question)? *He ignores it completely.*

There is, on the other hand, an important social reason for the impracticability of Aramis that the reader without any technical background will recognize immediately. Picture yourself riding, late at night and all alone, in one of these little Aramis cars. It stops at a station and two men get on, nasty-looking types with what look like ten-inch lengths of lead pipe bulging in their pockets . . . We rest our case. What has the sociologist Latour to say about such inherent problems of social interaction? The matter does not seem to have occurred to him at all.

Well, what, then, *does* interest Latour about Aramis? He seems very excited about the *semiological* aspects of the thing, the fact that "information" and "control" are such important *metaphors* for it. He is intensely aroused by the fact that the cars are to be connected by "information" but not "physically," as though he'd never heard of radio-controlled toys. He studies the evolving shape—even the color—of the prototype cars as aspects of the social representation of technology. In short, he indulges in all the ex-Gallic, jargonistic mumbo-jumbo about signification, and about social metaphor, that the devotees of cultural criticism have come to expect, without saying much of interest about the scientific or social reality. Eventually, his epistemological conceits emerge, decorated with the usual doodles and diagrams. It is hard to see what they have to do with Aramis—or any other episode in the history of science and technology—but they are very dear to Latour.

Some of the glaring gaps in Latour's analysis of the Aramis project are characteristic of his work as a whole. Mathematics is a symptomatic weak point of his. His discussion of Aramis avoids it completely, as we have seen, but, even worse, his discussion in *Science in Action* of the mathematical nature of scientific theories,[22] and the invocation of formal mathematics in order to express them, is naive and obtuse—he has a tin ear for mathematics. His account completely fails to resonate with the thought of mathematical scientists—a term that goes well beyond those formally described as mathematicians—and is deaf to how they reason with and persuade each other. The one reference to an actual piece of mathematical research manages to misunderstand an anecdote utterly.[23] The brief discussions of correlation coefficients[24] and Reynolds numbers[25] are mere occasions for sneering that completely avoid serious engagement with the deep, and enormously fruitful, concepts involved. Indeed, Latour fervently minimalizes and trivializes formalization, abstraction, and mathematization. His discussion of the matter is a series of flippancies, whose intended point is that the deep and surprising predictions about the real world that emerge from exacting logical analysis of abstract models are really no more than tautological parlor tricks.[26] Here, Latour's resentment of science seems to become overpowering. It should hardly need saying that this stubborn inability to deal accurately, comprehensively, and honestly with this central and most characteristic aspect of modern science effectively disbars the most grandiose claim of Latour's book—that it instructs the sociologically sophisticated "how to follow scientists and engineers through society."

We recall Latour's own imprecations against the anthropologist who failed to grasp the nuances of a tribal people's categories of "witchhood" and "nonwitchhood." His own evident failure, as a would-be "anthropologist of science," to grasp the categories in which scientists think and through which they judge and decide convicts him of a similar offense, and on a much larger scale. Latour's work is thus a very inadequate prop for any radical attempt to rethink scientific epistemolgy, or to indict science for unwitting relativism or perspectivism. Its appeal is almost wholly a matter of style, not of substance. It is a prime example of Feyerabend's cantankerous description.

Plutocrats

Cultural constructivist theories of science have lately infested the usually staid domain of the history of ideas. One well-known example is the work of Shapin and Schaffer, whose book *Leviathan and the Air Pump* has a wide circle of admirers. This work is rather more orthodox, on a superficial level, than Latour's. It is an intellectual history of some of the resounding disputes that surrounded the birth of "experimental" science—physics in particular—in the last half of the seventeenth century. What particularly concerns Shapin and Schaffer is the quarrel between some of the most prominent founders of the Royal Society—Boyle, Hooke, and their circle—and the philosopher Thomas Hobbes, author of *Leviathan*. This is the fulcrum upon which they attempt to push the case that, contrary to its flattering image as a uniquely wide-open and tolerant enterprise, welcoming of all new facts, information, and ideas that bear upon its investigations, modern science has been from the first the province of a tightly organized, well-insulated coterie, jealous of its prerogatives and hostile toward outsiders who intrude without the proper credentials. Moreover, this self-appointed scientific aristocracy is seen as organically connected to the ruling elite of Western society. Its views are derived, albeit subtly, from the dominant metaphors of that elite. By the same token, its prestige, authority, and epistemological monopoly are guaranteed by the power of the state and the social formations it principally serves. The argument between Hobbes and the adherents of the Royal Society is offered as an instance of this phenomenon:

> The restored regime [i.e., that of Charles II] concentrated upon means of preventing a relapse into anarchy through the discipline it attempted to exercise over the production and dissemination of knowledge. These political considerations were constituents of the evaluation of rival natural philosophical programmes [i.e., that of the Royal Society's experimentalists, as opposed to the a prioristic rationalism of Hobbes].
>
> Thus the disputes between Boyle and Hobbes became an issue of the security of certain social boundaries and the interests they expressed.[27]

The heart of the matter, as far as Shapin and Schaffer are concerned, is that the confrontation illustrates the degree to which Boyle and his friends were concerned not only with scientific issues, in the narrow sense, but also with the question of credentials. Their supposedly empirical rules, it is said, constituted a specific social practice. They were preoccupied with the question of who should count as a scientific authority, whose judgment was to be respected in scientific disputes, whose evidence was to be accepted as reliable, whose minds were to be acknowledged as sufficiently unpolluted by common prejudice that their observations could be taken at face value.

If we are to believe the Shapin-Schaffer thesis, worthiness to participate in learned discussion of experimental philosophy was closely correlated to rank, wealth, religious orthodoxy, and, in terms of Restoration doctrine, political reliability. This exclusivity was reinforced not only by the money, status, and political connections of many of the members of the Royal Society and their patrons, but in addition by their exclusive possession of the physical instruments of the new experimental method. The air pump of the title was not a common device. Only a handful existed during the 1660s, and thus the possibility of investigating experimentally the emerging theories of the weight and pressure of gases, now associated with Boyle's name, was limited to the corresponding handful of people who had access to one.

Hobbes, ever the gadfly and eager controversialist, was only too happy to point out this flaw in empiricism. The viciousness of the response to his challenge is to be explained not simply by the theoretical threat it posed to the self-assumed authority of Boyle, Hooke, Oldenburg, and the rest, but, as well, by Hobbes's dark reputation as atheist, philosophical materialist, and general subverter of the sanctity of authority. He was the natural target of distrust because of his lingering reputation as a duplicitous sycophant, willing to flatter either crypto-Catholic king or radical Protestant regicide as the opportunity of the moment suggested. Even more important was his enmity toward religious orthodoxy, and therefore toward the stability of a hierarchical society. The attempt to exile him from the realm of Natural Philosophy therefore must be seen as an act of political prophylaxis.

On this, the Shapin-Schaffer view, the nascent Royal Society was, from the first, the creature and deputy of a political and social viewpoint. The society's supposedly objective science is thus to be read, in large part, as a construction of its ideological commitments, which rejected simultaneously the republican sentiments and leveling enthusiasm of the most radical Puritans and the unconstrained absolutism of the Stuart monarchy. Shapin and Schaffer accept the idea that Hobbes was identified with *both* kinds of threat.[28] As a defender of absolutism, he could be read as the proponent of a government of unconstrained sovereign power. Yet his fierce independence, which devolved at times into a taste for rancorous disputation, was reminiscent of the intellectual licentiousness of the religious and social radicals of the Civil War period. Given this perspective, the scientific community led by Hooke and Boyle, which echoed the aspirations of a moneyed class that sought immunity from the whims of royalist autocracy while casting a suspicious eye on the tumultuous mass of the unpropertied, had no place for the likes of Thomas Hobbes.

It is not hard to transcribe this view to a contemporary context, as Shapin and Schaffer undoubtedly wish us to do. The analogies are clear. Modern orthodox science is also obsessed by "credentials" in the shape of formal training, academic degrees, and a long period of acclimation to the reigning "paradigms." It polices dissidence and safeguards its monopoly by an elaborate educational system and a forbidding insistence on "peer review." It flourishes with the connivance and support of the organized forces of wealth and authority as constituted in the state, in huge corporations, and in supposedly philanthropic foundations. It has exclusive control over the instruments of empirical investigation, some of which—like multibillion dollar particle accelerators and orbiting observatories—are far less accessible to the uninitiated than was Boyle's air pump. And it has its heretics.

Here is Shapin and Schaffer's last word on the general epistemic principle that their particular historical study is supposed to illustrate: "As we come to recognize the conventional and artificial status of our forms of knowing, we put ourselves in a position to realize that it is ourselves and not reality that is responsible for what we know."[29] So, in the end, we come back to the dichotomy—fallacious in that it posits total opposition between "reality" and "convention" where there is, in fact, intense and continuing interaction—so favored by Latour and other constructivists.

The questions raised by *Leviathan and the Air Pump* are serious and genuine. No intellectually astute history of the interplay between science and its supporting social matrix could afford to ignore them. The flaw, however, lies in attributing a deep and irretrievable source of error to what is ephemeral, local, and inconsistent in its operation. Let us examine the particular picture of seventeenth-century scientific life offered us by Shapin and Schaffer.

Were the panjandrums of the Royal Society really so rigid and intolerant in deciding who was to be accepted as a Natural Philosopher in good standing? Was it true that "the social order implicated in the rationalistic [i.e., a prioristic] production of knowledge threatened that involved in the Royal Society's experimentalism?"[30] It's hard to believe! Recall the roster of thinkers, Continental and English, who were heard with deep respect in the scientific and philosophical debates of the period: In addition to Boyle and Hooke, we have Descartes, a French Catholic; Spinoza,[31] a lapsed Jew doubtful of all religious orthodoxies; and the Royal Society's own Halley, undoubtedly an atheist. Above all, we have Newton himself—by no means a man of property, having little in the way of family connections, and a radical Protestant of fanatical intensity. Newton's singular religious views, be it recalled, prevented him from seeking ordination in the Established Church, a step he diplomatically avoided by becoming Lucasian Professor at Cambridge. New-ton was, in fact, an abjurer of the doctrine of the Trinity and thus from many points of view a *heretic*. He was a hostile, secretive, jealous recluse suffering intermittently from mental instability, an unrelenting enemy to the Stuart monarchy in its attempts to sponsor Catholic scholars at Cambridge (thereby opposing its attempts to exercise "discipline . . . over the production and dissemination of knowledge"). To top it all, he was probably homosexual.

Yet consider the celerity with which he was not only embraced but virtually deified by the English intellectual elite, once it became clear that his incom-parable mathematical skills had led him to those insights into the nature of physical reality that to this day remain staggering to comprehend. Consider, in particular, the rather touching story of the publication of the *Principia*. It will be remembered that Halley had to drag it out of Newton by main force (imagine a comparable situation involving a contemporary scientist). And Halley, a man of no wealth, put up his own money to see the work through press, taking his compensation in the form of copies, which he had to sell himself. Recall, once more, that Halley was, in fact, an *atheist*, while New-ton, on his own testimony, hated atheism above all things! Clearly, there is more to be said about rigidity and latitudinarianism, intolerance and liberty of opinion, in the seventeenth-century scientific community than that the Royal Society constituted a kind of thought police.

Consider again the question of Hobbes's banishment from the circle of the scientific elect. How accurate and complete is Shapin and Schaffer's analysis of the dispute between Hobbes and his foes in the Royal Society? From time to time, they advert to Hobbes's drawn-out fight with the Oxford mathemati-cians Ward and Wallis, as though its technical aspects were peripheral to their central thesis. They note the existence of the acrimony, and the readi-ness of the devout Wallis to bring Hobbes's ostensible irreligion into it; but they say nothing about the mathematical substance, claiming that it would carry them too far afield! But of course this is a *central* and highly illuminating question!

Hobbes, be it recalled, had little mathematical training in his youth. He took up the study of Euclidean geometry for the first time in his forties (mathematicians are often said, with some justice, to be washed up at the age of forty) and was an old man at the time of these controversies. Wallis, on the other hand, was, aside from Newton himself, the greatest English mathemati-cian of the seventeenth century. A partisan of Parliament during the Civil War (in anticipation of Alan Turing,[32] he served as code breaker for the

Puritan forces), Wallis was an ordained cleric, though of Presbyterian, rather than radical Puritan, leanings. He opposed the execution of Charles I, however, and migrated politically to a position of support for the Restoration settlement. Politics and theology aside, Wallis was a superb, creative mathematician, in contrast to Hobbes, who was, to put it bluntly, incompetent—utterly out of his depth in dealing with subtle mathematical matters.

The controversy between Hobbes and Boyle on the weight and pressure of air must be viewed to a considerable extent as an episode in the twenty-year wrangle between Hobbes and Wallis. It was Wallis who published the most pointed rejoinders to Hobbes, not Boyle himself. The animus between the old philosopher and the Oxford mathematician had arisen from Hobbes's futile criticism of Wallis's mathematics—particularly his great work on infinite series—which antedates the "experimental philosophy" dispute by a number of years. Subsequent to the attack on Boyle's physics, Hobbes once again turned his guns on Wallis's mathematics. But the most revealing, as well as the most comical, quarrel arose when Hobbes published his incorrect solutions to the ancient problems of "squaring the circle" and "duplicating the cube."[33] Wallis, of course, demolished the poor old philosopher's pretensions, and Hobbes compounded the sin, in the eyes of posterity, by being unable (or unwilling) to see the point of Wallis's refutation.

The relevance of these facts to the Shapin-Schaffer hypothesis is that this long and (to Hobbes's admirers) lamentable history provides a concrete and substantive reason, *in contrast to an ideological one*, for Hobbes's notoriety in scientific circles. So far as mathematics is concerned, Hobbes was simply dead wrong in these exchanges, as any competent mathematician would have seen. It is then no wonder that his authority to pass judgment on scientific matters was not well regarded, even if those matters had nothing directly to do with squaring the circle or the like. He was, after all, a strenuous advocate of a rational-deductive methodology *based on that of synthetic geometry*, as an alternative to the emerging experimental empiricism. Shapin and Schaffer emphasize this fact, but unaccountably fail to link it to the question of Hobbes's doubtful mathematical competence.[34] His grotesque failures as a would-be geometer, however, can hardly have been irrelevant.

Leviathan and the Air Pump would have been a rather different book had it addressed these matters directly. The image of Hobbes as brilliant and devastating iconoclast would have taken some hits, at the least. Moreover, Shapin and Schaffer would have put themselves in the position of conceding the existence of sound, objective reasons for deciding at least some scientific controversies—that between Hobbes and Wallis being an important case in point. Inevitably, they would have been led to concede that there are reasonably valid criteria for deciding the scientific competence of individuals, for distinguishing, in most instances, between worthwhile theorists and cranks. After all, in terms of mathematics, Hobbes was a crank. Such concessions, however, do not sort well with a relativist or conventionalist position, especially one grounded on a radically antielitist politics. Shapin and Schaffer sidestep the issues that might entail such admissions by insisting that all such disputes are ideological.[35]

Leviathan and the Air Pump is exhaustively and meticulously researched as a narrative of events and personalities during a short span of time. Nonetheless, the ideological perspectives of its authors make it an exercise in tunnel vision. To concentrate on the idea of empirical science as a manifestation of cultural and political imperatives is to omit important dimensions of the story, both human and philosophical. The effort of Boyle and his colleagues to put science on a solid experimental footing and to restrain the impulse toward a priori speculative systems was a project facing substantial practical difficulties at that stage. It is one thing to embrace "empiricism" in the abstract, quite another to find practical and reliable methods for developing and extending concrete knowledge. The early experimental philosophers were confronted with the necessity to minimize the effects of human fallibility and bias, and it is shortsighted to condemn them out of hand for addressing the difficulties in language that occasionally smacks of snobbery or political insecurity. The verdict of history must be that they succeeded magnificently in sketching the broad methodological outline by which the physical and biological sciences have attained their present scope and power. To put it another way, irrespective of the "social" grounding of their ideas, what more could they possibly have done, short of inventing the theory of experimental design and developing the techniques of mathematical statistics and error theory that underlie it?

Furthermore, it is false to read their rejection of Hobbes as a blanket denial of the value of speculative and deductive thought. Such reasoning was eagerly received when it was the product of genuine intellectual competence, as in the case of Huygens and, of course, of Newton himself. The singular genius of the period was to exploit new and powerful mathematical reasoning in the service of physical science *without* falling into the trap of contempt for mere experience. The authors concede that they have not quite come to understand how the experimentalism of Boyle was made to dovetail with the mathematical science of Newton; but this may well be because they have been celebrating the wrong hero. Hobbes, the mathematical dilettante and bumbler, simply does not belong in the same pantheon with Descartes, Huygens,

Newton, Leibniz, and Bernoulli. His misadventures are tiresome and, in the last analysis, uninstructive.

A final word about the rhetoric of the book: Once more, we find an argument designed to appeal to a certain kind of readership on grounds other than strict logic and evidence. To side with Boyle and the Royal Society crowd, as the book presents them, is to side with snobbish, purse-proud, rank-conscious plutocrats in their fear of the disorderly masses. If Hobbes cannot be construed as radical democrat (as indeed he cannot—his motivations are markedly authoritarian), then at least he can be made to stand for the voice-less and excluded masses—and the intellectuals without serious scientific training—to whom science is an inaccessible mystery, seemingly beyond human control. Thus we are forced in our reading of the book to see it as a parable, whose fulsome celebration of Hobbes conveys the implication that "philosophers" who are not professional scientists (for which we must read "historians" and "sociologists") should have the authority to pronounce, or even to prescribe, on scientific questions. As we have observed before, this kind of stacking of the emotional deck has great persuasive force on the academic left, irrespective of the soundness of the argument that encodes it.

Cultural Constructivism as a Political Code

We close by once more denying any covert program on our part to exclude historians, sociologists, or even anthropologists from the study of science and technology as social phenomena, whether on the grand scale or on the level of interpersonal relations. We cheerfully allow that such work might well incorporate a passionately held leftist (or, for that matter, rightist) point of view. The fact that much of the best of it comes from natural scientists such as Stephen Gould does not inevitably put similar effort by social scientists in the shade. Historians with a left-wing perspective, such as Marc Bloch[36] and Gar Alperovitz,[37] to take but two vastly different examples, have certainly made their respectable mark. But a political point of view is one thing; the pursuit of philosophical phantoms in order to give leverage to doubtful ideological claims is quite another. We insist on making the distinction. The central ambition of the cultural constructivist program—to explain the deepest and most enduring insights of science as a corollary of social assumptions and ideological agenda—is futile and perverse. The chances are excellent, how-ever, that one can account for the intellectual phenomenon of cultural con-structivism *itself* in precisely such terms.

The doctrine, whether nakedly asserted without much attention to historic and scientific detail (as in Aronowitz), or built on a minute, but overly restrictive, examination of the social and historical record (as in Shapin and Schaffer), is clearly designed to flatter a certain political perspective, and to assert the sovereignty of a certain kind of political alertness over the domain of history and philosophy of science. Even apart from "ideology," in the narrow sense, it functions politically (as universities understand these things) to redress the grievances of the social scientists, and to elevate their knowledge claims to the level historically enjoyed by physicists and chemists. Thus, the insights of the constructivists are ripe to be turned against them. One must scrutinize their precepts and their practices for signs that their theories are "value-laden" to a considerable, perhaps an unacceptable, degree. The evidence is there, we submit, and reveals far more about the nature of the constructivist program than that program has ever revealed about the nature of science.

CHAPTER FOUR

The Realm of Idle Phrases: Postmodernism, Literary Theory, and Cultural Criticism

If you're anxious for to shine in the high aesthetic line
As a man of culture rare,
You must get up all the germs of the transcendental terms
And plant them everywhere.
You must lie upon the daisies and discourse in novel phrases
Of your complicated state of mind,
The meaning doesn't matter if it's only idle chatter
Of a transcendental kind.

W. S. GILBERT, BUNTHORNE'S SONG FROM *PATIENCE*

The Ascent of Postmodernism

Future historians, composing a chronicle of the life of the mind in the United States during the period 1975–90, may well feel obliged to pay close attention to the role of academic humanists and social scientists. Assuming that they do, they will have to contend with the curious phenomenon of postmodernism, a stance that has inflected the thinking of hosts of scholars in these areas. Postmodernism flourishes chiefly in departments of English, comparative literature, art history, and the like; but anyone familiar with contemporary American universities is well aware of how far it has spread into such unlikely areas as sociology, history, political science, anthropology, and philosophy.

To give a concise statement of postmodern doctrine would be an almost impossible task. It is too variegated and shifty to allow easy categorization, and too willfully intent on avoiding definitional precision. There is even a risk of misleading in calling it a body of ideas, for postmodernism is more a matter of attitude and emotional tonality than of rigorous axiomatics. Nonetheless, as critics of postmodernism in one of its currently most vigorous forms—science criticism—we owe the reader some sense of how we understand the term.

Perhaps the easiest entry into this body of ideas (and prejudices) is to understand it as a negation—particularly as the negation of themes that have reigned in liberal intellectual life of the West since the Enlightenment. If we accept the notion that there is a generalized intellectual "project" of the Enlightenment, one that is intent upon building a sound body of knowledge about the world the human race confronts, then postmodernism defines itself, in large measure, as the antithetical doctrine: that such a project is inherently futile, self-deceptive, and worst of all, *oppressive*.

Contrasted to the Enlightenment ideal of a unified epistemology that discovers the foundational truths of physical and biological phenomena and unites them with an accurate understanding of humanity in its psychological, social, political, and aesthetic aspects, postmodern skepticism rejects the possibility of enduring universal knowledge in any area. It holds that all knowledge is local, or "situated," the product of interaction of a social class, rigidly circumscribed by its interests and prejudices, with the historical conditions of its existence. There is no knowledge, then; there are merely stories, "narratives," devised to satisfy the human need to make some sense of the world. In so doing, they track in unacknowledged ways the interests, prejudices, and conceits of their devisers. On this view, all knowledge projects are, like war, politics by other means.

It is fascinating that postmodernism, a point of view that must flirt continuously with nihilism, has become so conspicuously identified with radical scholarship and campus political activism on behalf of left-wing causes. As much as anything can be, postmodernism is the unifying doctrine of the academic left, having largely supplanted Marxism, except to the extent that the latter has been able to cover itself in postmodern dress. In a new and highly politicized area such as women's studies, for instance, virtually every scholar and student pays tribute to the supposed depth of postmodernist insight and the richness of postmodernist methodology.

The realm of cultural studies, only a few years old, and yet the virtual center of current left-wing theorizing, is to all intents the institutional embodiment of postmodernism. This is not to say that all feminists or all left-academic cultural critics are committed to the most caustic forms of postmodern skepticism. Some such scholars are at pains to make clear their doubts and reservations. Nevertheless, and however reluctantly, they almost inevitably find themselves aping the language and style of postmodernist prototypes, and drawing upon the manifestos of noted postmodernist thinkers, to lend authority to their own musings. As historian John Patrick Diggins notes in his comprehensive chronicle of the American left, "Entering the academic world, New Leftists would find in various poststructuralist theories ready-made answers to their defeat and disillusionment."[1] For "poststructuralist" here, read "postmodernist" (see Chap. 1, n. 2).

There is a paradox in all this. In scorning the Enlightenment, the postmodern left is clearly cutting away the roots, emotional as well as intellectual, that formed and sustained its most deeply held egalitarian ideals. In embracing the brittle skepticism of postmodern thought, would-be leftists are never more than an inch away from passivity, ineffectuality, and cynical despair. A criticism frequently advanced by opponents of postmodernism—justifiably, in our view—is that the doctrine, at its most virulent, is hardly distinguishable from the moral blankness, the *Viva la muerte!*, upon which fascism was erected in the first half of this century.

Yet the seductions of the postmodern stance are also obvious. In an earlier day, Marxism, in the form of a disciplined Communist movement, lured intellectuals by offering them the illusion of membership in a priesthood, an inner circle of initiates privileged to understand, by means of esoteric doctrine, the secret inner workings of the world, a coven of hierophants signaling to each other in an arcane jargon impenetrable to outsiders. It was the promise of numinous power, inherent in arcane doctrine and obscure lexicon, that convinced instinctive radicals that Marxist communism alone had the potential to purge the world of its indwelling evils. The melancholy chronicle of Communism in America, and its horror-laden history in those parts of the world where it has at one time or another actually held power, have by now demolished its intellectual prestige beyond hope of resurrection.

Nevertheless, if we examine the popularity of postmodernism with a view to understanding its appeal to the politically discontented, we see that psychological factors are at work echoing those that lured previous generations to Marxism-Leninism. Again, what is offered is the possibility of becoming an initiate, part of an elect whose mastery of a certain style of discourse confers an insight unobtainable elsewhere and authorizes a knowing (and often smug) attitude. The promise of power to mold the world in accordance with one's sense of justice is far more qualified and ambiguous than was the case with hard-line Marxism. All the same, that promise, however muted, is still there.

Resistance, subversion, and *transgression* are among the most popular postmodernist nouns, and the sense in which they are used clearly conveys the idea that bourgeois society, founded on racism, sexism, and the enforcement of rigid social roles, is under attack, its vulnerabilities being exposed. Moreover, the peculiarly quixotic view of the antagonism between "representation" and "reality" that is so thematic in postmodernist thought vouchsafes its practitioners an eerie absolution from having to measure their theories against the unyielding matrix of social fact. If one holds, as most postmodernists do, that "reality" is chimerical or at best inaccessible to human cognition, and that all human awareness is a creature and a prisoner of the language games that encode it, then it is a short step to the belief that mastery over words, over terminology and lexicon, is mastery over the world. As Diggins says, "to the extent that the Academic Left partook of various structuralist theories, reality eluded its vocabulary. Such terms as 'power and hegemony' and 'domination and discourse' marked a shift from labor to language in which text, speech, and other forms of communication came to be seen as more refined systems of control, with power ubiquitous and anonymous."[2] In the cold light of day, such a creed seems pathetic as well as futile, a desperate amalgam of solipsism and magical thinking. But the world of postmodern thought is well provided with devices for keeping out the cold light of day.

The idea that close attention to the words, tropes, and rhetorical postures of a culture gives one transmutative power over that culture finds acceptance for a number of reasons. First of all, it shifts the game of politics to the home turf of those who by inclination and training are clever with words, disposed to read texts with minute attention and to attend to the higher-order resonances of language. At the same time, it allows scholars of a certain stamp to construe the pursuit of their most arcane interests as a defiantly *political* act against the repressive strictures of society. This is exhilarating: it is radicalism without risk. It does not endanger careers but rather advances them. It is a radicalism that university administrators and even boards of governors have found easy to tolerate, since its calls to arms generally result in nothing more menacing than aphorisms lodged in obscure periodicals. It is, finally, a politics upon which the wear-and-tear of ordinary political life can have little effect. If something bad happens, one's doctrine is confirmed: if something good happens, it is vindicated.

One startling aspect of postmodernist thought is its belief in its own omnicompetence. It pronounces with supreme confidence on all aspects of human history, politics, and culture. If there is a prototype of postmodernism, a previous thinker whose sweep and ambition are mirrored in its swagger and whose corrosiveness is echoed in its skepticism, it is probably Nietzsche. Whatever one thinks of Nietzsche as philosopher and cultural critic, he is obviously a talismanic figure. From his nominal base in an obscure and hermetic discipline (classical philology), he reaches out with searing criticism of society and its follies, strips away its pretenses, and flays its complacency. Contemporary postmodern critics, themselves situated, by virtue of the extreme specialization that prevails in the training of academics, in scholarly pigeonholes, consider themselves similarly called upon to be philosopher-kings. Postmodernism is, among other things, a device for amplifying the special insights of a narrow area of literary criticism or rhetorical analysis into a methodology for making judgments of the entire cultural spectrum.

Necessarily, this entails considerable intellectual coarseness. The confidence of the postmodern cultural critic is the confidence of a generalizer who excuses himself from many of the usual obligations of erudition. Under this dispensation, a wide variety of disciplines may be addressed and pronounced upon without requiring a detailed familiarity with the facts and logic around which they are organized. A recent article by Heather MacDonald wryly analyzes this phenomenon, which, in its most impudent form, generates scholarly essays that seem to have as their subject everything in general and nothing in particular, and which, under the postmodernist regime, are equally suitable for symposia in literature, history, sociology, or feminist theory. She writes specifically about a recent forum devoted nominally to the history and analysis of twentieth-century art, many of whose participants turned out to have no particular knowledge thereof. This is not an anomaly—it comes closer to being characteristic of scholarly life among contemporary humanists. Notes MacDonald:

> Concurrently with its internal colonization of academic disciplines, [postmodern] Theor-ese broke down the institutional barriers between them. The growth of "interdisciplinary studies" in the university and the fascination in the non-academic creative world with "crossover" work are manifestations of the universalizing drive of Theor-ese. Its final triumph lies in the establishment of entire academic departments devoted solely to itself—"Departments of Critical Theory," "Units for Criticism and Interpretive Theory," and misleadingly-titled "Humanities Centers."[3]

American postmodernism is often accused, with considerable justice, of being little more than mimicry of a few European thinkers, mostly French, who rose to prominence in the midst of the bewilderment afflicting intellectual life when the protorevolutionary struggles in the late sixties in France, Germany, and Italy fizzled out without having produced any real impact on bourgeois society. The most recurrent and inevitable names in postmodernist circles are those of two French philosophers, Jacques Derrida and Michel Foucault. Derrida, founder of the deconstructionist school of textual analysis, has by example fostered many of the stylistic affectations that bespangle modern critical writing—puns, coinages, words made ambiguous by internal parentheses and other whimsical punctuation, facing columns of apparently unrelated text which, to the initiate, are supposed to comment on one another. This kind of writing, as much as anything else, has been responsible for the ambiguous reputation of deconstruction and related critical methods.

Derrida's deep epistemological pessimism has infected his disciples as much as have his stylistic eccentricities. Deconstructionism holds that truly meaningful utterance is impossible, that language is ultimately impotent, as are the mental operations conditioned by linguistic habit. The verbal means by which we seek to represent the world are incapable, it is said, of doing any such thing. Strings of words, whether on the page or in our heads, have at best a shadowy and unstable relation to reality. In fact "reality" is itself a mere construct, the persistent but illusory remnant of the Western metaphysical tradition. There is no reality outside the text, but texts themselves are vertiginously unstable, inherently self-contradictory and self-canceling.

On the face of it, this position would seem to offer little cheer to the would-be revolutionary or radical reformer. In the peculiarly constricted world of leftist literary intellectuals, however, it has come to be read as a road map for the continuation of a political struggle that seemed, by the late seventies, to have run out of steam. "Yet deconstruction had enormous value to New Left literary academics, explains Diggins. "Having lost the confrontation on the streets in the sixties, they could later, as English professors in the eighties, continue it in the classroom. A new nemesis haunted the Left. Everything wrong with modern society would be explained no longer by the mode of production but by the mode of discourse."[4]

Derrida and deconstruction (in the strict sense) have seen their prestige erode somewhat in the past few years. This is *not* due, for the most part, to philosophical or political rebuttal of Derrida's ideas (although, in contrast to literary critics, few serious philosophers have had much use for them). Rather, the reason is the adventitious exposure of two figures closely associated with Derrida—one as disciple, the other as philosophical forebear—as having behaved abominably during the heyday of Nazism. Derrida's chief American follower, Paul de Man of Yale University, was posthumously disgraced by the revelation of his pro-Nazi writings as a literary journalist in occupied Belgium. This, moreover, turned out to be just one episode in a life filled with dissimulation, opportunism, and betrayal.[5] At about the same time, new facts came to light concerning the enthusiasms of the influential philosopher Martin Heidegger for Nazi doctrine, enthusiasms that now appear to have been heartfelt, and that led Heidegger, as rector of his university during the thirties, to perpetrate unforgivable acts of repression.[6] Since Derrida had always claimed derivation of his thought from Heidegger, his own credibility as a liberatory thinker came under challenge.

Derrida's declining prestige was not, however, merely a matter of guilt by association. In trying to defend de Man and Heidegger, Derrida and those closest to him sent forth a stream of polemic and vituperation that stupefied many of its readers by its unreason and its resort to special pleading. What is more, Derrida fell into the ironic position of insisting that texts, *especially his own*, have quite determinate meanings which he, as author, was uniquely privileged to understand, and that history and facts were on his side. Thus, at a crucial point, the panic-stricken deconstructionists ran headlong from the implications of their own doctrine, which had loudly proclaimed the "death of the author" and had despised appeals to historical fact. [7]

If we turn to Michel Foucault, we find a more sympathetic, but still disturbing, figure. Foucault was, primarily, a philosopher of history, whose thinking led him to ever-deeper and more pessimistic considerations of the role of language and discourse in constructing the conditions of human existence. To Foucault, life is built around language, but language itself is not neutral. Rather, it is structured and inflected by the relations of power and domination in a society. In fact, language itself *creates* power and social authority. We are irremediably trapped in a linguistic web that determines not only what we can say but what we can conceive. All systems of thought, then, are artifacts of the prison-house of language and thus stand in a questionable relation to the real world.

Like Derrida, Foucault is a thinker whose appeal has been mainly to social theorists and literary intellectuals, rather than to philosophers, who are less swayed by the emotive aspects of his writing, and who tend to regard it as a kind of poetry, rather than as philosophy proper or sound history. For one thing, Foucault's epistemological relativism arises from a study of the presumably exact facts of social history, which his best work examines minutely. [8] Thus, despite himself, Foucault is ultimately tied to the postulate of a real world, definitely knowable in at least some of its aspects. Moreover, his reputation, too, has been attenuated of late, perhaps unfairly, by revelations of his deeply neurotic and self-despising personal life which, one cannot help feeling, dictated the tone of his speculations, as well as giving them their peculiar emotional force.

Notwithstanding these reversals, the influence of Derrida and deconstructionism, of Foucault and his ideas of consciousness and domination, remains strong. One particular aspect of their style that continues to command imitators is their assurance that they are capable of profound insight into anything and everything. This style of philosophizing had been in eclipse during most of the twentieth century, abandoned in favor of a technical mode of analysis that focuses with precise intensity on narrow questions and fine distinctions. But with Derrida and Foucault, among others, we see the rebirth of the philosopher as comprehensive sage.

The Conscription of Science as Metaphor

Science, arguably the dominant mode of thought in the contemporary world, has thus come under the scrutiny of Foucault, Derrida, and their followers. In the case of Foucault, skepticism is expressed in the form of doubts about the human *importance* of scientific truth, rather than on the possibility of achieving it. Nonetheless, his basic idea, that a mode of discourse is inevitably a code of power relations among the people who use it, has profoundly influenced other postmodern skeptics and has contributed importantly to the notion that science is simply a cultural construct which, in both form and content, and independently of any individual scientist's wish, is deeply inscribed with assumptions about domination, mastery, and authority.

For their part, Derrida and his epigones take a curiously ambivalent position toward science. On the one hand, scientific texts enjoy no special dispensation from the deconstructionist view of textuality. They are, it is asserted, just as indeterminate, as ultimately self-contradictory as any other text. The "privileged" status of scientific discourse is yet another illusion deriving from the conceits of Western metaphysics, and must therefore be rejected. Moreover, it has been put forth seriously that literary scholars trained in deconstruction or some related methodology are capable of a "deep reading" of scientific texts, a reading that reveals aspects of meaning and unconscious intent invisible to the scientists themselves. Later in this chapter we examine what some of these claims amount to.

On the other hand, deconstructionists, as well as other postmodern thinkers, have been eager to point out how modern science supposedly generates insights that confirm their own view of the universe. Kurt Gödel's celebrated incompleteness theorem is a constant point of reference. The argument is that this deep and startling result, which shows that no finite system of axioms can completely characterize even a seemingly "natural" mathematical object (that is, the set of whole numbers and its familiar arithmetic), can be taken to imply, in some sense, that "language is indeterminate." Mathematicians and logicians are dubious about such vague analogies, but many literary scholars are deeply impressed by them and recur to this particular example in paper after paper, even though it is doubtful that very many of them have any exact idea of what Gödel's result says, or any sense of how it is proved. They have fallen into the trap described by George Steiner, who understands, as few of the new "cultural critics" seem to do, that deep scientific ideas must be comprehended, first of all, on their own terms: "Having no mathematics, or very little, the common reader is excluded [from science]. If he tries to penetrate the meaning of a scientific argument, he will probably get it muddled, or misconstrue metaphor to signify the actual process."[9]

A further sense of Derrida's eagerness to claim familiarity with deep scientific matters can be obtained from the following quotation, which also gives one some sense of how seriously to take such claims: "The Einsteinian constant is not a constant, not a center. It is the very concept of variability—it is, finally, the concept of the game. In other words, it is not the concept of some thing—of a center from which an observer could master the field—but the very concept of the game."[10] The "Einsteinian constant" is, of course, c, the speed of light *in vacuo*, roughly 300 million meters per second. Physicists, we can say with confidence, are not likely to be impressed by such verbiage, and are hardly apt to revise their thinking about the constancy of c. Rather, it is probable that they will develop a certain disdain for scholars, however eminent, who talk this way, and a corresponding disdain for other scholars who propose to take such stuff seriously. Fortunately for Derrida, few scientists trouble to read him, while those academics who do are, for the most part, so poorly versed in science that they have a hard time telling the real thing from sheer bluff.

This is not, we assure the reader, an isolated case. In various other Derridean writings there are to be found, for example, portentous references to mathematical terms such as "differential topology,"[11] used without definition and without any contextual justification. Clearly, the intention is to assure readers who recognize vaguely that the language derives from contemporary science that Derrida is very much at home with its mysteries. An even more egregious and unambiguous example of the same sort of pretentiousness occurs in a piece by a young scholar writing in the important postmodern journal *October*: "The discourses of philosophy, linguistics, and sociology must be supplemented in a truly psychoanalytic account of AIDS by concepts drawn from the discourse of mathematics, principally post-Euclidean geometry, which provides for topological mappings based on a non-Euclidean concept of space."[12] Scientists who are genuinely familiar with the terminology invoked by declarations of this sort have no choice but to regard the whole business as a species of con game.

This kind of pretentiousness is not limited to Derrida and his clones. It seems to have become a habit with many postmodern thinkers. Jean Baudrillard, for example, tells us that "there is no topology more beautiful than

Moebius' to designate the contiguity of the close and the distant, of interior and exterior, of object and subject in the same spiral where the screen of our computers and the mental screen of our own brain become intertwined with each other as well."[13] This is as pompous as it is meaningless; but it is well contrived to impress readers whose knowledge of mathematics is superficial or nonexistent. Jean-Francois Lyotard is another celebrated postmodernist thinker—he is chiefly responsible for the popularity of the term—who has pontificated about science at great length. Lyotard lets us know that "the games of scientific language become the games of the rich in which whoever is wealthiest has the best chance of being right."[14] Not all Lyotard's readers, even among nonscientists, are eager to accept his scientific authority, however. In his book *A Blessed Rage for Order: Deconstruction, Evolution, and Chaos*, Alexander J. Argyros takes note of Lyotard's propensity to play a similar game:

> Lyotard's postmodernism is not to be understood as ideological or theoretical fiat, but, we are led to believe, if only by implication, as the consequence of new developments in the natural and mathematical sciences. Therefore, Lyotard enlists such allies as Gödel, Thom and Mandelbrot in his campaign to reduce ethics to paralogy . . . I think Lyotard's appropriation of mathematics and science is biased and tendentious in general.[15]

Even if we stick to mathematics alone, it is not hard to find other examples of postmodern thinkers whose urge to pontificate on science far outruns their competence to do so. The recent compendium ZONE 6—*Incorporations*[16] is replete with examples. This is a volume of meditations on science, technology, and culture by a throng of well-known postmodernists. In perusing it, we find papers by Gilles Deleuze, Gilbert Simondon, Peter Eisenman, Alluquère Roseanne Stone, Frederick Turner, and Manuel de Landa, wherein they attempt to make references to deep mathematics—some of them at length, others just in passing.[17] In each case, there are amateurish errors or efforts to pass off mere verbal tinsel as mathematical knowledge. Biology is similarly ill served.

We do not claim that all of the named writers are hostile to science: many of them, indeed, profess to admire it greatly. Moreover, a number of them are at best equivocally "postmodern," under the meaning of the term as we have defined it. Nonetheless, whether deliberately or by inadvertence, they help to set the stage for a kind of hostile science-criticism which assumes that a grounding in the postmodern critical style, with its formulary, its litany, its rhetorical gimmicks, provides by itself sufficient intellectual leverage for insight into the workings of science, for criticism of it, and avoids the necessity of actually learning it.

How much has science itself been affected by these goings-on among the humanists? To this point, in the "hard" sciences—mathematics, physics, chemistry, and most of biology—the effects have been minimal or indiscernible. The same holds for applied science and engineering. Despite sweeping postmodernist claims of "paradigm shifts" and radical breaks in the reigning episteme, scientific practice in the more rigorous disciplines goes on as usual, driven for the most part by the internal logic of the subject and the unyielding contours of reality. The alarums and excursions that have shaken the halls of English and comparative literature departments have reached scientists, even those strictly within the academy, only as vague and amusing rumors.

In the social sciences, however, the effects have been drastic. The notion of "cultural critic," in its postmodern form, embraces a certain kind of sociologist as well as a certain kind of literary scholar. They publish in the same journals and appear at the same symposia, speaking the same language and sharing the same attitudes. According to the eminent cultural anthropologist Robin Fox, his own discipline has been permeated by jargon, philosophical dogma, and political attitudes drawn from the world of postmodern literary criticism:

> English literature departments are reconstituting themselves as cultural studies departments and are trying to take over the intellectual world. It's a heady time for them and a scary time for science . . . My own interpretation is that lazy minds are happiest with the mere voicing of opinion, or with the easy task of dressing this up to make it look plausible. In modern literary criticism they have found the perfect model of this, along with a new doctrine of extreme relativism that says that everything is only opinion anyway, to justify it. Thus the otherwise odd vision of thousands of social science children cavorting after the Pied Piper of Lit. Crit. and discourse analysis.[18]

Cultural anthropologists, Fox reasons, were particularly susceptible to this invasion because "it makes a good excuse to dodge the rigors of science—the demand for verification and falsification—and promotes the relativism with which the social sciences have always sympathized."[19] Moreover, those whose politics inclined toward the left were all too happy to have a rationale for reconstituting their discipline as part of a social movement to champion the oppressed races, castes, genders, and sexual outcasts of the earth, freed of any need to analyze their situation "objectively." In Fox's view, however, many of the peoples whom this strategy is designed to help are, in the end, poorly served: "Science, with its objectivity . . . remains the one international language capable of providing objective knowledge of the world. And it is a language that all can use and share and learn . . . The wretched of the earth want science and the benefits of science. To deny them this is another kind of racism."[20] It is difficult to judge whether Fox overstates the extent of the damage—or understates it. The necessary census has not been taken. There certainly has been damage, and plenty of it. We hope that we shall not be obliged to compose a similar lament for polymer chemistry or biophysics in the near future.

The Political Temptations of "Theory"

Professors in the humanities are not, by and large, any more feeble-minded than the general run of humanity, nor are they particularly feckless in the affairs of day-to-day life. Moreover, despite the hopes of readers of the *National Review* or the *American Spectator*, left-wing political opinions are not especially inconsistent with high intelligence either, nor do they lead to a generalized susceptibility to muddled thinking. Why, then, has so large a proportion of the left-wing professoriate in literature and adjacent disciplines been so ripe for seduction by the potpourri of views—deconstructionist, Foucauldian, and otherwise—traveling under the catchall term *postmodernism?* Deconstructionism in its pure form would seem to be an unlikely candidate for such popularity. It is a uniquely disenchanted and crepuscular philosophy, carrying the reek of a decadent mandarinate that has seen everything too often. To toy with ideas in such an idle and self-vitiating fashion would seem to confess a lack of interest in bringing about salutary change in human affairs. For its part, Foucauldian analysis, despite the tender-heartedness of some of its instincts, seems equally to lead to resignation and quietism. If consciousness is such a prisoner of power—and Foucault seems much more gloomy than Marx in this respect—then hopes for a break with the oppressive past must be futile indeed. Notes Alan Ryan, Princeton professor of politics:

> It is, for instance, pretty suicidal for embattled minorities to embrace Michel Foucault, let alone Jacques Derrida. The minority view was always that power could be undermined by truth . . . Once you read Foucault as saying that truth is simply an effect of power, you've had it . . . But American departments of literature, history, and sociology contain large numbers of self-described leftists who have confused radical doubts about objectivity with political radicalism, and are in a mess.[21]

The well-known Marxist literary critic Frank Lentricchia voiced similar doubts with respect to Paul de Man (this before the revelation of de Man's shabby political history): "De Man, unlike Schiller or Wordsworth, has no desire to employ the literary in the redemptive work of social change . . . his talk of 'critical crisis' is academic in the most debilitated sense of the word; it can only interest professors of literary theory."[22] Though less disdainful of Foucault, Lentricchia is in the end disenchanted with that thinker's beatification: "Foucault's theory of power, because it gives power to anyone, everywhere, provides for a means of resistance but no real goal for resistance . . . In this version, the economic version of exploitation seems insignificant."[23] An even more sweeping uneasiness is expressed by Bogdan Denitch, influential political scientist and co-chairman of the Democratic Socialists of America: "Politics of identity and mechanical imports of French intellectual fashions have trivialized and decentered attempts to build genuinely broad coalitions that could provide an arena for a resurgent left."[24]

Even those among leftist intellectuals who have in part accepted the stance or methodology characteristic of postmodernism are left with a degree of unease. Alexander J. Argyros, in the statement of purpose that begins his book, flatly asserts: "Since it is essentially a negative methodology, when deconstruction is called upon to address concrete issues, such as political ones, its penchant for eliding commitment and its resistance to postulating scales of value render it ineffectual at best and reactionary at worst."[25] In the important theoretical journal New Left Review, Elizabeth Wilson, defending herself against a charge of abandoning the rationalistic legacy of the Enlightenment, writes: "As someone who still finds Marxism highly relevant in the present world I absolutely reject, however, any attempt to align me with, the likes of Rorty, et al."[26] And, in a similar vein, the radical-feminist philosopher Kate Soper proposes exploiting postmodern skepticism without being overwhelmed by it, and advocates combining "alertness to the deficiencies and crudities of much traditional value-discourse with alertness to the self-defeating quality of the attempt to avoid all principled positions in theory."[27] As will be evident from a day's skimming the unbound periodicals in any university library, such cautionary voices are in general not being heard. The impulse to embrace postmodernism and to adopt the velleities of its chief figures in one's polemical language has run powerfully among left-wing intellectuals whose academic anchor is in the humanities.

A curious apology for this infatuation can be found in a paper of Kate Ellis, a radical-feminist literary scholar.[28] Her argument for embracing deconstruction, or some partially non-Derridean variant of it, is, roughly, that in instructing (or rather, indoctrinating) her women's literature students in the virtues of a radical feminist critique of society, she finds herself obstructed. Her problem is the tendency of those students to construe situations in fiction, and presumably in life, with reference to a narrative model emphasizing redemption (in the bourgeois-liberal mode), self-realization, and autonomy, by seeing those as the outcome of a sufficiently strong will and the ability to make the right choices. In Ellis's view, feminism requires a more strongly *destabilizing* view of things, which deconstruction fosters: "It means that no one person or group has the power of totally constitutive speech, and that no subject position can guarantee the truth of the speaker."[29]

As is usual with rationales for deconstruction, the linear logic of Ellis's argument is an implicit rejection of the very position it argues for. Beneath that there lurks a still more curious paradox, for upon analysis, Ellis's rhetoric reveals an underlying cast of mind flagrantly inconsistent with the cool pose of deconstruction. What is undeniable is her strict and unassailable moralism, as steadfast as that of any Sunday-school teacher. For Ellis, gender oppression and class oppression are *absolute* evils; all her theoretical moves are made with the intent of abolishing them. Whatever persuasive force can be found in her piece derives from her appeal to these values, whose epistemological standing for her is, of course, beyond question. They are so much a part of her that she is hardly conscious of them; she would scarcely allow them to be regarded as a mere casual consequence of her "position in the discursive field" or some such. The odor of unassailable rectitude that pervades what is supposed to be a case on behalf of untrammeled relativism is what makes this essay a little ridiculous—and just a bit admirable. Such emphatic ethical commitment, when all is said and done, puts to flight the formal skepticism being conscripted on its behalf.

As is exemplified by Ellis's work, the postmodern stance and its attendant philosophical buzzwords have become obligatory on the academic left, save for unrepentant die-hard Marxists. There is no one, overriding reason for this; a number of mutually reinforcing factors seem to have come into play. First of all, postmodern philosophy, in its guise as literary theory, flatteringly concedes a high degree of power to the skills and habits of mind of literary critics. The practice of close, exegetical reading, of hermeneutics, is elevated and greatly ennobled by Derrida and his followers. No longer is it seen as a quaint academic hobby-horse for insular specialists, intent on picking the last bit of meat from the bones of Jane Austen or Herman Melville. Rather, it has now become the key to a full comprehension of the profoundest matters of truth and meaning, the mantic art of understanding humanity and the universe at their very foundations. At a stroke, the status of literary studies as a genteel backwater of the world of affairs is reversed; and the image of the sophisticated critic as a new Dr. Faustus, conjuring secrets from the remotest circles of heaven and hell, is set in its place. Like all great con men, the high priests of deconstruction and the like are flatterers.

Secondly, postmodernism, whether chiefly derived from one philosophical source or drawing eclectically on a flock of them—Lyotard, Baudrillard in addition to Derrida and Foucault—is, in its skepticism about everything save itself, an incarnation of the anti–Philosopher's Stone. Everything it touches is drained of value, authority, validity, and even the right to stand for what it has always stood for and to be understood as it has always been understood. Thus, in the game of intellectual *subversion*, which is always important to the academic left (though the wider world goes on much as usual), it is felt to be the instrument for dethroning the proudest symbols and most sublime achievements of Western—that is to say white, patriarchal, violent, imperialistic, capitalistic, greed-ridden—civilization. Notes Vincent Pecora, a literary critic of emphatic left-wing sympathies but scornful of deconstruction and its political consequences: "To many of Derrida's critics, the deconstructive rejection of humanism and the Enlightenment has seemed mere nihilism. But it is precisely this anti-'Western' stance that has been the key, I think, to the influence Derrida's work has had on a broad spectrum of the academic left."[30]

Everything by which that civilization contrives to hold itself in high regard—Shakespeare and Dante, Descartes and Kant, Locke and Jefferson, Newton and Einstein, Mozart and Beethoven—wilts under the deconstructive gaze (at least in the minds of those doing the gazing). It is a heaven-sent device for avoiding close argument and the analysis of particulars. Once a postmodern critic has at hand a license to read every proposition as its opposite when it suits his convenience, analytic skills of the more traditional sort are expendable and logic is effaced in the swirling tide of rhetoric. Once it has been decided that determinate meaning is chimerical and not worthy of slightest deference from the well-honed poststructuralist postmodernist, the entire edifice of hard-won truth becomes a house of cards. Once it has been affirmed that one discursive community is as good as another, that the narrative of science holds no privileges over the narratives of superstition, the newly minted cultural critic can actually revel in his ignorance of deep scientific ideas. That this is a canny political act is accepted as an article of faith, no matter how much it seems to elevate wishful thinking over hard social fact.

The feeling that the postmodern critique is inherently political in a fashion helpful to the left is made evident in the recent rise of what has come to be called "cultural studies" on the campus. This term covers a multitude of free-form speculations about social institutions current and past. It is a recombination of social history and sociology, practiced largely by scholars whose background is in literary studies, when it is not in women's studies or something of the sort. It combines a pugnacious vindication of the demotic and popular cultures with a truculent interrogation of anything that issues from the high culture of the elite or from the dominant attitudes of the bourgeoisie. In that sense, it is a Foucauldian project on its face. The role of the skepticism and relativism of the deconstructionists is also clear; if no text is "privileged," no narrative tradition closer to ethical, aesthetic, or historical truth than any other, then there are no grounds for regarding the traditional venues of humanist scholars—high literature and high art—as sacred ground. Thus, it becomes permissible for professors of English to inquire solemnly into what are by tradition (and in fact) trivial matters, and to festoon those inquiries with the abundant neologisms of the postmodern lexicon, giving thereby further assurance that the subject at hand, be it rap music or professional wrestling, has deep implications for theory.

Philosophical Revenge

While many scientists in and out of the academic community hold progressive and leftist—sometimes emphatically leftist—views on a variety of questions, the postmodern stance has made little headway among them and seems, when they become aware of it, to evoke indifference or amused contempt. To the extent that contemporary theory, as it is understood by the humanists, is likely to influence scientists at all, the effect will probably be to drive them further from active political engagement along the lines hoped for by the left. This fact rarely deters the postmodern left from pursuing its favorite will-o'-the-wisps; if anything, the humanist radical is persuaded by the opposition or indifference of scientists that he must be on the right track. We suspect that this phenomenon is partly rooted in intellectual trends that were manifest in university life three or four decades ago.

At that time, the no-nonsense logical positivism adumbrated in such influential books as A. J. Ayer's *Language, Truth, and Logic* was widely discussed and supported. It is safe to say that some version of this viewpoint—with Popperian addenda—is still embraced, at least tentatively, by most working scientists who have reflected at all (as most have) on the issues of knowing and truth. As it was made known to the academic community, however, positivism, while flattering to physical and biological scientists, was devastatingly hurtful to the *amour propre* of traditional humanists, and hardly more comforting to social theorists. This philosophical doctrine imposes severe tests of meaningfulness on all sorts of propositions. Statements in the language of academic, as well as everyday, discourse that seem, on their face, to be making some kind of factual assertion about the world are, in the harsh glare of positivism, often dispossessed of such pretensions.

The propositions of science, by and large, escape humiliation, while those of the humanities, including such venerable philosophic areas as ethics and aesthetics, emphatically do not. Thus, while statements about the emission spectra of planetary nebulae are perfectly meaningful for the positivist, the assertion that Racine is superior to Corneille (or Schubert to Mendelssohn, or that the Napoleonic Code is ethically inferior to Anglo-Saxon common law) collapses into meaninglessness. The latter is understood as an example of "emotive" utterance, to which truth-value cannot properly be ascribed.

Given that humanists—and, in particular, literary scholars of the traditional sort—have always labored just as hard, examined the relevant data just as minutely, and argued as exhaustively in reaching their judgments as physicists and mathematicians do in reaching theirs, the news that their conclusions cannot, in principle, even be wrong (in the sense that the contrary proposition is right) was a sour revelation indeed. The fact that scientists tended to accept it more or less complacently cannot have been much comfort to professors of English and art history.

A further source of unhappiness was the reaction of a good part of the social-science community, which responded to the logical-positivist critique (or to its vulgarization) with various attempts to introduce quantitative methods, mathematical models, "replicability," and "falsifiability" into sociological work. Many of these attempts were brutally reductionistic and flew in the face of common sense, obtaining results that were either painful elaborations of the obvious or, even worse, procrustean absurdities. As the acerbic Stanislav Andreski puts it:

> The recipe for authorship in this line of business is as simple as it is rewarding: just get hold of a textbook of mathematics, copy the less complicated parts, put in some references to the literature in one or two branches of the social studies without worrying unduly about whether the formulae which you wrote down have any bearing on the real human actions, and give your product a good-sounding title, which suggests that you have found a key to an exact science of collective behavior.[31]

This sort of thing, while for the most part unimpressive to the scientists, tended to convince many humanists—and a good part of the social-science community as well—that a craving for methodological respectability— "scientism" or "physics envy" as it was sometimes called—must lead to a sterile (and politically reactionary) view of human affairs, denying ineluct-able truths about the human situation. [32]

Thus it probably came to pass that when the brutally skeptical views of the postmodernists began to gain currency some years later, many humanists, and many social scientists as well, were quick to lay hold of them as instruments of revenge. [33] If the carefully crafted opinions of literary experts were to be consigned to epistemological limbo by analytic philosophy, those experts and their academic progeny now had in hand—or so they thought—an instru-ment that could drag down the scientists and other pursuers of "objective" knowledge with all the rest. This view accounts at least in part for the paradox that on embracing postmodernism, humanist scholars have in many in-stances cruelly repudiated the accomplishments of their own disciplines, even to the point of denouncing their own earlier work.

However varied the reasons for the embrace of postmodernism in the universities, it is clear that the phenomenon is almost wholly associated with the self-described political left. As far as the ideological right is concerned, the situation presents it with welcome opportunities for polemical sallies, counterblows that avoid the necessity for justifying the illogical or evil prac-tices of their own heroes and of whatever world (usually of the recent past) they like to think of as the best of all possible worlds. With postmodernism the target, conservatives easily move the discussion onto the loftier plane on which the relativist caprices of Derrida, Lyotard, and the rest are the princi-pal focus. Roger Kimball's *Tenured Radicals* adopts this strategy in part, but its underlying politics are relatively transparent, compared to arguments that seem never to stray in the slightest from disinterested philosophical inquiry. [34] It is often hard to read the writer's political position from such critiques, and some of them, in fact, issue from impeccably left-wing thinkers.

For the first time in modern American history, right-wing theorists seem on the point of establishing themselves upon the ethical and philosophical high ground, thanks to the postmodern contortions of the left. This fact, however, has little penetrated left-academic discourse; the entanglement of would-be progressive intellectuals with the conceptual freak show of postmodernism continues to isolate and neutralize them, at least outside the hothouses (i.e., academic departments and conferences) in which they flourish. One will cheer or deplore this fact as one's political tendencies dictate.

Cultural Studies: Playing Intellectual Hooky

We can now come to the question of the relation between the postmodern styles pervading so much of current humanist thinking and the traditional scientific disciplines—mathematics, physics, chemistry, biology, engineering. It would have been idle to hope that the ambitions of postmodernism would be satisfied by a revision of the standard modes of analyzing literature and the arts, and by a new methodology for thinking about sociology and social history. The mentality of postmodernism has an emphatically *totalizing* component, even as it pretends to denounce the totalizing propensities of whatever it wishes to attack. The centrality of science to the contemporary world, its crucial role in shaping the material conditions under which we live as well as many of the assumptions we bring into our discussions of the world, guaranteed that sooner or later postmodern dialecticians would feel bound to turn their guns on it.

It may be argued that in revolutionizing literary criticism, postmodernism will have created a valuable legacy, although many people (including students) who simply love literature and look to academic criticism for relevant inspiration and deeper insight about it have been cruelly disappointed. Still, the analysis of social questions may have benefited from postmodern intellectual strategies, however susceptible to subjectivism and giddy pontifications they may be. In the area of the hard sciences however (and we hold to the usage, anticipating the jeers of Derridean or Foucauldian skeptics), it has by now become clear, after a few short years, that criticism and analysis informed by postmodern attitudes has been, by and large, an irrelevant botch.

We could not wish for a more straightforward example than the following paragraph, appearing in the recent volume *Cultural Studies*, a massive tome clearly intended to be a sourcebook and text for the hordes of anticipated students in this newly delivered academic hybrid. The quotation is taken from the essay "New Age Techniculture" by Andrew Ross, professor of English at Princeton, editor (with Stanley Aronowitz) of the fashionably postmodern-leftist journal *Social Text*, and glamorous cult-figure in the movement.[35] Despite its length, it is worth quoting in full:

In this respect [i.e., the distinction between authentic science and pseudoscience] it is worth drawing an analogy between the demarcation lines in science and the borders between hierarchical taste cultures— high, middlebrow, popular—that cultural critics and other experts involved in the business of culture have long had the vocation of supervising. In both cases, we find the same need for experts to police the borders with their criteria of inclusion and exclusion. In the wake of Karl Popper's influential work, for example, falsifiability is often put forward as a criterion for distinguishing between the truly scientific and the pseudoscientific. But such a yardstick is no more objectively adequate and no less mythical a criterion than appeals to, say, aesthetic complexity have been in the history of cultural criticism. Falsifiability is a self-referential concept in science inasmuch as it appeals to those normative codes of science that favor objective authentication of evidence by a supposedly objective observer. In the same way "aesthetic complexity" only makes sense as a criterion of demarcation inasmuch as it refers to assumptions about the supposed objectivity of categories like the "aesthetic" refereed by institutionally accredited judges of taste.[36]

So much, then, for three thousand years of struggle to develop a systematic method for getting reliable information about the world! So much for the notion that refutation by experience is good grounds for abandoning a theory, or at least taking it in for major repairs! To see whether this petulant paragraph says anything, we have to strip away the irrelevancies concerning the relative nature of aesthetic judgments and resolutely ignore the dreaded Culture and Science Border Patrol in order to get at it. Doing so, we seem to be left with something like: "Science backs up its claims, whereas pseudoscience doesn't, but I don't care about the difference." Perhaps, however, we are being uncharitable. If we work on it for a while, it is just possible to construe Ross as meaning to say, "The empiricist philosophy by which science proceeds cannot be justified by an appeal to empiricism. We can't solve the Problem of Induction by appealing to inductive inference."

Welcome to freshman philosophy! True, there is a serious point here, one with which most scientists are quite familiar. *But it does nothing to elevate pseudoscientific nonsense to the epistemic dignity of genuine science.* In our mind's eye we envision this paragraph leaping from the occult–New Age section of the bookstore, where, no doubt, it graced some puffery about channelers or healing crystals, to the spanking-new cultural studies section where, thanks to Professor Ross and the editors of *Cultural Studies,* it takes on a new life as a contribution to learned discourse.

Ross's piece in *Cultural Studies* is largely incorporated into his *Strange Weather: Culture, Science and Technology in the Age of Limits.* This work studies contemporary popular subcultures that vulgarize standard science while, to some degree, challenging its authority. Ross is for the most part in sympathy with these enthusiasms, which include the New Age movement, "alternative" health care regimens, science fiction (especially of the cyberpunk, feminist, or gay variety), computer hacking, and visionary radical ecologism. He celebrates them as possible nuclei of resistance to a monolithic, global, capitalist techniculture—a monster, as Ross would have it, sustained by modern science and sanctified by its canons of validity. *Strange Weather* is

Ross's magnum opus, and it enlarges his claim to be expert on matters of science and technology. It does so, however, on the basis of an argument that is thin and irresolute, unwilling even to try to formulate a clear and consistent case against the putative scientific world view, but reflecting a desperate unhappiness that science is such a powerful force, materially and intellectually, in contemporary life.

Predictably, the sociologist Bruno Latour is one of Ross's gurus, and the ideas of Stanley Aronowitz are called upon as well. Equally predictably, Ross parrots all the New Age mystifications of quantum mechanics—without, however, displaying any but the vaguest understanding of physics in general or quantum mechanics in particular. This is a book that is content, in the main, to posture, rather than to argue. It is driven by resentment, rather than by the logic of its ideas. Ross doesn't know what he wants to do about science. He doesn't like it, but—he wistfully allows—he might be persuaded to like it if it were to change into something "that will be publicly answerable and of some service to progressive interests."[37] "Of some service to progressive interests" seems reasonably clear, if frighteningly Stalinist in tone and at root. One infers, however, that a "publicly answerable" science is the sort of thing that the common man or woman can do pretty much at will, the sort of thing that involves not too much by way of hard work or thought, of deep analysis or difficult concepts. Above all, that old demon, *rationalism*, must be banished. "How," asks Ross, "can metaphysical life [i.e., New Age] theories and explanations taken seriously by millions be ignored or excluded by a small group of powerful people called 'scientists'?"[38]

Strange it is that a well-known scholar at one of the world's most distinguished universities should write a lengthy book upon a subject about which he knows, evidently, virtually nothing. It is stranger still that he can boast of his ignorance in the very first words. "This book is dedicated to all of the science teachers I never had. It could only have been written without them." Such hubris, such eagerness to put on an antic disposition and yet be taken seriously, speaks volumes on the canons of acceptable scholarship under the postmodern dispensation. In the end, however, it is less appalling than Ross's confident assumption that he and his coterie know what's best for the human race politically and socially, an assumption that presumably licenses this silly attempt to declare Western science ripe for overthrow.

Note well that the nonsense Andrew Ross propounds is a predictable, in fact a near-inevitable, consequence of applying to questions of scientific validity the sophomoric skepticism—and the visceral urge to champion the demotic whenever it comes in conflict with "official" culture—that characterizes the postmodern stance as absorbed and refracted by left-wing academics. Note also the way in which Ross's manifesto serves the impulse to inflict retribution on those recent modes of philosophy that are flattering to the claims of science and dismissive of literary criticism and the like as mere opinion-mongering. Clearly, resentment is in the saddle here, and it is a disastrous methodology, one that propounds ostensibly deep thoughts on shallow aspects of science and culture, while generating shallow thoughts on their deep aspects.

Chaos Theory: A Brief Guide

We are obviously not able to consider here each and every intellectual curio that arises from the now widespread effort of postmodern theorists to bring science under their scrutiny. A few of the more redolent examples must suffice to illustrate the general tendency. For the sake of unity, and to reserve discussion of other sciences for later chapters, those we consider here have to do with a certain recent development in the mathematical sciences—so-called "chaos theory"—that has drawn an unusual (for a mathematical subject) amount of public interest. Quite naturally, it has been a proving ground for postmodern critics eager to try their apparatus in the venue of modern scientific thought, and eager to justify their philosophic maxims by appeal to ostensible "paradigm shifts" in science. This tactic is not so rudely dismissive of science and scientists as Andrew Ross's frontal assault. Nonetheless, in positing the emergence of a "postmodern" science which, it is claimed, illustrates the validity of the postmodern weltanschauung, these "analyses" in effect derogate the reliability and accuracy of standard science, and snidely disparage those scientists—that is to say, the vast majority of *all* scientists—who have been oblivious to this ostensible revolution in thought.

Any but the briefest description of chaos theory would be out of place here. The term refers to developments in pure and applied mathematics, particularly to a branch named dynamical systems theory: the study of systems that change with time. Typically, these are deterministic; that is, the state of the system at one instant completely determines its state at all subsequent moments. The *locus classicus* of dynamical systems theory is the great work of Isaac Newton on celestial mechanics, that is, the theory of how stars, planets, moons, asteroids, comets, and so forth move under the influence of gravity.

Chaos theory essentially addresses this conundrum: knowing that a system is in theory deterministic is by no means equivalent to having an effective means for *predicting* its behavior as a function of an initial condition, namely, the state of the system at one particular time. The optimism of late eighteenth- and early nineteenth-century mathematicians, astronomers, and physicists that practicable methods of computation would become available for making such predictions in *all* reasonably simple cases turns out to have been premature, although it inspired brilliant work sufficient to handle *many* such problems, *including most that have to be dealt with in engineering and day-to-day science.* As it happens, however, there are very simple systems, in the sense that they involve a small number of parameters and a very straightforward "law of evolution," where prediction becomes essentially impossible beyond a short period of time. What's more, there are significant "qualitative" as well as "quantitative" aspects to this inability. For instance, the trajectories of evolution of two systems that start out "microscopically" close, as far as their respective initial conditions are concerned, can diverge wildly, not only in a numerical sense but in their geometric aspects. Thus, to make matters more concrete, an astronomer may find that it is impossible to make a good prediction about the qualitative behavior of a planetary system because, first of all, the methods at hand for solving the relevant differential equations are far too inaccurate, and furthermore, even if the first difficulty could be got round, a tiny error in determining the initial condition (which is of course inevitable) can result in a gross *qualitative* error in characterizing the long-term behavior of the system.

Dynamical systems theory is a very geometric subject, as modern mathematicians understand the term, and, in consequence, many of these bizarre phenomena can be illustrated, with the aid of computer-generated graphics, by weird and beautiful pictures. This alone accounts for much of the public interest in these developments. (It accounts as well, we must admit, for much of the popularity of the subject among mathematicians themselves, not to mention legions of computer-users with skill at graphics.) Nonetheless, despite their didactic value, the accessibility of such pictures may have the effect of deceiving the intelligent layman into believing that he grasps the subject better than he really does. To be undiplomatic, a solid understanding of what is really involved requires a considerable amount of formal mathematical knowledge.

The foundations of the subject were laid at the end of the nineteenth century by the great mathematician Henri Poincaré, and in a sense modern chaos theory represents a resumption of this work after a long hiatus. The reasons for this slumber are as follows:

1. The attention that Poincaré's work should have attracted from physicists and mathematicians was understandably diverted by the stunning developments in theoretical physics that took place at the beginning of the twentieth century (special and general relativity and quantum mechanics). These quite naturally absorbed the lion's share of intellectual energy of those best placed

to follow up Poincaré's implications. (It is worth noting in passing that, even had Einstein never been born, Poincaré would almost certainly have come up with relativity on his own at about the same time.)

2. The possibility of developing chaos theory as a mathematically consistent subject depends on a huge body of foundational mathematical work in such areas as topology, differential geometry, and the theory of computational complexity, most of which was done long after Poincaré's day.

3. Mathematical theories cannot grow without a host of specific examples on which the mathematician must rely to sharpen and modify his intuition before setting out to erect a systematic mathematical structure incorporating them. So far as chaos theory is concerned, most of the paradigmatic examples cannot be worked out by ordinary pencil and paper computations; nor can geometric pictures easily be drawn by relying solely on naive intuition. These examples were forthcoming only after the development of high-speed electronic computers in the 1950s and 60s and the subsequent refinement of computer graphics techniques.

The best-known book on the subject is James Gleick's *Chaos—The Birth of a New Science*. While commendably accurate on the underlying mathematical principles and their relevance for a host of scientific questions, Gleick's book, perhaps inevitably, overdramatizes the history of the subject in trying to make its protagonists fascinating. In point of fact, there is nothing, on the level of personal idiosyncracy, that can be said to distinguish specialists in chaos theory from other mathematicians and theoretical physicists. They are not pointedly more heretical in temperament. They just happen to work on dynamical systems theory, as opposed to low-dimensional topology or geometric measure theory or Hopf algebras. Moreover, chaos theory, for all its beauty and scientific relevance, is *not* the dominant theme in contemporary mathematics, for the simple reason that nothing is. Mathematics is stupendously vast and varied, and every year results appear in one specialty or another that are just as delightfully surprising and involve just as great intuitive leaps as those of chaos theory. So far as physics and the other mathematical sciences are concerned, chaos theory is certainly a helpful source of new techniques and insights, but it cannot by any means be said to put everything else in the shade. If one insists on calling the development of chaos theory a "paradigm shift" in the Kuhnian sense of the term, it probably does no harm, as long as it is kept in mind that within the scientific community there is not much sense of foundations being overturned. A more apt metaphor is that a bright light has been turned on, better illuminating what we already knew, making visible some fascinating fine detail, and revealing promising paths for further investigation. There has been very little in the way of culture shock.

We point out that some contemporary popular myths about chaos theory are corrected in the book by David Ruelle,[39] one of the founders and accomplished masters of the subject. For those who want a brief and clear exposition of the basic notions, we recommend also Harmke Kamminga's essay in, of all places, *New Left Review*.[40] (This may reassure some on the left who misunderstand the polemical intentions of this book. On the other hand, even the most conservative reader will be able to get through Kamminga's piece without elevation of blood pressure.) Kamminga wisely chose to write her exposition "in consultation with a number of experts," and she observes prudently that "chaos and nonlinearity may be in danger of being seen as the solution to everything. Uncritical use of the notion of the 'butterfly effect' and glib assertions that 'life is a strange attractor' threaten to turn chaos theory into a new mysticism." Ruelle would certainly agree, as would most mathematicians and physicists. This has not forestalled the emergence of a fad among historians, social theorists, and literary intellectuals, a number of whom are given to studding their essays with knowing references to chaos theory as a way of dressing up truisms about the complexity of life, art, and human experience.[41]

Chaos as Nonsense I: Steven Best

Kamminga's piece and the publication of it in a staunchly ideological journal are ample evidence that left-wing political commitment is not, of itself, inconsistent with a healthy and productive interest in various aspects of modern life, science included. However, there are in contrast examples of attempts to deal with the same subject that rely for their doctrine and methodology on the arcana of full-blown postmodern theory. The results are grotesque. They illustrate both the megalomaniac pretentiousness and the utter impotence of postmodernism under full sail as it attempts to engage itself with the world of honest science.

Steven Best is a fledgling philosopher and co-author of *Postmodern Theory: Critical Interrogations*, a well-received and approving account of the trends in current intellectual life grouped under this term. He has also written a curious and turgid essay, "Chaos and Entropy: Metaphors in Postmodern Science and Social Theory," for *Science and Culture*, a journal that focuses on the relation between science and society from a left-wing, often quite explicitly Marxist, point of view. However, Marx is very much in the background so far as Best is concerned; his heroes are the usual suspects—Foucault, Derrida, Lyotard, Baudrillard, *und so weiter*. His version of postmodernism is in the standard mold: "Postmodern social theory vigorously rejects every key axiom of modern philosophy and sociology: it renounces foundationalism and representational epistemologies. Postmodernism stresses the relativity, instability and indeterminacy of meaning; it abandons all attempts to grasp totalities or construct Grand Theory."[42]

In his paper, Best attempts to describe and (more or less) to serve as advocate for something he refers to as "postmodern science." This, one is given to understand, has as its mission the overthrow, rather than the fulfillment of "modern" science: "Like postmodern social theory, postmodern science sees modernity and modern reason as inherently repressive." In other words, hidden beneath the relativistic pose, there lurks a stiff-necked moralism. Postmodernism in general and postmodern science in particular have come to liberate us from repression. This mission is especially crucial since only postmodern science can save us from the ecological catastrophe into which modern science is driving us: "Postmodern science draws the conclusion that a new, postmodern paradigm is necessary, one which is philosophically sophisticated, scientifically complex, ethically sensitive, spiritually aware, and ecologically sane."[43]

The reader may be inclined to characterize these dicta as a Very Grand Theory indeed, which would seem to subvert *ab initio* their self-proclaimed postmodernity, within which there is supposed to be no Grand Theory. Letting this pass, however, we ask *what* is thought to constitute "postmodern" science. Best gives us this categorical description of it: "Within its own discipline, postmodern science has three main branches of influence: thermodynamics . . . quantum mechanics . . . and chaos theory." That, now, is a very curious amalgam for something forthrightly claiming to be postmodern: it covers a period in the history of physics from the 1820s to the present day, but mysteriously omits both special and general relativity. (Perhaps, not having Jacques Derrida's credentials in these latter subjects, Best thinks it best to steer clear of them.)

One would think that the much-trumpeted emergence of self-conscious postmodernism in the two decades past would have produced a generation of explicitly postmodern scientists proud to identify themselves as such. Strangely enough, the examples Best comes up with are unimpressive. Noted chemist Ilya Prigogine is among them, of course (his name keeps coming up in postmodern discourses with depressing frequency); but a realistic view of Prigogine's science would have to come to terms with the fact that serious contributions have not been forthcoming for a couple of decades, and that he has slipped into habits of speculation that involve him in very shaky science and even shakier mathematics. Best's other paragon of contemporary postmodern scientific thought is Jeremy Rifkin, author of *Entropy*, *Algeny*, and

Beyond Beef, among a series of books and pamphlets devoted to imminent environmental catastrophes. While it is possible that Rifkin's high-pitched rhetoric performs some service in alerting a sluggish public to the existence of ecological problems, it is widely felt, even by those scientists most passionately committed to environmentalism, that Rifkin's unrelieved alarmism rests on ill-founded and unscientific theorizing, and that his distortions and fantasies damage the political cause he seeks to inspire.[44]

In contrast to such dubious champions, undisputed masters of the most startling ideas in current scientific theory—Hawking, Witten, and Guth for instance—are very much out of Best's picture. This is hardly surprising, given Best's evident incompetence at understanding the scientific and mathematical ideas he tries to cite in his favor. His understanding of chaos theory in particular is shallow and confused, and apparently arises from a botched reading of popularizations like those of Gleick and Kamminga. Certainly, he fails to take heed of Kamminga's warning (cited above); and in his account, chaos theory does indeed become a new mysticism.[45] This is a grave charge; but here are some examples to substantiate it.

Best asserts, for instance, that in the systems studied by chaos theorists, the inability to determine initial conditions exactly frustrates prediction because "errors made at that level will be exponentially magnified in subsequent calculations." Certainly, this is often true for these so-called nonlinear systems; but then again it is equally true for the most classical of *nonchaotic linear* systems as well. For instance, the differential equation of exponential growth, $y' = ky$, $k > 0$, which is usually studied in elementary calculus courses, exhibits the behavior prototypically, and it has no hint of chaos. As any competent freshman can see, if the initial condition is approximated erroneously by even the smallest amount, the subsequent error will grow exponentially with time. On the other hand, for a *nonlinear* equation like $y' = ky^{1/2}$ this will *not* happen; over time, initial errors are amplified at a less than exponential rate.

Best's physics is similarly shaky. He solemnly intones: "The dialectic between order and disorder also suggests a revaluation of the Law of Entropy, no longer viewed simply as system decay and breakdown but as creations of new forms of order."[46] Unfortunately for the gravamen of his argument, this realization represents no breakthrough inspired by chaos theory. The formation of the orderly arrangement of a snowflake, for instance, from an unordered collection of water molecules is, in fact, an entropy-increasing process, a fact that is quite well understood in classical thermodynamics and is, again, taught in elementary courses.[47]

This error, absurd as it is, is merely symptomatic of an even deeper igno-

rance. Best begins an ill-advised attempt to inform the reader of the difference between "linear" and "nonlinear" mathematics with the clause, "Unlike
the linear equations used in Newtonian and even quantum mechanics."[48]
This is a howler; for, whereas the fundamental equation of quantum mechanics (the Schrödinger equation) is what is technically known as a linear partial
differential equation, the Newtonian laws of celestial mechanics are expressed by a decidedly *nonlinear* system of ordinary differential equations
(which is why such classical conundrums as the Three-Body Problem have
been such a headache since the seventeenth century).

Best is likewise in deep trouble when it comes to understanding what he
offhandedly describes as the "Newtonian paradigm." Newton's equations of
planetary motion are, of course, nonlinear, for which reason some of the most
interesting examples in chaos theory occur in classical celestial mechanics.
Thus the theory represents, in a major sense, the triumphant reemergence of
Newtonian mechanics, *not, as Best would have it, its overthrow.* What we have
in chaos theory is a recommitment to taking seriously some deep old issues,
such as the "structural stability problem."

Best's paper is rife with similar errors, historical as well as narrowly scientific. Yet his deepest misunderstandings are not of the kind that can be
remedied by a quick refresher course in elementary math and physics. They
arise, when all is said and done, from the metaphysical hubris of postmodernism as such. As a thinker, Best is ill equipped to draw inferences of any kind
from his contemplation of chaos theory (or quantum mechanics or thermodynamics) simply because his grasp of these matters is so rudimentary and so
tied to secondhand paraphrases. That he plunges, notwithstanding these
difficulties, into vaporous pontifications is evidence of the conceit with
which postmodernism infuses its acolytes. Nothing is excluded from the
sweep of their judgments; the sententious generalities that constitute the core
of their doctrine are held to excuse them from the necessity of actually
learning the particulars of the disciplines they criticize. The inevitable result
is the philosophical styrofoam of Best and those who theorize along the same
lines.

Chaos as Nonsense II: N. Katherine Hayles

Among these is the literary scholar N. Katherine Hayles, whose specialty is
the relations among science, literature, and contemporary literary theory.
She too is committed to the idea that chaos theory is somehow pardigmatic of
the postmodern condition, but unlike Best she tries to illustrate this, not in
reference to some presumed political ideal, but rather by arguing for deep

parallels and assonances between the mathematics of chaos and the theoretical practices of textual critics loyal to the tenets of postmodernism. (Best, for his part, advances a similar claim in passing.[49]) Her recent book, *Chaos Bound*, is devoted to promulgating this odd hypothesis.

Hayles's underlying assumptions seem curiously Hegelian. The cultural moment, she reasons, has brought forth chaos theory simultaneously with Derrida's *Of Grammatology* and de Man's *Allegories of Reading*, and hence, some unspecified mechanism of the zeitgeist must be responsible for both developments. This is a bizarre thesis. Why should the theory of dynamical systems be more closely related to the gyrations of literary exegetes than it is to major league baseball or Jane Fonda's workout tapes? Aside from the irrelevant fact that both kinds of theorizing take place, for the most part, on university campuses, there is no ground for positing any kind of conceptual relationship. Hayles's arguments, such as they are, are based on subjective and shoddy analogies, leaky metaphors, and (not unusual for work immersed in postmodern theory) flat and unsupported assertions. She is one of those who are eager to tell you, earnestly and at length, precisely why a raven *is* like a writing desk—especially if a publication can be got out of it. Her comparisons—like that of Derrida to the mathematician Mitchell Feigenbaum,[50] to take but one example—are strained and arbitrary and informed by a logic that would make everything a metaphor for everything else. Why not compare Derrida to Charles Manson, or the Feigenbaum number[51] to Roger Clemens's earned-run average? Why not compare the "unexpectedness" of chaotic phenomena to the surprising twists and turns in Haydn's string quartets, icons of the Enlightenment though they be? It would make easily as much sense. What Hayles does is not analysis. It is name-dropping.

Even philosophers who see some parallels between deconstructive literary theory and the mathematics and physics of chaos theory are loath to push the comparison as far as does Hayles. Alexander Argyros, commenting on Hayles's work, notes: "I suspect that this apparent compatibility may be implying to literary theorists that chaos is a validation of deconstruction. My own view is that such a claim is, for the most part, wrong . . . While it is certainly true that deconstruction and chaos are both interested in highlighting non-linearity, to claim that they are fellow travellers is, I believe, to make an unwarranted assumption."[52] (Of course, there is the further question of what "non-linearity" means to a mathematician, as contrasted to what it might signify to a postmodern literary theorist. We address this point below.)

In trying to grasp what chaos theory is and how it relates to other aspects of mathematics and the physical sciences, Hayles falls into the same kinds of amateurish errors that plague Best's paper. It is clear, over and over again, that she really doesn't know what "linearity" means in a mathematical context. In that her book is so much longer, her mistakes are correspondingly more numerous. They are all the more embarrassing as well because she genuinely attempts to give an expository description of some deep mathematics and physics—fractal geometry, nonlinear dynamics, Gödelian incompleteness, information theory, and so forth. Naturally, the usual doodles are present—fractals, Cantor sets, the Lorenz attractor, bits and pieces of the Mandelbrot set (no diagram of the Peano space-filling curve, more's the pity)—but in this context, they are no more than intellectual tinsel, since they are unilluminated by genuine understanding or exposition. Hayles repeatedly if unwittingly illustrates the everlasting soundness of Pope's axiom—a little learning is a dangerous thing.

Unlike most of the works we have examined, *Chaos Bound* is not primarily concerned with leftist political agenda. Nonetheless, Hayles can't resist dropping portentous hints about the transformative political significance of both postmodern literary theory and nonlinear topological dynamics, especially when they are viewed as manifestations of the same putative eructation of the guts of the culture. As well, there are the predictable genuflections to the feminist-critique-of-science mafia, especially to Donna Haraway. Mercifully, however, the explicit political claims are muted. Or perhaps it would be better to say that they are diluted by a sea of muddy abstractions. Hayles provides a chapter entitled "The Politics of Chaos," filled with ruminations on such themes as tensions between local and global, contingent versus universal, laden with the sense that these are vibrant with political significance. Of course there is the tendency to conflate the mathematical terms *local*, *global*, and *universal* with the same words as they occur in poststructuralist discourse—an impermissible tendency in our opinion, and one that rests on Hayles's shallow understanding of how mathematics really uses such terms. The deeper trouble, however, is that the word *politics* itself is used so abstractly that one has no sense of what all this introspection is supposed to signify for those aspects of human existence usually covered by the term.

One might argue that Hayles's analysis, in contradistinction to most of the critiques of science emanating from the academic left, has at least the virtue of regarding science as, on the whole, liberatory and politically progressive. But this approbation comes at the cost of such a distended misreading of science, in equal measure grotesque and condescending, that it is hardly distinguishable from hostility. In any event, Hayles's subsequent work[53] reverts to the tone of orthodox radical feminism and rails at physics (fluid dynamics in particular) as deriving from a worldview deeply tainted by sexist *imagery*. We shall not comment on this latest exercise in self-righteous hermeneutics, except to observe that it is tendentious and strained to the point of absurdity.

It would be an endless task to compile a detailed list of Hayles's solecisms. A very few examples will have to suffice. On one page, for instance, we find that "The special theory of relativity lost its epistemological clarity when it was combined with quantum mechanics to form quantum field theory. By midcentury all three had been played out or had undergone substantial modification." This will come as a terrible shock to physicists! Special relativity and quantum mechanics are as solidly confirmed as it is possible for physical theories to be. While there may be some lingering doubts whether general relativity is quite the right model on the cosmological scale, the special theory has always triumphantly passed every empirical test. However physics develops in the future, any modification must subsume rather than displace special relativity, just as special relativity subsumed Newtonian mechanics. The story is much the same for quantum mechanics, with the additional element that theorists (John Bell for instance) are in the habit of deriving wildly counterintuitive conclusions from the quantum mechanical formalism, only to have these confirmed in the laboratory as soon as the experimentalists can think of a way to test them! As for quantum field theory, one mathematical aspect of the great project to provide a truly unified framework for both relativity and quantum mechanics, this is an ongoing project that engages the deepest and liveliest intellects in physics and mathematics. "Played out?" The best that can be said for Hayles is that she confuses the fact that physics is very much a continuing discipline, and, therefore, has fascinating foundational problems left to solve, with some kind of philosophical and spiritual exhaustion. If anything is played out, it is the postmodernist's pretension to have something interesting to say about physics.

Hayles is similarly at sea when it comes to philosophy. On the *same page* a few lines further on we find that "logical positivism had its heyday in the closing decades of the nineteenth century," which is rather like saying that Babe Ruth's career was at its height in the closing decades of the nineteenth century. Logical positivism was, of course, the philosophical school devised in the late twenties and early thirties of this century by the so-called Vienna Circle, a group of philosophers, mathematicians, and physicists who sought to accommodate the mind-stretching, and then still-recent, developments of relativity and quantum mechanics. In the English-speaking countries, it gained great influence through the work of A. J. Ayer, and the fifties and sixties were, as previously noted, its *real* heyday. Nor is it a dead letter, even today: its critics, such as Rorty and Feyerabend[54] are the ones who now seem to be in at least partial retreat. In charity, perhaps we should assume that Hayles has confused Carnap[55] with Comte.

All this is embedded in a discussion of scientific work that might be read as attempts to find something like absolute grounds for knowledge. The *Principia Mathematica* of Bertrand Russell and Alfred North Whitehead is cited, reasonably enough, as such an attempt, but so is Einstein's special relativity—at least that seems to be Hayles's view of the matter. This ignores the stunning and immediate philosophical implication of relativity, apparent from the very first, namely that such quantities as size and duration, which common sense had always held to be absolute properties of things and events, are in fact relational, not absolute. The real point is not that relativity provides "an overarching framework within which observations from different inertial systems could be reconciled," but that relativity theory discovers the previously unsuspected fact that some such reconciliation might be necessary!

Hayles then cites the Gödel incompleteness result as the deathblow to the Russell-Whitehead program (although, of course, there is no corresponding reference to any analogous demolition of relativity, since there hasn't been one). This is intended to figure the movement away from post-Enlightenment ideals of "universal" knowledge to postmodern skepticism, which regards knowledge merely as "representation" conditioned by the local culture. Hayles seems unaware, however, that Russell's own skepticism, expressed in such works as *Our Knowledge of the External World*, is in some ways close to postmodern ideas of "representation" (albeit Russell is far keener and far, far less windy than the postmodern heavyweights), while Gödel, *qua* philosopher, was a fervent believer in absolutes! "Gödel turned out to be an unadulterated Platonist, and apparently believed that an eternal 'not' was laid up in heaven, where virtuous logicians might hope to meet it hereafter."[56]

It becomes clear that Hayles's cultural constructivist prejudices, her convictions that even the most abstract scientific ideas are closely tied to the zeitgeist, are responsible for such distortions. Her egregious remark about logical positivism is embedded in the following context:

If we think of these projects as attempts to ground representation in a non-contingent metadiscourse, surely it is significant that the most important work on them appeared before World War I. Einstein published his papers on the special theory of relativity in 1905 and the general theory in 1916; the *Principia Mathematica* volumes appeared from 1910 to 1913; and logical positivism had its heyday in the closing decades of the nineteenth century. After World War I, when the rhetoric of glorious patriotism sounded very empty, it would have been much more difficult to think language could have an absolute ground of meaning.[57]

Her point seems to be that the *Principia* and the special theory are products of the halcyon times of pre–World War I Europe (a conceit that leads to the gaffe on logical positivism). This ignores the fact that both efforts are the result of extreme intellectual *crises*, albeit crises of which the general culture, and even the rarefied cultures of philosophers and the literary intelligentsia, were entirely unaware. The *Principia* sprang from Russell's discovery of the set-theory paradox that bears his name, which rendered unsatisfactory the prior work of Cantor and Frege on the foundations of mathematics. Relativity derived from Einstein's realization that the mathematics of Maxwell's equations raised serious questions about classical notions of absolute time and space. These "crises" were known only to a handful of mathematical scientists. The idea that something in the ambient culture—whether under Hayles's interpretation or some competing version stressing the subterranean tensions that led to World War I—generated these magnificent works is thus wildly implausible.

Correspondingly, Hayles's idea that such intellectual projects as the *Principia*, relativity, and logical positivism would have been far more difficult after the war (ignoring even the fact that logical positivism was one of the most *characteristic* such postwar projects) is a febrile delusion of doctrinaire reading. The silliness of the cited paragraph is perhaps most apparent when one considers that both Einstein and Russell were highly conscious of the emptiness of patriotic rhetoric long before the war, even as they were in the midst of their great work on the foundations of physics and mathematics respectively. The lesson to be learned, then, is that cultural constructivist theories of science deserve to be treated with the gravest suspicion, whether they derive from sociology, Foucauldian historicism, or deconstructive literary theory.

All the strange pronouncements upon which we have focused occur, as we note, on one page. There is nothing particularly special about that page. This book is stuffed with similar solecisms, which makes reading it a painful experience. Yet the work is published by a distinguished university press and has garnered Hayles a substantial degree of recognition, including an endowed chair at a major university, a Guggenheim Fellowship, the presidency of the Society for Literature and Science, and the chairmanship of the literature and science committee of the Modern Language Association; so we ought not to conclude that this is some kind of crackpot tract of the New Age movement (although the word *crackpot* unkindly leaps to mind when one has to read it). This is very much in the academic mainstream, as commandeered by the votaries of postmodernism.

The point, finally, is not to berate Hayles—or Best, for that matter—for mathematical subliteracy. That, in itself, is nothing like a disgrace. Hundreds of millions of bright, able, and accomplished people share this minor affliction (and quite a few mathematicians are weak, to say the least, on postmodernist thought). But when such solecisms as we find in these writings are confidently put forth as *scholarly discoveries*, with every assurance that something profound is being uttered, one must wonder about the system—and the ideology—that nurtures and rewards them. Whence, we must ask, does such grossly misplaced intellectual self-confidence come? The smug, hermetic, self-referential atmosphere of politicized academic postmodernism obviously has a great deal to do with it. In this milieu, there is not much thought given to simple scientific accuracy. The caution and scrupulousness that working scientists are conditioned to expect are swept aside, because, in the final analysis, postmodernist work is in great measure prophetic and hortatory, rather than analytic: it announces and cheers on a sweeping "paradigm shift" within our civilization, a change that is supposed to liberate us all.

We suspect that the reader who has followed this brief survey will be left with a few questions. First of all, why is the technical question of mathematical "linearity" versus "nonlinearity" so intriguing to supposed experts in culture and literature? Of course, as we have noted, the success of Gleick's book on the emergence of mathematical chaos theory and its scientific applications has left much of the literate but scientifically inexpert public with a somewhat distorted sense of the overall configuration of the mathematical sciences, of the enormous compass of contemporary mathematics, both pure and applied, and of the relative importance of various ideas within that field. The very accessibility of Gleik's work, and subsequent efforts in the same line, have thus had the unintended consequence of calling forth portentous pronouncements from "cultural critics," whose knowledge of the relevant science is largely limited to these necessarily oversimplified accounts.

Beyond this, however, there is a deep confusion of categories, and a surprisingly naive sense that the use of the same English word in widely separated contexts assures that there are deep thematic similarities. To a paid-up member of the postmodern academic left, the word *linear*, for example, carries negative connotations. It suggests relentless sequentiality, unbending purposefulness, singlemindedness, the triumph of the instrumental—in other words, the mentality that is held to underlie the predicated Western ethos of conquest, domination, objectification, and rigid delineation of oppressive categories via "binary oppositions."[58] Inevitably, *nonlinearity* is seen by contrast to have liberatory implications. It suggests many-sidedness, multiculturalism, diversity, polymorphism, the effacement of boundaries. Thus the revolution for which the postmodernist yearns, realistically or otherwise, is one in which the "linear" regime of late capitalist society will be supplanted by a "nonlinear" ethos, in which multiplicity reigns in the cultural and sexual realms, and in which all sorts of boundaries may freely be crossed.

It should—but obviously does not—go without saying that the mathematical notion of linearity, or its absence, in regard to functions, differential operators, dynamical systems, or whatever, while technically indispensable, has nothing whatsoever to do with such sociocultural questions. Of course, anyone is free to read pictures of fractal geometry and the like *subjectively* as emblems for a revolution in sensibility—or in politics, for that matter. The point is, however, that this is utterly subjective; it is poetry of the most idiosyncratic sort. Postmodern cultural transformation is no more inscribed in the mathematical peculiarities of nonlinear dynamical systems than Nazi doctrine is to be read in the geometric configuration of the swastika. To hold otherwise is to revert to the magical, emblematic thinking of premodern (rather than postmodern) times. It certainly doesn't deserve the name of scholarship.

It is also useful to consider the sense in which these theoretical extravagances of would-be philosophers of culture are hostile to science and to scientists. Obviously, there is some subtlety here. Some of these critics seem, after all, on the face of things, to be *celebrating* science, or at least some of its recent achievements. They see certain new themes in the sciences as harbingers of a desirable cultural change. Hostility is there, however, and its presence becomes clearer when we take note of the moralizing undertone. What is really being asserted is that there is a "modern" science, linked to "phallogocentric" thought and the mechanisms of capitalist-racist-patriarchal domination—in other words, the science that William Blake, in an earlier era, decried as "single vision and Newton's sleep." By contrast, there is supposed to be an embryonic "*post*modern" science that points to the overthrow of the old order. This theme can be traced in the continued insistence that the "chaos theory" postmodernists think they are talking about is "post-Newtonian" (even though it is perfectly clear to the mathematically literate that Newtonian themes are *central* to these new developments, whether they address Newtonian celestial mechanics or the fractal geometry of the basins of attraction of the roots of a polynomial that appear when Newton's method is applied in the complex plane).[59] The "Newton" that postmodern cultural critics are trying to escape is Blake's figment, not the preeminent mathematician and physicist of the same name.

We conclude that hostility to science is, after all, an inextricable element of these postmodern philosophical excursions. It takes the form of the "good guys (persons?) versus bad guys" scenario that the critics impose relentlessly on the history and sociology of science. It is mirrored in the remarkable arrogance with which postmodernists address these issues. Virtually all of them claim to discern important intellectual themes and political motifs in past and current science, themes and motifs that are quite invisible to the scientists themselves. These supposed insights rest, as we have seen, on a technical competence so shallow and incomplete as to be analytically worthless. Their arrogance, then, is comparable to that of "creation scientists" in addressing evolutionary biology, or to that of Galileo's persecutors within the Inquisition in their response to his cosmology. We probably don't need to fear for the safety or intellectual freedom of the sciences on the basis of these bizarre lucubrations: but that is not the issue. What does concern us is that these intellectual misadventures are so well received in nonscientific academic circles, especially on the left, and that they provide the route to publication, tenure, reputation, and academic authority for a growing body of would-be scholars.

We must hope that the painful bolus of postmodernism will pass through the costive bowels of academic life sooner rather than later. Pass, of course, it will eventually.[60] Keeping the hard sciences from contamination should not be impossible, provided that the scientists' resistance to jargonistic snow jobs is as high as it ought to be. We do worry about that, however. In the meantime, unfortunately, the postmodernists will be out there trying to dominate every intellectual conversation. Have they not imbibed the wisdom of the sage?

And everyone will say
As you walk your mystic way
"If this young man expresses himself in terms too deep for *me*,
Why, what a very singularly deep young man
This deep young man must be!"

CHAPTER FIVE

Auspicating Gender

> Would not physics benefit from asking why a scientific world view with physics as its paradigm excludes the history of physics from its recommendation that we seek causal explanations of everything in the world around us? Only if we insist that science is analytically separate from social life can we maintain the fiction that explanations of irrational social belief and behavior could not ever, even in principle, increase our understanding of the world physics explains.
>
> SANDRA HARDING, *THE SCIENCE QUESTION IN FEMINISM*

Feminist Success

American universities have adopted feminism. History, literature, art criticism, psychology—all have had to come to terms with a militant, sometimes angry challenge to their settled ways of doing business. In its obvious form, feminism has concentrated on educational opportunity and careers, demanding an end to practices that have excluded women, and strong remediation, which includes not only affirmative action but also the establishment of women's studies programs and women's centers. On the conceptual level, it has forced a reevaluation of scholarly practices and opened neglected questions of the history, status, and particular interests of women. It has resurrected the work and built the reputations of some women artists and thinkers whom history and male indifference had discounted.

The natural sciences take their share of the heat. In point of opportunities for women, the traditional recruitment and apprenticeship system has been unfair and exclusionary. Strenuous pressure for change has been the predictable result, as women claim their right of equal access to any vocation, no matter how long tradition has regarded it as a province of the male intellect. Until recently, however, the substance and the cognitive style of science per se had not been the target of much feminist complaint. The main demand was for a fair chance at careers, in and out of academic life—a just claim, unproblematical in its philosophic standing if not immune to vexations. Aspiring women chemists and physicists were not insisting upon a female thermodynamics; women mathematicians did not struggle to relate the Mittag-Leffler theorem to gender.

Lately, however, a new academic industry has sprung up: feminist criticism of science. This criticism is not limited to the discovery and censure of discrimination, although, needless to say, those are recurrent elements. The new criticism is far more sweeping: it claims to go to the heart of the methodological, conceptual, and epistemological foundations of science. It claims to provide the basis for a reformulation of science that reaches deeply into its content, its ideas, and its findings. The key process of this critique is insistence that inasmuch as science has until now been a male enterprise, it is ipso facto biased by unacknowledged assumptions derived from the patriarchal values of Western society. On the other hand, the argument continues, a body of insights, attitudes, and sympathies corresponding to the suppressed female culture has been unable to penetrate official science, depriving it therefore of alternative points of view and condemning it to distortion. Such a position has been promulgated with extraordinary success in the humanities and social sciences, even in legal education and research. (Economics has seemed, for some reason, relatively resistant.) The only surprise about the assault on science is that it was so long delayed.

This literature grows with astonishing speed. In the university where one of us works, the reserve reading section of the undergraduate library, alone, lists at the time of writing 143 items on "science and feminism"—during the quiet of summer vacation. Not only are the books being produced; increasingly, college courses adopt and are built around them. The best-known critics are accepted as legitimate historians and philosophers of science, in circles far wider than their feminist peers. They receive generous academic emoluments, large grants, distinguished lectureships, well-subsidized visiting positions, and tenured professorships at leading universities. In at least one discipline, women scientists, inspired by these analysts, convened a conference from which men were excluded on the ground that the particular relevance of female life-histories needed to be brought to bear on their research, free of male interference.[1] In that perennial horse-trade of curriculum revision there now arise without fail proposals to redesign science instruction so as to accommodate the ways of knowing available to women.[2] By any reckoning, therefore, although it is in its early stages, the feminist attack upon science has attained a position of respect and influence.

The central argument varies from one critique to the next, depending upon which of the sundry standpoints within feminism the critic represents.[3] Nevertheless there are broad agreements. The firmest of these is that feminist insight and practice must, *by definition*, improve the range and depth of scientific theory, and must by definition eliminate errors arising from unconscious commitment to patriarchal assumptions. Thereby, the validity of science, as well as its scope, are to be enlarged. On the other hand, the influence of postmodernist theorizing is not absent: many feminist tracts accept and defend the notion that there is no "objective" science, merely a variety of "perspectives," one of which—patriarchal science—has been "valorized" and "empowered" so as to preclude until now the possibility of a feminist science. On occasion, finally, feminism joins hands with New Age attitudinizing, yearning for the rebirth of a prelapsarian golden age, wherein the human race knew and worshiped a goddess-nature, without artificial categories, tortuous cerebration, and the elaborate physical devices of male technology. The editors of a collection called *Women, Knowledge, and Reality* explain,

> We point out that the practices of science have a broad variety of gender implications, ranging from the structure of laboratory work to the most fundamental scientific concepts. We examine androcentric bias in its myriad forms in theories, models, and experiments. We ask whether science can serve what Sandra Harding calls "emancipatory ends" for gender, race, and class. Flowing from the critique of the gendered character of science, we raise many epistemological questions about objectivity, about rationality, about the possibility of a value-free science, and about the ways in which beliefs and knowledge are related to social experience.[4]

Cultural constructivism is the underpinning of all these attacks, even when they are made by self-styled empiricists. All the familiar and some original forms of relativism are found in the copious literature and in the classrooms where it is taught. Most of the analyses insist that a feminist or women's science is—or *should* be, or *will* be—different from and much better than the kind we have now. The announced goal, upon which feminists of the most disparate schools agree, is a science transformed, purged of sexist, racist, classist, homophobic taint.[5] The self-assigned task of feminist critics and their growing band of followers is to administer the purgative. The earlier, less controversial goal of uncovering past and present discrimination, of bringing to light neglected contributions of female scientists, has been subsumed under this enormously more ambitious project: to refashion the epistemology of science from the roots up.

The favorable reception these polemics get in universities is due in large part to their origins in a morally unobjectionable ambition: to recognize and rebuke misogynist practices that have plagued Western science as they have most Western (and, indeed, non-Western) institutions. The record of science, until recently, is—in its social aspect—tarnished by gender-based exclusions (and, as well, of course, by class snobbery, anti-Semitism, racialism, and vulgar nationalism). At times, baseless paradigms in medicine and the behavioral sciences have been pretexts for subordinating women. Pseudoscientific doctrines of innate inferiority and moral frailty have been used to discount female capacity for achievement and to confine women to subservient roles. All this is beyond dispute and generally recognized in intellectual circles, even those of the most conservative bent.

Inevitably, there is not only a thirst among women (and not just militants) for justice and for reparations, but more broadly an atmosphere that allows a truly remarkable latitude to feminist intellectuals. To put it bluntly, the reigning posture is that the weight of men's historical misdeeds is so great that it is bad form, in fact indecent, for male academics to object, even to the most aggressive and speculative announcements of their feminist colleagues. As a result, "women's studies" (like "multicultural" programs generally) has almost everywhere a sacrosanct status, an unprecedented immunity to the scrutiny and skepticism that are standard for other fields of inquiry. Feminist criticism of science (and of culture in general) has become, to borrow a favorite item of lit-crit palaver, "privileged" within the academy.

Sexist Discrimination Today

What are the realities of discrimination against women in science today, at least in the American universities? We take a position that is not likely, in the climate described, to endear us to a majority of our colleagues in or out of the sciences, or to the political and administrative avant-garde. It is that sexist discrimination, while certainly not vanished into history, is largely vestigial in the universities; that the only widespread, *obvious* discrimination today is against white males. These days, nearly half of all medical students are women, and the ratio is accurately reflected in the proportion of women among residents in training and among younger practitioners. A similar situation is found in biomedical research. The numbers are comparable, furthermore, in psychology and anthropology. In the "harder" sciences— mathematics, physics, chemistry, computer science—the proportion of women is much smaller, but it is growing. One of us teaches in a major university mathematics program where half of the graduate students are women. In engineering the numbers of women are still quite low, but there too the trend is upward. The multitude of affirmative action and equal opportunity strategies in use at universities across the country makes it unlikely that any young woman's scientific ambitions and talent will go unencouraged. *Science*, the premier and most widely read general science journal for scientists and engineers, dedicates a *second* annual issue (16 April 1993) to "Gender and the Culture of Science." It comes to no conclusions on the weighty issues to be mined below; but it *is* deeply and positively concerned to encourage every possible effort to recruit women to science. Within recent memory, no social and political movement other than feminism has had such an approving, and massive, exposure in *Science*.

The overwhelming majority of active scientists neither practice nor condone discrimination. Their attitudes toward political and social questions involving gender and women's rights place the majority at the feminist end of the spectrum. Special programs for recruiting women (and minorities, as defined ad hoc) into the sciences have solid support from working scientists, granting agencies, and all the national scientific societies; and so do other programs intended to undo the discouragement young women scientists are said to meet in the wider society. The pieties of affirmative action are supported, not merely tolerated, although enthusiasm for them is tempered in practice by an abiding respect for meritocratic standards (a respect shared by most accomplished women scientists). We would be greatly surprised if, in any current, major American university search for a scientific leader—professor, department chairperson, dean—there were not some mandatory step in the process at which the search committee *must* justify the absence of women (and of minorities), if such is the case, from its short list.

Whatever the actual sex ratio in various fields, the attempt to explain lower numbers of women by the indictment of contemporary male scientists as sexist malefactors is naive—or worse. The truth is *much* more complicated. That doesn't prevent, of course, the charges of male evil-doing being made, published, and honored, even by males. Facts and logic are one thing. Guilt feelings are quite another. Of course these remarks violate the feminist metaphysics according to which every institution of this society is irremediably sexist, and every male, even the most sympathetic, ineradicably guilty by association with it. Some positions, even among persons brought up in the logophallocentric West, are well beyond the reach of rational argument. Feminist fundamentalism shares that distinction with other dogmatisms, such as religious fundamentalism; and when all is said and done, similar mentalities give rise to both.

Recent feminist theorizing about the sciences therefore contains heavy doses of dogma. The claims are immensely strong and usually counterintuitive; it would seem that a correspondingly powerful case, built upon incontrovertible evidence and bound together with iron logic, would be required to make them credible. *We would have to be shown that there are palpable defects, due to the inadequacies of a male perspective, in heretofore solid-looking science and that the flawed theories can be repaired or replaced by feminist insights.* The issue before us is knowledge, scientific knowledge specifically, and the extent to which the prevalent feminist critique, as agent of methodological or conceptual change, is relevant to its advance. To examine the issue we need (and are led by feminist authors to expect) not just stories of past or present discrimination, but examples of scientific knowledge informed, reformed, enhanced by feminism. As far as we are aware, there are as yet no examples. It's that simple.

A principal reason for the absence of examples, of even the effort to identify examples, may lie in the conceptual roots of feminist science-criticism. Not surprisingly, given the academic backgrounds or venues of many of its champions, the feminist critique is overwhelmingly concerned with metaphor, rather than with the logical content and analysis of scientific results. But scientific results, we must insist, are not simply metaphors. If they survive, they do so because they work,[6] for a large number of people of hugely varied backgrounds and interests, that is, they function as distinctly more than images or figures of speech, and their values are, likewise, more than merely figurative. What we learn from these critiques is unexceptional and also not in the least surprising: that men have traditionally dominated the upper reaches of science (and nearly everything else); that their idiom, especially in informal speech, has reflected that dominance.

This is not a trivial fact and we don't mean to diminish it. But it is not new. Why it is a fact remains a question of modest interest (if properly asked); but the fact itself provides no useful gauge of quality, or utility, or even of representational value in science. Incessant linguistic criticism has not yet produced a single revision of the body of serious science. Feminist cultural analysis has not yet identified any heretofore undetected flaws in the logic, or the predictive powers, or the applicability of mathematics, physics, chemistry, or—much complaining to the contrary notwithstanding—biology. Of course, a superficial reading of the feminist literature, and indeed of cultural criticism in general, leaves quite the opposite impression. But, as seems to have been forgotten in the current rise of antifoundationalism, assertion is not evidence. We cannot in any practical sense *prove* that no old science has been effectively dismissed, nor any new science produced, under the influence of feminism. Proving such a negative would require inspection of every potential counter-example. We can, however, following our adopted practice, examine some characteristic products of feminist science criticism, for which there has been fulsome praise and from which there come typically strong claims. We can show, in a progress from light to heavy samples, why we think the products and the associated claims fail to stand up to honest evaluation.

Feminist Algebra

Our initial sample is drawn from feminist criticism of mathematics, an area where one might imagine it hardest to draw connections between the content of the field, abstract as it is, and social and sexual attitudes of the circumambient society. One therefore expects analysis of some depth and subtlety, analysis of a kind it would be unjust to demand in a feminist examination of, say, the behavior of obstetricians. The example we have before us—"Toward a Feminist Algebra," by Maryanne Campbell and Randall K. Campbell-Wright[7]—is, however, remarkable for its absence of subtlety and for an ideological fervor more appropriate to an old-time camp meeting than to "analysis" of any kind. What passes for the idea behind this piece is that women and other disempowered groups are discouraged in the study of mathematics because most of the concrete problems they encounter—"word problems" or "narrative problems" of the "if-a-man-and-a-half-makes-a-dollar-and-a-half-in-a-day-and-a-half" variety—refer to situations that are sexist, racist, class-bound stereotypes. Thus the authors would doubtless condemn the "man-and-a-half" problem because it encodes the assumption that men work, and it therefore implies slyly that women don't, or shouldn't.

It may seem to innocent readers, if any such remain, that we are putting words in the authors' mouths; but no: they disapprove of a particular problem in which a girl and her boyfriend run toward each other (even though the girl's slower speed is carefully explained by the fact that she is carrying luggage) because it portrays a *heterosexual* involvement. They object to a problem about a contractor and the contractor's workers (sex undeclared), because they assume that the student will envision the workers as male. On the other hand, they offer for our approval a problem about Sue and Debbie, "a *couple* financing their $70,000 home."[8] Their general maxims call for problems "presenting female heroes and breaking gender stereotypes" and "analyzing sex similarities and differences intentionally" and "affirming women's experiences." All this, mind, is to be done in an *algebra* class.

The underlying pedagogical theory is a commonplace, and it is shaky. It holds that a proper social context stated or implied in little problems of this sort is crucial in making "disempowered" students comfortable, enabling them to solve such problems, or at least to give them serious attention. Thus, women (dare we say girls?) will do better if problems involve powerboat races between Hortense and Maxine, rather than Fred and Algernon, and black kids will be more inspired if Johnny is allotting the money he has saved up for Kwanza, rather than for Christmas. The empirical basis for such an assumption is, as we say, dubious in the extreme. Generations of Jewish kids have done quite well at these problems, despite having to concern themselves with Johnny's Christmas money, rather than Menachem's Chanukah gelt; and in recent decades an even greater cultural dissonance has done little to trip up vast numbers of young algebraists of Chinese, Korean, or East Indian background.

Nonetheless, such alterations would seem to be at worst harmless. In themselves, they are unobjectionable, although they will almost certainly turn out to be futile. Concentrating on such matters ignores the nub of the teacher's difficulty, which is precisely to train students to *ignore* the superfluous context of such problems in order to extract their mathematical and logical essence. When one is doing such problems correctly, Sue and Debbie's sexual arrangements are neither here nor there—at least until one gets the answer. An excellent—and famous—counter-example to the educational psychology propounded here can be found in Lewis Carroll's books *Symbolic Logic* and *The Game of Logic*,[9] which, in whimsical Carrollian fashion, teach the student to work in the efficient realm of abstraction by presenting concrete situations that are delightfully absurd.

"Toward a Feminist Algebra," as we have thus far characterized it, may appear to be no more than an overly solemn formulary for teaching simple mathematics; and the reader may be forgiven for wondering why it should be invoked as an example of feminist criticism of serious science. First of all, the authors insist they are speaking of *college* algebra. In less perplexing times, we might have been able to point out that the material they so designate is properly high school, or better, junior high school algebra, the term *college algebra* being reserved for a study that begins with the theory of matrices and vector spaces, and continues with abstract algebraic systems—groups, fields, rings, and the like. Today, however, this simple material is routinely taught in many universities, with only the barest whisper that it is remedial work. That is a minor point, however.

More importantly, this paper insists that it is making deep and serious points about the mature *science* of mathematics as it has developed in the past few hundred years. It strongly suggests, moreover, that the changes proposed in the wording and "social context" of simple exercises will somehow induce more women to become mathematicians. That latter is the worthiest of goals. However, even if we grant the pedagogical efficiency of feminist-approved terminology, and concede that it might help some reluctant young women to handle simple algebra, the fact remains—and it is a fact—that anyone beyond the age of twelve or thirteen who has real difficulty with such problems, no matter what the social connotations of their wording, is simply not destined to be any kind of mathematician. A young lady who makes a game stab at "Maude and Mabel" problems but balks at "Joe and Johnny" versions of the same is almost certainly without the knack for abstraction that is an indispensable ingredient of mathematical talent.

If the problematical assertions of "Toward a Feminist Algebra" went no further than this, it would be a little silly, but not ridiculous. However, there is a far more solemn proposal being made here, one that insists that serious mathematics, "higher" mathematics, is saturated with sexist ideology. It is this, rather than concern with pedagogical efficiency, that really prompts the authors' obsession with the gender-hermeneutics of speed-and-distance or principal-and-interest problems. One understands their predicament. It is possible—though tendentious—to impute sexist intent to a simple algebra problem in which "Peter is meeting his girlfriend Melissa at the airport."[10] It's a rather more difficult trick to find sexism if the problem reads "Prove that a 1-connected closed 3-manifold is homeomorphic to S^3."[11] Nevertheless, "Toward a Feminist Algebra" is determined to have us believe that one attribution supports the other.

The paper is not without further arguments in this direction, but they are equally weak. There is some reliance on the clichés of cultural constructivism —mathematics is the product of a certain social order, thus contaminated by its ideological transgressions, especially sexism, and so on. Appeal is made to feminist thinkers such as Sandra Harding and Evelyn Fox Keller (of whom more below). This is reinforced by mechanistic applications of postmodern literary dogma: "Mathematics is portrayed as a woman whose nature desires to be the conquered Other."[12] (Language like this makes it difficult to forget that one of the authors is in an English department.)

This strange notion of mathematics as the willing victim of date rape is, we must admit, a new one on us—and one of us has been earning a (marginal) living at it for thirty-five years! Such peculiar ideas are supposedly bolstered by a purported examination of the language in which mathematics is couched. "If you torture the data it will confess" is cited as a typical example of violent mathematical rhetoric.[13] The trouble is, we have never heard anything remotely like it spoken! True enough, such denounced terms as "brute force" and "grinding it out" *are* common mathematical slang—but Campbell and Campbell-Wright neglect to inform us that these are universal terms of *disparagement*; grinding the answer out by brute force is what one does if one is *not* clever enough to think of something efficient and elegant. (Campbell-Wright should know better: he is supposedly a mathematician). Easily the most absurd part of this indictment is the insistence that terms such as *manipulate* ("manipulate an algebraic expression"), *attack* (a problem), *exploit* (a theorem) are evidence that mathematics at all levels is a foul nest of aggression, violence, domination, and sexism.

All this reveals a mind-set we shall encounter again and again in feminist science-criticism. Metaphorical language is scrutinized microscopically for evidence that the science in which it occurs is tainted by sexist ideology. "Metaphor" is one of the subheads of the Campbell/Campbell-Wright essay, and it labels a section beginning: "Metaphor plays a central role in the construction of mathematics."[14] No! It does *not*—certainly not the kinds of metaphors this paper alludes to. One of us, speaking as a mathematician who has seen an awful lot of mathematics "constructed" and has constructed some himself, can testify to the *uselessness* of metaphor in mathematical invention, although *analogy*—a rather different notion—can be of some help. Mathematical intuition is something much more mysterious than metaphor.[15] True enough, mathematicians have their informal slang—what clan does not?—but that is quite another matter.

Metaphor mongering is the principal strategy of much feminist criticism of science. It is invoked to accomplish what analysis of actual ideas will not. "Toward a Feminist Algebra" is a particularly childish example of this, although we shall see others, more sophisticated, shortly. The worst thing about this paper, however, is not its shoddy theory of mathematical epistemology. It lies, rather, in the fact that the ultimate aim of the authors is *not* really to advocate devices for improving the mathematical education of women and other disempowered classes. Rather, one finally discovers, the purpose is to justify the use of mathematics classrooms as chapels of feminist orthodoxy. The purpose of the carefully tailored feminist language and imagery is not primarily to build the self-confidence of woman students, but rather to convert problems and examples into parables of feminist rectitude. It is, at bottom, not different from an imaginary Christian fundamentalist pedagogy requiring that all mathematics problems illustrate biblical episodes and preach evangelical sermons. Campbell and Campbell-Wright really want mathematics instructors to act as missionaries for a narrow, self-righteous feminism. *That* is far more disturbing than bad philosophy of mathematics! Sermonizing—Christian, Muslim, Buddhist, or feminist—is not the function of science instruction. It is a strange world in which two would-be pedagogues can advocate such a program, in the belief, doubtless justified, that some of their colleagues will take it seriously.

Haploid Hermeneutics

Our next example, less blithely outrageous than "Toward a Feminist Algebra" but no less reliant, in the end, on metaphor mongering, comes from biology. No fewer than nine co-authors, calling themselves "The Biology and Gender Study Group,"[16] have collaborated to show us "what feminist critique can do for biology." At least one of the authors is a well-known practicing biologist: Scott Gilbert has written a distinguished and best-selling textbook in developmental biology (which one of us uses in the classroom) and writes on the history of biology as well. If any group can speak to the improvement of science via feminist epistemology, it is presumably one like this.

We take the liberty of summarizing some of the claims of the BGSG (taking care to segregate those from the abundant self-congratulation) in *The Importance of Feminist Critique for Contemporary Cell Biology*. The introduction sets forth the standard claim of feminist theory with respect to biology:

> Gender biases do inform several areas of modern biology and . . . these biases have been detrimental to the discipline. In other words, *whereas most feminist studies of biology portray it—with some justice—as a privileged oppressor*,[17] biology has also been a victim of the cultural norms. These masculinist assumptions have impoverished biology by causing us to focus on certain problems to the exclusion of others, and they have led us to make particular interpretations when equally valid alternatives were available. (Emphasis added.)[18]

Without equivocation, we are here promised evidence, from people who work in the scientific discipline they are talking about, of (1) bad science that would be better if it were opened to feminist insight, and (2) good science ("equally valid interpretations") made possible or generated by such insight. This seems to guarantee a measured assessment of scientific results, rather than the usual linguistic, cultural, psychoanalytic, or political auspication. The hope proves, alas, to be bootless. The article starts by referring to seed-and-soil cosmological myths from Aristotle to the eighteenth century. That Aristotle harbored opinions we would nowadays dismiss as sexist is probably true; but that is hardly something new and it has no relevance for contemporary biology, in which Aristotle is no authority figure. We come next to "Sperm Goes A'Courtin'," which turns out to be an indignant attack on a book published in 1890—not a technical book at that, but a popularization by Sir Patrick Geddes and John Arthur Thomson dealing with sexual physiology. This was published soon after the discovery of syngamic fertilization (sperm-egg fusion), but before the discovery of chromosomal sex-determination. It contains much talk about *metabolism*, about the anabolic (building-up, nutritive, vegetative) qualities of eggs vis-à-vis females and the catabolic (breaking-down, active) qualities of spermatozoa vis-à-vis males. These notions are related by the BGSG to the masculine British ritual of the hunt (which we might have thought to be concerned with class, rather than sex), and back to an Aristotelian emphasis on nutrition (of course most eggs do contain nutritive yolk, while the male gamete does not), all this with appropriate feminist aspersions.

We readers are supposed to join in the derision of Geddes and Thomson, even though they were in fact of no importance to the science they wrote about. It is hard to see why derision is called for. Metabolism (not just nutrition) was then in the air within the fields of biology and medicine. It was and remains an important concept, and there was in those days new knowledge about it. The science that was to mature as "physiological chemistry" and, later, as biochemistry was being built. Moreover there was the hard fact that sperm *are* motile: they swim actively; eggs do not. Everybody in those days related everything to these jots of new knowledge, just as everybody who is anybody in biology today relates everything to the nucleotide sequences in DNA[19] (and just as everybody who is anybody in cultural criticism relates everything to "overdetermined figures"—or to Madonna; everybody in literary criticism to "the indeterminacy of the text"; everybody in feminism to a universal and ineluctable "gendering"). The language in this ancient book is innocuous, except to a perfervid eye. One wonders why the authors bothered to belabor it.

The next example offered is a paragraph from a paper of C. E. McClung, the American cytologist who discovered chromosomal sex determination. The authors begin their critique of a McClung paper (of 1901!) by suggesting that he had been unduly influenced by Geddes and Thomson. "Using a courtship analogy wherein the many spermatic suitors courted the egg in its ovarian parlour, McClung . . . stated that the egg 'is able to attract that form of spermatozoon which will produce an individual of the sex most desirable to the welfare of the species.' He then goes on to provide specific gender-laden correlation." What gender-laden correlation is that? The authors quote a long paragraph from McClung that includes such presumably offending passages as "the ovum . . . reacts in a way best to subserve the interests of the species. To it come two forms of spermatozoa from which selection is made in response to environmental necessities."

It is a mystery that the authors find any of this "gender laden," unless they mean that he attributes *too much* power to the egg (as he does, justifiably in the context of his time). Of course the words refer to *sex*. That's what the research (and the writing, which is not the same thing) were about. Of course we know that McClung wasn't quite right about what the egg does at fertilization (which demonstrates the continuous, self-correcting character of science); but this is not simply because he attributed to it a stereotypical female passivity. The most sexist, most gender-laden statements here are the paraphrases concocted by the *authors*.

McClung's notions of the egg as a sperm selector are of minor relevance to the assessment of his contribution to reproductive biology.[20] What is important is that, following up earlier work of Thomas H. Montgomery and others, McClung was the first to understand (from microscopical research on grasshoppers, as it happened) the general meaning for sex determination of the "accessory chromosome." This led to E. B. Wilson's presentation of the theory relating the sex chromosomes, sex determination, and sex-linked inheritance.[21] McClung's discovery was one of the foundations of modern knowledge of sex determination and sexual development. His cytological observations were rock-solid.

Where are those questions, unasked because of bias, that we were promised by the BGSG? Where are the "equally valid alternatives?" So far this critique is nothing but the all-too-familiar metaphor mongering, likely to convince no one who is not already excited by the idea that Francis Bacon and Isaac Newton were advocates of rape.[22]

The section entitled "The Sperm Saga" attacks statements that treat the spermatozoon as a victor in conflict. It is concerned with "narratives" in which, this time, the egg doesn't choose a suitor but is instead the prize, for which the victor has competed with hordes of others. The authors excoriate textbook authors who perpetuate "this epic of the heroic sperm struggling against the hostile uterus." There is a long quotation from a once-famous introductory biology text (by William R. Keeton), a quotation whose language is far more moderate than that used by the authors to attack it, and whose factuality, whatever adjectives are used, is unequivocal (e.g., that in mammalian reproduction large numbers of spermatozoa die before reaching the cervix).

The BGSG next moves to narratives of fertilization they characterize as "I alone am saved" (referring to the successful spermatozoon). Here they disparage an obscure manual for expectant mothers, objecting to such "images" as sperm lying in wait for the ovum, an "army" of spermatozoa, "penetrates," "electrifies," and the like. The authors identify such images with "a kind of martial gang-rape," thus exceeding by an order of magnitude the floridity of language of the manual's unfortunate author (who surely never had rape in mind). The BGSG sees fit, moreover, to compose this vehement burlesque: "The fertilizing sperm is a hero who survives while others perish, a soldier, a shard of steel, a successful suitor, and the cause of movement in the egg. The ovum is a passive victim, a whore, and finally, a proper lady whose fulfillment is attained."

An imaginative, deconstructive reading of the original, trivial text! The criticism is riper by far than the criticized. There can't be many places in the literature of science where the unfertilized egg is called a proper lady, much less a whore. Note well that the argument has nothing to do, so far, with the scientific results supposedly under discussion. Nothing said or implied about the observable events of conception in the text under attack was wrong at the time it was written. No useful alternative science of feminist origin is suggested. Only the desirability of a different metaphorical conceit is implied; yet the BGSG doesn't trouble itself to give an example of how the thing might have been said free of sexist images. (A deep foreboding afflicts us as we try to imagine the sanitized language.) There is nothing in this section of the paper that is a genuine case of research opportunity missed, or a genuinely alternative interpretation of the data slighted due to sexist bias; no improvements of biology are offered. We are still in the realm of quibbles about incidental metaphors in secondary books.

Cherishing the Ovum

The next section of *The Importance of Feminist Critique for Contemporary Cell Biology* looks at first more serious. It concerns the activity and passivity of egg and sperm. Here the BGSG comes up with what is meant to be an example of the benefits in store for biology once it has been enriched by feminist awareness. They propose that until recently, textbooks emphasized the passivity of the egg. They then cite with admiration a popular article on fertilization published in 1983 by G. and H. Schatten. This account (entitled, quite reasonably, "The Energetic Egg") discusses electron-microscopic data that imply an active role of the egg surface in attaching the sperm head and drawing it inward. Following this are other items of information, in light of which the egg's role in fertilization and development is anything but "passive." We are supposed to conclude, then, that without the habits of thought provided us (and the Schattens) by feminist insight, we would still be mired in the thought of the egg as a fat, immovable female-vegetable of a cell, and of the sperm as a steely bearer of glad tidings, a swift warrior.

But—and we are aware that each time we say it we lose a few more friends—this is nonsense. Reproductive biologists of either gender who spoke that way would be considered by their colleagues, and doubtless by their mates, as overdue for deep psychoanalysis. And this is nothing new; the same would have held in the nasty old nineteenth century. Artificial parthenogenesis was demonstrated and announced in 1899, by Jacques Loeb. Artificial parthenogenesis is metabolic activation of the egg, and in some systems the initiation of embryonic development, by a stimulus *other* than the spermatozoon. It was sensational news at the time, not only to biologists but also for the lay public. Loeb's unnatural, asexual stimulus was to change the chemistry of the seawater in which sea urchin eggs were suspended. Loeb did not have the benefit of feminist lustration such as that of the BGSG to guide his thought, his experimental methods, or—more importantly—his language. [23]

That eggs of many species are capable of developing partly or completely with an appropriate stimulus other than the penetrating spermatozoon is taught in every embryology course and in most courses of introductory biology. One of us has demonstrated it to several generations of college students. Parthenogenetic development is a normal option in some species (although not, so far as we know, in ours!). There is no parthenogenetic development from *sperm*. Classical developmental genetics was for a long time preoccupied with a proper attribution of phenotypic (visible) outcomes of development to the maternal or paternal gene-set. Among the most important (and epistemically fruitful) kinds of gene mutations affecting animal development have been the so-called "maternal effects." This is standard science, whose recognition and promulgation owes nothing to feminism or masculinism, or to any other ideology.

As long ago as 1964, one of the means was discovered by which the egg—that huge, complex, and unique cell—knows what to do in the critical, early hours following fertilization. The key is the "maternal messenger RNA," a population of stabilized genetic messages made and distributed in the egg cytoplasm during maturation in the maternal body. *There is no such population in the spermatozoon.* That story too is standard biology, treated well and comprehensively, in fact, in Gilbert's textbook[24]: these RNA molecules, acting alternately or simultaneously with others copied from the embryo's nuclear DNA, are responsible for laying out the plan of future body development. Eric Davidson's authoritative monograph on genes in early development, the standard work now through three editions, concerns itself almost exclusively with the egg. [25] There is not a trace in it of the kind of "imagery" in which the metaphor mongers delight.

There is thus a vast and serious science of what the egg does—actively—relative to the sperm. It has emerged over the last thirty years, independently of feminist or any other kind of cultural criticism. The important contributions to it have been made by women as well as men, and women are among today's leaders of the field. None of the debates that have raged in it have had anything whatsoever to do with gender or metaphor. The large literature in which the new knowledge is embedded has, as far as we know, no significant component written in sexist language or driven by gendered images; and even were the academic Bowdlers to find such, it would have no effect upon the data or the facts they represent. In short: no case has been made in this—entirely representative—field of biology for opportunities overlooked due to sexist bias, or for valuable alternative interpretations having arisen out of feminist (or any other brand of cultural) criticism. Not, that is, if by "interpretations" we mean efforts of objective analysis that relate empirical data, predictions, and experimental designs to one another by means of logic. Considerations of space and balance prohibit further analysis of this work by the BGSG. We would urge anyone among our readers who has done the sort of college courses taken by premedical students, and who may remember a little organic chemistry, to search out in it the amazing unmaskings of such nasty masculinisms as "nucleophilic attack."

If there is anything of substance in the BGSG critique, it is that writers on science—especially popular writers—ought to be careful with figures of speech and should be vigilant in avoiding the pathetic fallacy.[26] This stricture is more appropriately addressed, however, to feminist critics, the BGSG included, than to working scientists. What is saddest about this well-meaning group—we do not impugn their sincerity or their desire for equality between the sexes—is the contrast between the decency of their intentions and the triviality of the results. "By using feminist critique to analyze some of the history of biological thought," the authors assert, "we are able to recognize areas where gender bias has informed how we think as biologists. In controlling for this bias,[27] we can make biology a better discipline . . . We become what biology tells us is the truth about life. Therefore feminist critique of biology is not only good for biology but for our society as well." Wishful thinking is the customary name for this such "analysis."

The Spin on Sexuality

No thoughtful observer of original science and its popular expositions denies that the language of interpretation is influenced by the tastes of the interpreter. The further the interpreter is from the original work, moreover, the more likely those tastes are to color the statement. It is a commonplace among relativists of all kinds to ignore or dismiss the self-correction process by which good science survives and bad science—*that which is not verifiable by others of different tastes and tendencies*—vanishes in due course. In many cases, among students, for example, the ignorance is excusable as such. In others it is a deliberate dismissal, and that is inexcusable.

There is always much ado among feminist science-critics, as must now be clear, about tendentious interpretive language. These critics are governed by the impulse to take language very seriously, even when it is clearly metaphorical or simply whimsical. Their censoriousness applies just as strongly to the offhand self-deprecations of the late, brilliant Richard Feynman as to the turgid metaphors of Francis Bacon (who did little of scientific value and goes unread by the vast majority of scientists). The temptation to construe colloquialisms as tokens of deep epistemological error has been a ceaseless element of feminist criticism, and one of the most fatuous.

Nevertheless, granting that women have been discriminated against in science, that their contributions have often in the past been undervalued, one can justify a certain watchfulness. However negligible the power of inappropriate metaphor may be to shape the *ultimate* body of scientific knowledge, there is no great harm in sensitizing people to it. The less so, then, is there harm in a keen alertness to consciously politicized interpretations of science for public consumption, that is, in the popular media. If there were any longer anything like a hegemonic, white, masculinist slant to popular natural and social science, denunciation of it would be proper.

The shoe, however, is now on the other foot. Anyone who gives prime-time television a passing glance (we hope none of our readers give it more) is familiar with today's universal spin, for example, on women's careers. Who has not looked sidewise at the screen and seen a beautiful young woman (political correctness in the media does not yet frown upon "lookism"), high heels, lipstick and all, leaping about with her 9-mm. Beretta held, two-handed, in the approved barrel-up manner, dodging around corners, stalking a murderous criminal? Who has not seen her straddle and handcuff the oaf, toward the end of the show? Who has not seen the impenetrably tough, young woman lawyer face down a crooked male judge in court, and then, as a sop to story line and the *connectedness* of women, make a lonely phone call to her mother, or her sister, late that night? Who, for that matter, hasn't seen the new, standard children's books, in which Mama Bear, like Papa Bear, goes to work or runs a honey-packing business? Who is so asleep as not to have noticed that Dagwood of the funnies, always an amiable idiot, is today a bigger jerk than ever, now that Blondie is a successful businesswoman while he remains under the thumb of his boss?

If there is any spin on popular science, it is today of that kind, and not of the "females of small brain and weak body" kind. As a foil to such hand-wringing as that of the Biology and Gender Study Group, we cite for convenient example a special issue of *Discover*, the most (deservedly) successful slick magazine of science published in the United States. This issue[28] was devoted to the science of sex, that science to which feminist critics give their closest attention. Its handsome, four-color cover sets forth the questions to be explored inside, including:

Tales of a Sperm	The Hermaphrodite Paradox
The Aggressive Egg	Why Do We Know So Little about
Evolution of the Orgasm	Human Sex?

As is usual in popular science writing, the articles vary in detail and fidelity to the original science and in the depth of understanding displayed by the writers (some but not all of whom are practicing scientists, not journalists). Some of the articles in this issue are, as usual, perfectly sound. Some are preposterous. Without exception, however, they embody and contribute to a certain spin on matters of gender. It is that the new scientific knowledge of sex(uality) reveals the rigidity and narrowness of all earlier formulations. The articles and sidebars are replete with imagery in which metaphor mongers might well delight; they offer speculations looked upon with great favor by feminists.

Among them, for example, is a paean to the Bonobo, "a rare species of chimplike ape in which frequent couplings and casual sex play characterize every social relationship—between males and females, members of the same sex, closely related animals, and total strangers. Primatologists are beginning to study the bonobos' unrestrained sexual behavior for *tantalizing clues to the origins of our own sexuality*" (emphasis added). This comes from the folk who sneer at sociobiology for extrapolating from animal to human behavior! This article, one among siblings, is replete with photographs of the charming primates engaging in the described sexual acts. "Maiko is on top, and Lana's arms and legs are wrapped tightly around his waist. Lina, a friend of Lana's, approaches from the right and taps Maiko on the back, nudging him to finish. As he moves away, Lina enfolds Lana in her arms, and they roll over so that Lana is now on top. The two females rub their genitals together, grinning and screaming in pleasure."[29] What fun! "In reconstructing how early man and woman behaved," we are told by the author, Meredith F. Small, "researchers have generally looked not to Bonobos but to common chimpanzees." The burden of what follows is that this was a *very* bad mistake. The spin is that we have undeniable evidence already for that fondest of constructivist, feminist fantasies: an original, genderless human society of perfect sexual and behavioral equality, full of great games and no domination.

David H. Freedman, writing on the "aggressive egg," takes an unwitting page from the writing of C. E. McClung, whom the Biology and Gender Study Group scourged for gender-laden imagery. Freedman's article, near its beginning, describes the spermatozoa as "a *wastefully* huge swarm . . . flop[ping] along, its members bumping into walls and flailing *aimlessly*." Then, once they are in the vicinity of the female gamete, "the egg *selects* one and reels it in, pinning it down in spite of its efforts to escape. It's no contest, really. The gigantic, hardy egg yanks the tiny sperm inside, distills out the chromosomes, and sets out to become an embryo" (Emphasis added). Look on and smile, Beretta-wielders!

The spin is conscious. Its purpose is to highlight the small size, the prodigality, the incompetence, the misdirectedness of male gametes, and the purposeful, dominant object that is the female. The author's real heroine, subject of the article, is anthropologist Emily Martin, who is reported to have set out to undo popular notions of warrior sperm and damsel-in-distress eggs. "Can biased metaphors be eliminated from science?" Freedman (and Martin) ask. "Martin doesn't think so . . . The goal shouldn't be to clean the imagery out . . . but to be aware that it's there." The implication (Martin's and Freedman's) is that until now nobody had heard of metaphor, or of the distinction between metaphor and the underlying facts of science.

Not unexpectedly, this special issue of *Discover* includes higher-level commentary; this by the well-known Anne Fausto-Sterling, who insists elsewhere that nature is a peaceful *socialist* collective, rather than (as in the capitalism-stained imagination) "red in tooth and claw."[30] It is entitled "Why Do We Know So Little about Human Sex?" Her answer, which *assumes* that we know much less about that subject than we ought to, is as firm as is the denial, in her polemical book on the subject, of significant biological differences between men and women.[31] It is that Senator Jesse Helms and Congressman William Dannemeyer, driven by a deep-seated fear of homosexuals and supported by constituencies of like-minded conservatives, have blocked funding of the necessary research. This, she explains, is but the latest chapter in a saga of repression, by which the efforts of brilliant investigators, from Krafft-Ebbing to Kinsey—and after—have been marginalized.

While we have no immediate objection to the blaming of politicians for inserting themselves into the peer review process, if and when they do, it seems to us quite unsporting not to praise male college presidents, deans, vice-presidents for student affairs, and affirmative action officers. Hundreds —literally—of those go daily far out of their way to provide positions (tenure-track and otherwise), honors, and research support for such work, and for political-cultural critics who write about it. Their support is worth *much* more in dollars than Jesse Helms could possibly subtract. But then again, "sporting" is an encomium of male provenance.

Discover's spin is heavy: science, unleashed, opened to the insights of the marginalized, can give us endless treasures of *knowing*, like those so colorfully presented in the magazine. It could give us new hope for true sexual freedom, were it not for the obstructionism of (male) politicians and their retrograde, homophobic supporters. Well, let a hundred flowers bloom, as the hero of an earlier left-wing enthusiasm was wont to observe. Let us not pretend, though, that masculinism prevails any longer in the metaphors, let alone in the substance, of science. That substance, we are forced to report, seems stubbornly resistant to spin applied in either direction. You may choose to see spermatozoa as a mindless mob, each one a chromosome set strapped to a hot-rod; or you may see them as streamlined engines of delivery for the indispensable paternal genes, without which there can be none of the combinatorial advantages of sex. The less of such "seeing," the better the science. Neither trope matters in the slightest when serious scientific issues are on the line. What matters is the evolutionary consequence of sex in reproduction, from microbial to human. What matter are the remarkable details of devices for sexual genetic recombination. What matters is the physics, the chemistry, and the information content of the sperm and of the egg, and what happens when all those become the property of a new, single, diploid cell—the zygote, the beginning of an individual multicellular life.

Dealing with Physics

Our epigraph, two sentences from Sandra Harding's influential book on feminism and science, stands here as a signal of the bewilderment in store for the innocent but literate explorer of the feminist critique. It represents not only the characteristic animus, but also a style of argument frequently employed in justifying such hostility. Harding's success in establishing a reputation as a major thinker on science and epistemology would be incomprehensible in an age less determined to celebrate difference at all costs. To "difference" we might have added "heterodoxy," except that it is our own views that are now unorthodox, at least to those outside of science. The success of her "justificatory strategies"[32] may be deserved in a society whose most celebrated thinkers tend to be TV anchorpersons and talk-show participants. Hers is a discursive style crafted for a certain constituency, designed to gratify without challenging, to offer the emotional rewards of rebellion without the actual work and dangers of combat.

Let us examine her approach to the important question of physics. What can the quoted statement possibly mean? What can be said in defense of it? Let's have a look at the first sentence: "Would not physics benefit from asking why a scientific worldview with physics as its paradigm excludes the history of physics from its recommendation that we seek causal explanations of everything in the world around us?" First of all, the sentence is freighted with assumptions. There is supposed to be a regnant worldview, with physics as its paradigm. Given that only a vanishingly small fraction of the world's five billion souls know anything about physics, it cannot be that physics is any sort of demotic weltanschauung. If we limit our attention to what likes to think of itself as the Developed World, the situation is hardly different. Let us then narrow the focus to the microscopic subworld of professional intellectuals. At least half of those—shall we say most humanists, most historians, a good fraction of sociologists, a surprising number of philosophers—know virtually nothing about physics either. Such ignorance is not a particular virtue, but neither is it a vice: it has not precluded remarkable intellectual achievements. Humanists have not, traditionally, felt the need to apologize for this gap in background, *nor should they* (although a grain or two of embarrassment might well be felt for the propensity of a few colleagues to pontificate on the deep meanings of physics without having bothered to learn it).

We are left, presumably, with natural scientists and some of the social scientists who have been trying, with variable success, to introduce into their disciplines a measure of what is usually thought of as scientific rigor. It is to some degree justified to assert that this community takes physics for its paradigm, out of acknowledgment that physics is the most successful of empirical sciences. To say that, however, is to gloss over the important debates that engage scientists in all fields. How and to what extent *should* physics serve as a model? There is the issue of reductionism in the strict sense, the distinct question of whether a given discipline *ought* to rely on a logico-mathematical model for its theoretical structure, the question of the degree to which constructs appearing naturally within such a structure should be reified. These questions are as subtle as they are important. Leading thinkers in all fields of science confront them ceaselessly. Thus to speak of physics as a "paradigm" is to vulgarize the situation.

We are assured that this putative paradigm enjoins us "to seek causal explanations of everything in the world around us." Can this be so, even for the most conceited of physicists? Would it be possible to dredge out of the laboratory even one scientist who thinks it possible, or desirable, to find a causal explanation of the fact that Mozart wrote music that sounds like Mozart and not like Ditters von Dittersdorf? Then too, the statement seems to encode a naive notion of causality. As used by scientists, causality is hardly an unexamined idea. There has been painstaking attention to it from almost everyone who takes physics seriously. In fact the concept is employed with the closest attention to fine distinctions. The relationships among "causal explanation," "predictability," and "verification" are particularly subtle. But Harding, as polemicist and cheerleader, is indifferent to such questions.

So much for the assumptions of the sentence. What of its key assertion? Does the "paradigm" exclude the history of physics from its otherwise sweeping injunction to seek causes? From almost any point of view this seems to be nonsense. Has physics been inattentive to its own history? No: physicists seem in fact to be obsessed by it.[33] History is not some parochial hobby: great achievements have derived from it. To take the best-known example, the meditations that led Einstein to relativity had their origin in a serious examination of what—from the nineteenth-century point of view—might have seemed a work of purely antiquarian interest: Galileo's speculations on the invariance of natural law from one inertial frame to another.

Does physics, then, have an explanation for the history of physics? In one very strong sense it does: the history of physics as a collection of ideas is largely explained by the objective nature of the phenomena it describes and schematizes. Thus Kepler's laws of motion are explained by the fact that, to a high degree of precision, the planets move as predicted by those laws. This seeming tautology will leave relativists and cultural constructivists feeling quite out of sorts; but, as explanations go, it is supremely solid and convincing.

Does this mean that the "paradigm" rules out attention to other factors, social ones in particular, in explaining the history of physics? The answer, to judge by the attitudes and interests of physicists and other scientists, is surely "No" again. They would love to know why modern science—that endlessly fruitful marriage of empiricism with mathematical thinking—took root and flourished in seventeenth-century Europe, rather than in second-century Rome or tenth-century China, which would have been equally likely nurseries. Scientists welcome the sort of "social" explanation that examines minutely and honestly the intellectual, attitudinal, and—to be frank—the moral preconditions of culture that encourage and sustain the practice of science. More is the pity that academic opportunism turns some historians of science into ideologues.

Dare we go on now to read Harding's second sentence? "Only if we insist that science is analytically separate from social life can we maintain the fiction that explanations of irrational social belief and behavior could not ever, even in principle, increase our understanding of the world physics explains." What about the assumptions underlying this statement? They are obviously dead wrong if we take science, not unreasonably, to include such fields as behavioral psychology, cultural anthropology, and atmospheric chemistry. We give her the advantage, then, and take the statement to refer to "pure" physics. There she is quite right. In the view of most physicists, it has been analytically separate from social life. There have been some egregious exceptions, such as Philip Lenard, the Nazi physicist who, in denouncing Einstein, declared: "German physics? one asks. I might rather have said Aryan physics or the physics of the Nordic species of man. The physics of those who have fathomed the depths of reality, seekers after truth, the physics of the very founders of science. But, I shall be answered, "Science is and remains international." It is false. Science, like every other human product, is racial and conditioned by blood."[34]

We should be grateful that most physicists found this nauseating, as well as unpersuasive; and we think they would find Harding equally unpersuasive. They consider the analytic separation quite appropriate. Arguments to the contrary, whether from Engels, Lenin, the Institute for Creation Research, or Stanley Aronowitz, have been undone by the absence of competent attention to the substance and logic of physical ideas. This flaw applies to Harding's arguments as well; she doesn't know anything about physics. (In fairness, she doesn't claim to.) How then can she decree with any assurance what physics may or may not be analytically separate from?

We shall try to see. She evidently wants us to accept that explanations of irrational social belief can and do increase our understanding of the *physical* world. This is an amazing notion. We would need, at the least, some examples in order to judge. None are provided (unless "the world physics explains" means something other than the usual and agreed subject matter of physics). She wants to talk, presumably, about social attitudes, political prejudices, discriminatory myths, subconscious sexist assumptions. To the best of our ability to make out her program, as set forth in the book from which the quotation is taken, it is as follows:

1. Western society is in some measure rooted in assumptions about the differences in worth and ability between men and women.
2. These are irrational and harmful beliefs.
3. Such beliefs permeate all our social institutions and all aspects of our belief systems.

4. Therefore even physics is biased and distorted by the ineluctable influence of these irrational beliefs, and therefore

5. analysis of this root-and-branch unreason will lead eventually to clarification and rectification, even of the recondite world of physics.

Points (1) and (2) are unexceptionable; (3) is partially, but only partially true, and it is altogether too categorical; (4) is substantially *untrue* as regards the actual content and methodology of physics—there is no plausible body of evidence for it—and, consequently, (5) represents no more than the triumph of hope over logic.

As to this last point, we surmise that Harding herself feels its force. Elsewhere in the book, we find this remarkable admission: "If it is reasonable to believe that physics should always be the paradigm of science, feminism will not succeed in 'proving' that science is as gendered as any other human activity, unless it can show that the specific problematics, concepts, theories, language and methods of modern physics are gender-laden."[35] Fair enough! This compels, then, a detailed examination of the logical and conceptual structure of contemporary physics, a close analysis that must reveal, more sharply than by mere analogy, precisely how sexist attitudes inflect, indeed *distort*, the body of ideas now generally recognized as giving a cogent view of the physical world at the level addressed by physicists. Alas: nothing so interesting emerges. As we have said, Harding doesn't seem to *know* much physics. What we get instead is a fog of evasions and excuses for *not* tackling physics directly, culminating in the amusing "solution" Harding proposes to the problem she has posed: "If physics ought not to have this [paradigmatic] status, then feminists need not 'prove' that Newton's laws of mechanics and Einstein's relativity theory are value-laden in order to make the case that the science we have is suffused with the consequences of gender symbolism, gender structure, and gender identity."[36]

Not many persons with real experience of contemporary science, taking the time to examine such arguments as this, are likely to adopt them (except, perhaps, as articles of faith).[37] If this argument—and, in a kind of fairness, we must note that it is not quite so incoherent as is the succeeding section, which purports to demonstrate the illusory character of "pure" mathematics—represents the analytic method feminists want to urge upon physics, then there is no chance the proffered gift will be accepted, whether or not the potential converts are women.

The reader unacquainted with radical feminist epistemology, and especially with the flow of trendy commentary from secondary authors (as in the public media and in various theoretical organs of postmodern feminism), may judge that too much pressure is being placed here upon small samples: the innocuous BGSG's metaphor mining, Harding's text—that these remarks are an unfair burden upon a few sentences. But these are not atypical, nor are they her most redolent remarks. Her stirring assertion to the effect that Newton's *Principia Mathematica Philosophae Naturalis* is a "rape manual" may well have won her a lasting admiration in doctrinaire feminist circles and even a place among physicists. We pity coming generations of freshmen physics students who, titillated by this famous remark, will spend long hours thumbing through that magisterial work, looking for the dirty bits.

Lately, Harding has been enlarging her notional armament. In her latest work, she has boarded the multicultural gravy train. She now propounds a doctrine labeled "strong objectivity." The idea is that once proportionate numbers not only of women but of blacks, Native Americans, Latinos, gays, lesbians, and other disadvantaged groups join the ranks of science (whilst retaining full and unapologetic pride in *ethnos*), science will become more open, more inventive, and above all—more objective. We share with enthusiasm the hope that the ethnic and sexual demography of the sciences will come to resemble that of the human species as a whole. But the idea that physics is in for a major conceptual upgrade because multiethnic perspectives will be brought to bear upon it is sheer fantasy. Recall that since the end of the eighteenth century various groups at one or another time regarded by European Christians as lesser breeds have come increasingly to be represented in science. These groups include, *inter alia*, Jews, Indians, Arabs, Pakistanis, Chinese, Japanese, and Koreans. As individuals, many of them have made contributions of the first rank and of enormous influence, and many have been honored appropriately. To claim that their ethnicity left a particular stamp on the content of their achievement is to revert to the odious ethnic essentialism of Professor Lenard.

Expanding the pool of scientists will produce more and perhaps better science; but it will not create African science or gay science any more than it will women's science; nor will some new multicultural science arise upon the ashes of the old white European male science. We are happy to leave the last word on Harding's book and its admirers to the philosopher Margarita Levin, whose meticulous dissection of feminist science criticism was done half a decade ago:

> One suspects that feminists themselves sense the emptiness of their enterprise. Those confident of their product do not strain to oversell it, yet much of feminist scholarly writing consists of wildly extravagant praise of other feminists. A's "brilliant analysis" supplements B's "revolu-

tionary breakthrough" and C's "courageous undertaking." More disconcerting is the penchant of many feminists to praise themselves most
fulsomely. Harding ends her book on the following self-congratulatory
note:

> When we began theorizing our experience . . . we knew our task
> would be a difficult though exciting one. But I doubt that in our
> wildest dreams we ever imagined we would have to reinvent both
> science and theorizing itself to make sense of women's social experi-
> ence.

This megalomania would be disturbing in a Newton or Darwin: in the
present context it is merely embarrassing.[38]

Science and "Science Studies"

That the outpouring of feminist science criticism, a part of the larger genre of
"science studies," is revolutionary and of fundamental importance is a given
for the academic and intellectual left. Such works as Harding's *The Science
Question in Feminism* have already attained the stature of classics; likewise
their writers are widely acclaimed as members of a new wave in scientific epistemology. Nor are they reluctant to drape themselves in that mantle, as
becomes evident to the reader of an interview given by Donna Haraway, one
of the greats of the business. What is impossible to discover, unfortunately, is
exactly what contribution to epistemology has been made, or is being made.
The honored achievement seems to be not philosophy but cage rattling.

The interview referred to, for example, is introduced by the journal's
editors with a panegyric to Haraway, which does not fail to mention that she
was trained as a biologist and that now, having moved, presumably, up in the
world, she teaches in the history of consciousness program at the University
of California at Santa Cruz. To clarify any possible misunderstandings and to
set the stage for an exposition of her contributions to (presumably) "meaning," the editors summarize Haraway's position: "Her mission—which she
claims not to have (simply) chosen but seems to thoroughly enjoy—is to melt
frozen categories, rearrange the landscape, and generally mess up everything
on the map. *She continually insists on the socially constructed and politically
contested nature of facts, theory, practices, and power* (emphasis added).[39]

Fair enough: we can expect to see some social (or cultural) constructivism
in what follows, and some sharp "messing up" of arguments that fail to take it
into account, or to accord it proper respect. No such luck. In the interview
proper, Haraway is asked "How is this related to your recent efforts to *explode*

the idea of social constructivism, to question the assumption that once you've asserted that scientific knowledge is socially produced you've said it all?" (emphasis added). Whoa! How can she *insist* on "the socially constructed and politically contested nature of facts, theory, practices, and power" and at the same time be engaged in *exploding* social constructivism? We confess that, though we try to be careful readers, we are evidently not careful enough, because we can't make consistent sense of her assertions. Her reply to the question begins on an admirably firm note: "Plainly the social constructivist argument has its limitations, because it ends in relativism. The whole thing is set up within a very conservative philosophical tradition. And that's very problematic." *What* is very problematic, a very conservative tradition, or relativism? What follows seems to favor the former:

> I find that the most enlivening work tries to sidestep that set of philo-sophical traps, that part of our analytical inheritance. I'm also trying to sidestep it, by saying that nature, at every level of the onion, is artefactual—that is, made—but not just by us. All the actors aren't human, all the actors aren't machines.
>
> Of course the obvious binary objection to that is no, the world's not made, it's given. It's not a product, it's a gift. It pre-exists our actions upon it; it is matrix to our action; it is resource to our instrumentality. Those are exactly the traditional modes of Western philosophy. How in the world can we sidestep them, crack them open, deconstruct them?

Advanced thought of this sort cannot be expected to be merely *linear*, of course, so we must make the effort to follow the zigs and zags as they appear. The *resultant* of these vectors seems to be, roughly, that an opposition be-tween a "made" world (or a "made" reality) and a "given" one is itself not real: it has simply been taken as such by Western philosophy, from which we had best escape. Reverting to the original question, then, Haraway's response seems to be tending, not toward "exploding" social constructivism, hence relativism, but rather toward a denial that the underlying question (i.e., How do we know?) is necessary. This is hardly a firm position from which to conduct epistemological sorties; but, well, it *is* a position. In what follows, unfortunately, Haraway seems abruptly to become a full-fledged relativist:

> There are people in the world who don't have our problems, who inherit other ways of imagining, who really are discursively different, and that means materially different. But even Western discourse isn't homogeneous—it's incredibly eclectic. Hermeticism, for example, sees

the world as alive, says that matter is active . . . but it's available, and lots of people are using it in various ways. Certainly ecofeminist discourse relies on it.[40]

If Haraway judges hermeticism to be "available," then presumably she judges it as valid; and indeed the tone of the discussion makes it clear that it, and ecofeminism, are to be considered perfectly adequate alternatives to "post-Enlightenment scientific discourse." Let the confused reader not worry: we are confused, too; and so were the five thoughtful people, all strongly sympathetic to feminism, to whom we applied for clarification. They don't know, and we don't know, whether Haraway is for or against relativism; but a vote indicates that whatever "limitations" she sees in relativism, they are not fatal "at every level of the onion." For this "discourse"—and the quoted passages are entirely typical—on what is, after all, a quite fundamental question of scientific epistemology, we have been able to find only one *signifier*: it is Peter Mayle's term, invented originally to describe certain goings-on in Provence, especially in the season of wine tasting: "delusions of adequacy."

"Justificatory" Strategies

A 1989 volume edited by Anne Garry and Marilyn Pearsall, *Women, Knowledge, and Reality*, prepared for use in the teaching of philosophy and women's studies, includes a section on philosophy of science. It is, specifically, *feminist* philosophy of science, the book's purpose being, one judges, to convey in compact but authoritative form the new acumen brought to the philosophical subdisciplines by feminist thought. The editors chose well. The three contributors on science, Sandra Harding, Evelyn Fox Keller, and Helen E. Longino, are unquestionably among the leaders of the field; their papers, reprinted or reworked from earlier contributions, represented at publication the cutting edge of feminist thinking on the subject and are, furthermore, good summaries of the work of each writer. Here—finally—we may hope to find those sharp epistemic analyses that expose the philosophical shortcomings of science, along with comprehensible guides to such alternative ways of knowing the material world as patriarchal science has overlooked, by its nature or as a result of conspiracy.

We are giving her, perhaps, an unfair measure of attention, but Harding is among the chosen here, too. Her title, "Feminist Justificatory Strategies," is revealing and in character. The "strategies" discussed are to find means—any means, as we have already seen—by which to advance feminism and to defend a foregone conclusion: that science is a social construct, sharing the deficiencies of the society in which it has been assembled. She identifies three principal strategies in current use. One is feminist empiricism, by which she seems to mean ordinary empiricism disciplined by feminism so as to broaden problem choice and to open science to alternative hypotheses. By her account, these elements are absent from "ordinary" empiricism: "Missing from the set of alternative hypotheses nonfeminists consider are the ones that would most deeply challenge androcentric beliefs, ones that emerge to consciousness and appear plausible only from a feminist understanding of the gendered character of social experience."[41]

How this absence affects empiricist *procedure* is not clear, unless some definition of empiricism other than the usual is implicit or is being attempted. One pines for a concrete instance in which the "feminist understanding" that produces "alternative hypotheses" challenging "androcentric beliefs" has had, or at least promises to have, impact on some particular—any—scientific problem—high-temperature superconductors, protein folding, the population biology of herrings—anything! But alas, such hopes are idle.[42] But it doesn't matter: Harding observes immediately, in any case, that "feminist empiricism is ambivalent about the potency of science's norms and methods to eliminate androcentric bases. While attempting to fit feminist research within these norms and methods, it also points to the fact that without the assistance of feminism, science's norms and methods regularly failed to detect these biases." Feminist empiricism is, in short, empirical science done (with ambivalence) by feminists. Other empiricisms are wrong; but even this one may not be right. This conclusion is, presumably, one of the "justificatory strategies" for feminism. It is certainly consistent with the radical relativism displayed by less sophisticated voices in the movement.

Harding's second strategy is named "the feminist standpoint." Why this is distinguished from the first "strategy" is hard to determine. The feminist standpoint is, as far as we can determine, the standpoint of feminism. Harding describes it as a consequence of gender: "Women's distinctive social activities provide the possibility for more complete and less perverse human understanding—but only the possibility. Feminism provides the theory and motivation for inquiry."[43]

There follow lines about political struggle.[44] The point, if there is one, is that male empirical science cannot, *even in principle*, be rectified by importing the more enlightened styles of problem selection and hypothesis choice available in feminism. Only science done from an entirely feminist standpoint has a chance to be true. So much for the second strategy. But Harding's heart may now lie elsewhere. It is with the altogether more global problem of "whether there should be feminist sciences and epistemologies at all"—with the problem she identifies as having been brought to the fore by recent discoveries of postmodernism. This is, of course, the usual problem, for relativists, of *truth* (a problem because they would be out of a job if they allowed not only—as they always do—that there is no necessary truth in what others say, but also in what *they* say). The postmodernist-feminist way, then, lacking only explicit nods to Derrida, Foucault, et alii, is to be understood as follows: "Feminist claims should be held not as an 'approximation to truth' . . . but as *permanently partial instigators of rupture*, of rents and unravelings in the dominant schemes of representation. From this perspective, if there can be 'a' feminist standpoint, it can only be what emerges from the political struggles of 'oppositional consciousness'" (emphasis added).[45]

How one is to pursue a career as a permanent partial instigator of rupture is of course a mystery. Assuming that this doesn't involve being a straw boss on a loading dock, it may mean that one is to be some kind of epistemological Merry Prankster (although, since high feminism is in play, there is probably strict rationing of merriment). In any case, it doesn't seem to have much to do with taking ideas seriously—yours or anyone else's, scientific or otherwise.

Harding declines to choose among the three strategies because (1) she sees some merit in each one, and (2) each remains incomplete. She recommends for now that all three should be followed. What is needed, in order to bring modern science up to snuff, is—of course—more feminist research on the epistemology of science. But it must be said that "justificatory" strategies such as these cut no ice. They will not convince skeptics (among whom must be, not only scientists, female and male, but also most philosophers of science), since that which is to be justified is a postulate as well as the theorem. Elementary logic may have been superseded in postmodern theorizing; but most people who make a living from intellectual work still depend on it.

Our academic inner city is described by Kenneth Minogue as that locale "in which it is possible to combine theoretical pretension with comprehensive ineptitude," and for that reason it "has become the natural habitat of the ideological enthusiast."[46] Evelyn Fox Keller and Helen E. Longino, however, are anything but inept, and their writing appears at first laudably modest. Their aims are much less exotic than Harding's "permanently partial instigation of rupture"; but they, too, are iconic voices, defending ideology in the academy. In Harding's classification (although not necessarily in their own) these two might be feminist empiricists, attempting to build upon the achievement of existing science (which Keller and Longino generally acknowledge) a better, more comprehensive structure, one that will incorporate feminist truths.

Our sampling of feminist science-criticism would be incomplete without at least a glance at what it contains. That will be a positive change from the follies we have encountered, from time to time, so far in this chapter.

An die Natur

Forgive us a preparatory word about Johann Wolfgang von Goethe, who lived a long, honored, and uniquely productive life and nevertheless died a disappointed man. His scientific work, he believed, was at least as important as his poetry. Yet at no time, outside a circle of sycophants, did it receive the awed respect he believed was due. He had fought a battle against the (long-dead!) Newton over color and light, proposing in *Zur Farbenlehre* his own, very different theory; he had produced an encompassing theory of plant morphology; and he had immersed himself in comparative vertebrate anatomy. All was for naught, in respect of the hoped-for high honors. The reasons were examined with great elegance in an important lecture, given at Oxford fifty long years ago by the neurophysiologist Charles Sherrington, celebrating the bicentenary of Goethe's birth.[47]

Sherrington, making a devastating critique of Goethe's science, found the war with Newton instructive: it was a war against abstraction and the use of experiments and apparatus in the attempt to penetrate the inherently impenetrable "fundamental phenomena" of nature, a war against the use of mathematics to describe (or, in Goethe's opinion, to obscure) such phenomena. Goethe had attempted, with little success, to grasp the infinitesimal calculus of Newton and Leibniz. He had tried and failed to replicate Newton's decomposition of white light with a prism. Goethe's prisms were always "cloudy"; to the same cloudiness he attributed Newton's "errors."

Independently of such trials Newton was, in Goethe's view, self-evidently wrong. Light was for Goethe a fundamental entity of nature, one of those that cannot and must not be decomposed. No prisms allowed, and no calculations. His battle, grounded in a *Naturphilosophie* that had already been displaced, insisted upon *a closeness to and a feeling for nature*, an absorption in the object; a modesty and simplicity in the encounter with phenomena. Rhapsodizing on *Die Natur*, he wrote that "we are in her and she is in us . . . Life is her fairest invention, and Death is her device for getting more life. She sows wants because she likes movement: the game she plays with all is a friendly game . . . Those who will not partake of her illusions she punishes as a tyrant would punish. Those who accept her illusions she presses to her heart. To love her is the only way to approach her."[48] Goethe's was a self-limiting empiricism, a romantic oneness with reality doomed to blockade at the level of the obvious—that to which he gave the ringing name *Urphänomen*. It was an incapacity, as far as physical reality was concerned, and an unwillingness, to see beyond the immediate. Closeness to, identification with, the object, the substitution of ideals for logic and abstraction, of unfettered intuition for analysis, was a transcendent characteristic of romantic natural philosophy. Taken as principle, that characteristic was a root cause of its failure to produce useful science.

It is then an irony that in the search for defining characteristics of women's "ways of knowing" the most important and distinguished feminist writers propose that just this—closeness to, interaction with, the investigated object, this resistance to abstraction—is in fact the essential one. Keller, responding to critics of her view that most scientific practice is gender-valued, insists upon distinguishing between gender and sex; but she admits also that the definition of gender is unstable. It is nevertheless gender she identifies as *causing* the unique style of which women doing science are capable, a style rejected by masculinism: "What then are we to make of the fact that so much of what is distinctive about that [feminine] vision and practice—its emphasis on intuition, feeling, connectedness, and relatedness—conforms so well to our most familiar stereotypes of women? And are, in fact, so rare among male scientists?[49]

Keller's echoes of Goethe can sometimes be uncanny. Like Goethe, she decries scientists for "torturing nature" in order to extract its secrets (although her examples of "torture"—the high energies to which matter is subjected in particle accelerators, for instance—are hardly likely to engage the attention of Amnesty International, or even People for the Ethical Treatment of Animals).[50] Elaborate experimentation and intense abstraction make her uneasy. They are male paradigms.

Keller and Longino, one must assume, are among the writers who reject the claim that the known divergences of early human development *cause* structural and perhaps functional differences in the brains of males and females. They are, in conformity with one of the few widely held feminist principles, anti-essentialists. For them there are no important differences due to gene action and physiology between men and women, other than in their reproductive systems. Such differences as are so obviously *there*, except for urogenital anatomy, must therefore come from *gender*, which is a socially constructed, not a biological category. Hence so basic a cognitive distinction as the claimed tendency toward objectification and abstraction in men and its diametric opposite in women must also be socially constructed: an artifact of a gendered culture. Yet it is an exclusive connectedness with nature (for Keller if not for Longino), with the object of study, the real thing, that she seems to identify as the most characteristic capability of women in science. Because women have been excluded from androcentric science, that contribution is missing and science is therefore one-sided, epistemologically incomplete. Keller, like other feminists, insists that the social construction of science is well-established:

> Recent developments in the history and philosophy of science have led to a re-evaluation that acknowledges that the goals, methods, theories, and even the actual data of the natural sciences *are not written in nature*; all are subject to the play of social forces . . . Social, psychological, and political norms are inescapable, and they too influence the questions we ask, the methods we choose, the explanations we find satisfying, and even the data we deem worthy of recording. (Emphasis added.)[51]

Those "recent developments" turn out to be not so recent. Chief among them is—of course—the work of Thomas Kuhn, whose studies of theory choice led him to conclude that major upheavals of scientific theory— "paradigm shifts"—are conditioned not only by the officially recognized cognitive processes of textbook science, but by various social factors as well as personal whim and aesthetic considerations.[52] Only the most superficial reading of this work and of subsequent commentary by Kuhn on his critics can lend support to strong forms of relativism, a position that Kuhn is at pains most energetically to deny.[53] He is a firm believer in scientific progress and in the power of science to "solve puzzles," while harboring doubts only about the permanent representational value of any regnant paradigm. Moreover, he clearly believes that the dominant factors in theory choice are, indeed, the ones traditionally celebrated by scientists: logical economy, explanatory parsimony, and the capacity to synthesize once-disparate theories into a conceptual unity. Even that clarification of Kuhn's position, an epistemological progenitor of the more iconoclastic social-constructivist critiques, is hardly a recent development, and it hardly justifies Keller's hopes for epistemic reform, especially when one considers that Kuhn's work, well known as it is, is regarded with considerable skepticism by a majority of contemporary philosophers of science.[54]

No serious thinker about science, least of all scientists themselves, doubts that personal and social factors influence problem choice and the acceptance of results by the scientific community. Few serious thinkers about science, however, outside the camps of feminists and social constructivists, argue that the stable results of science, those that have been subject to empirical test over time and have survived, *are not written in nature*! Most know that whatever the underlying calligraphy, self-correcting science is the best translation of it we have.

Keller knows science too well—she, too, was trained in it—to mean all her animadversions upon "male" science to be taken literally. She argues, rather, that the psychosexual development of males endows them with certain combinations of values, for example, objectivity with autonomy or analysis with domination, that leave androcentric science incomplete, blind to categories of question and theory that cannot be imagined in the cognitive context of such mixtures. She cites, but to her credit with no great conviction, such unimpressive examples of supposed bias as inadequate attention from the medical profession to contraception and to menstrual cramps; an overuse of male rats by psychologists in studies on learning (justified by the short estrus cycle of females—an avoidable complication); supposed misrepresentation, via improper language, of the facts of primate sexual behavior.

"Not written in nature" means, on this view, presumably, that a subtle but pervasive cognitive bias distorts science because most of its practitioners have been men; that psychoanalytic theory has established the origins and the dimensions of that pervasive bias:

> Our early maternal environment, coupled with the cultural definition of masculine (that which can never appear feminine) and of autonomy (that which can never be compromised by dependency) leads to the association of female with the pleasures and dangers of merging, and of male with the comfort and loneliness of separateness. . . .
>
> Central to object relations theory is the recognition that the condition of psychic autonomy is double edged: it offers a profound source of pleasure, and simultaneously of potential dread. The values of autonomy are consonant with the values of competence, of mastery. Indeed competence is itself a prior condition of autonomy and serves immeasurably to confirm one's sense of self.[55]

Without attempting the quixotic journey to understanding of why she believes, for example, that "the values of autonomy are consonant with the values of competence, of mastery," one can deduce that Keller's psychoanalytic reveries have yielded a chain of proposed cognitive relationships, at one end of which is autonomy and at the other, aggression. All these characterize the male (gender), or—since she is not so naive as to deal in dichotomies—the male end of a spectrum of cognitive styles. And finally, for the long-awaited examples of epistemic failure brought about by such male psychosexual development, she cites the "central dogma" of molecular biology,[56] the idea of a "master molecule"—DNA—encoding and directing the destiny of the living cell, its aggregates, and the organisms those aggregates produce. For the example of alternative science, the holist, interactionist science done

by those whose psychosexual development has been *female*, she refers (denying as always that it is quite as simple a thing as "women's science") to her
studies on the life and work of Nobelist Barbara McClintock,[57] who eluci-
dated the developmental genetics of maize.

The claim is not only that McClintock's science is (or was, before it was
honored) qualitatively different from that of male geneticists, thus condemning it to a long term of obscurity, but that McClintock's story is diagnostic:
"What I am suggesting is that if certain theoretical interpretations have been
selected against, it is precisely in this process of selection that ideology in
general, and a masculinist ideology in particular, can be found to effect its
influence. The task this implies for a radical feminist critique of science is,
then, first a historical one, but finally a transformative one."[58]

The suggestion is utterly unconvincing. There are no developments in the
history and philosophy of science that *prove* a social (masculinist, ideological)
construction of the final product of empirical science. The best evidence of
such a negative is the set of plain truths that science (in the guise of, say
penicillin) works just as well for Australian aborigines (male and female) as it
does on Englishmen (and women); or that certain fundamentalist Christians,
firmly convinced in their social conditioning that the world is just a few
thousand years old, may nevertheless die in earthquakes, whose underlying
tectonic processes require intervals thousands of times that long. The psychoanalytic theorizing that identifies the traditional scientific mentality as
a deformed hybrid of autonomy, objectification, loneliness, and aggression
is no firmer than any other theorizing done in the psychotherapeutic
community—which is to say that it is unreliable. Even if it had a kernel of
truth, as some such impressionistic proposals do, it could not possibly establish a *clear* connection between the ideal of objectivity and a desire for
domination. Quite the contrary: ask a random sample of a hundred reasonably articulate people what they mean by an "objective" view of X; and they
will say something about suppressing one's prejudices about X, stepping outside oneself, about modesty, and the like.

The idea of DNA as "master molecule" is not, even among the most
radically reductionist biologists, a literal one: only secondary writers think *ex
DNA omnia.* "Nucleocytoplasmic relationships," the system of chemical interactions by which the information encoded in DNA (not only in the
nucleus, but in the mitochondria as well) comes to be expressed as the
structure and chemistry of the entire cell, has been for more than forty years a
central concern of cell and molecular biology. DNA as "master molecule" is
shorthand for "initial information source," nothing more; it carries no implication of "dominance." And finally, McClintock's splendid work *was* (al-

most, but not quite) overlooked for some years, but there is no convincing mark on it of femininity, as McClintock herself was the first to insist. Her closeness to the experimental material, her willingness to "listen to it," is characteristic of the work of some scientists and less so of others. There are *no* data suggesting that women scientists display the characteristic, in general, more often than do men. Moreover there is no lack of abstraction in McClintock's work: it is solidly grounded in the abstractions of formal genetics. She simply saw things that others didn't see. Her finest work supports the still-powerful Popperian idea that good science consists in framing hypotheses so that they are refutable and then designing experiments to do just that. Good experimentalists *must* be close to the experimental object in order to make effective designs; otherwise they are unable to identify and exclude intervening variables.

Keller's position rests upon unsupported speculations about psychosexual developmental differences between men and women; and its examples of consequences of such differences, in science, are questionable at best. Yet on such a basis, an already large body of writing has come to be taken with utmost seriousness, indeed to be celebrated, especially among cultural critics. What is most curious, though, is the acceptance of a form of essentialist doctrine by these anti-essentialist feminists. Keller presumably does not believe that the proposed cognitive differences between men and women are inborn. But so pervasive and fundamental a pattern, if it exists, fails of essentialism only in the feminist denial that *genes* are in any way involved. In all other respects the doctrine does not differ, formally, from the most essentialist belief in an inherent mathematical excellence of boys as contrasted to a verbal precocity of girls—and their weakness in spatial relations. To such beliefs (for which there is at least *some* empirical support) feminists usually respond with rage.

The proposal that a new and better science will emerge from an interactionist, holist, nurturant "feeling for the organism" with which women are supposed to have been endowed—by their nature or by their gender—is in its epistemic effect precisely Goethe's old argument against experiments, mathematics, and abstraction, and for *die Natur*. It is Goethe's—and Wordsworth's and Whitman's—Romantic idealism in this year's Paris original. It is not likely to affect future science any more than Goethe's sputterings against Newton affected the science we have today.

Context Ineluctable

"Can there be a feminist science?" This is the question that an articulate Helen Longino sets out to answer in the affirmative. Contra Keller, she begins her inquiry by dismissing as question begging the strategy of simply pointing to what feminists (or women) in the sciences do. She quotes, with astringent favor, Stephen Gould's[59] argument against that strategy in his review of Ruth Bleier's *Science and Gender*:

> Gould . . . brushes aside her connection between women's attitudes and values and the interactionist science she calls for. Scientists (male, of course) are already proceeding with wholist and interactionist research programs. Why, he implied, should women or feminists have any particular, distinctive contributions to make? There is not masculinist or feminist science, just good and bad science. The question of a feminist science cannot be settled by pointing, but involves a deeper, subtler investigation.[60]

For Longino, the important question, among the several deeper, subtler ones to be investigated, is why the original question is to be entertained at all. "What sort of sense does it make to talk about a feminist science? Why is the question itself not an oxymoron, linking, as it does, values and ideological commitment with the idea of impersonal, objective, value-free inquiry? This is the problem I wish to address in this essay." Her attempt is not overly respectful of other feminist strategies. Unlike Harding, she rejects the notion that established science is in some systemic way wrong about the phenomena it chooses to investigate. Either "feminist science" is indeed oxymoronic, seen against "the standard presuppositions about science" (i.e., value-freedom), or the products of science are generally in error. *The latter, she admits with laudable honesty, is nonsense.* The former needs investigation before one can decide; and, of course, it is important for a feminist to decide. The sticking point is those standard presuppositions, the most important of which is the freedom of science from values. Longino sets out to examine values relevant to science; and she claims to have established a clear dichotomy: there are two sets of values, distinguishable in that one—the *constitutional* values—refers to practice, to scientific method, while the other—the set of *contextual* values—belongs to the social and cultural context in which science is done. "The traditional interpretation of the value-freedom of modern natural science amounts to a claim that its constitutive and contextual features are clearly distinct from and independent of one another, that contextual values play no role in the inner workings of scientific inquiry, in reasoning and observation. I shall argue that this construal of the distinction cannot be maintained."[61]

The succeeding argument reports (and purports to explain), a claimed demise of logical positivism on the ground that scientific hypotheses are not—as a minimal form of positivism would require—simple generalizations of data statements. Hypotheses contain *language* that is nowhere in the observations from which hypotheses come and to which experiments, testing those hypotheses, are addressed. Hypotheses contain, in short, assumptions (language!) extraneous to data and not testable through the agency of data. Those assumptions may be, usually are, value laden; and some of those values may be (Longino does not insist that they *must* be) contextual values. Even when contextual values are hidden in hypotheses, however, and hence in the theories sustained by confirmation of hypotheses, "there is no formal basis for arguing that an inference mediated by contextual values is thereby bad science." It might be, or it might not be. Each case, as Longino sees it, has to be investigated.[62]

The important point of this argument is that contextual values cannot be eliminated. Science is not even in principle free of social and cultural values. Ergo, "feminist science," science informed by the social and cultural values of feminism, is not an oxymoron.

> The conclusion of this line of argument is that constitutive values conceived as epistemological (i.e., truth-seeking) are not adequate to screen out the influence of contextual values in the very structuring of scientific knowledge . . . The conceptual argument doesn't show that all science is value-laden (as opposed to metaphysics-laden)—that must be established on a case-by-case basis, using the tools not just of logic and philosophy but of history and sociology as well . . . [But] *it is not necessarily in the nature of science to be value-free.* If we reject that idea we're in a better position to talk about the possibilities of feminist science.[63] (Emphasis added.)

This is not the place to argue through the philosophical case, which is largely a summary of Longino's earlier (and later) work. It is sufficient to note that, unlike the arguments of other feminist critics of science, this one does not refer endlessly and pejoratively to science as "Western" or "androcentric." It suggests, with a justice to which we (and most working scientists) would readily agree, that "contextual" values can, unknown to the investigator, influence the design of experiments and the interpretation of data. It would be difficult to reject a suggestion made in terms so qualified as to imply not only that a feminist science has not yet been identified by anyone but that it might not ever be identified.

Longino endorses and promises specifics, analyses that really attend to what scientists do and how they think in detail. What remains for her to show, however, is that this kind of analysis can lead where she says it will; that some nontrivial, long-lasting outcomes of science *are* in error because of contextual values, and that science done in a new context—feminism—can be both important and correct. Without such demonstrations, her argument is so tentative as to deprive it of any force.

Once again we are disappointed. Among all the pressing possibilities for a searching-out of error in important and problematic science (e.g., the Big Bang theory, the origin of life, the search for extraterrestrial intelligence, dark matter in galaxies, computability, cellular signaling processes, the defeat of AIDS, the origins of human language) the most examined case of distorting male contextual values is that perennial feminist whipping boy, biological and behavioral differences between the sexes. Longino reports on her earlier work with Ruth Doell (treated later also, in Longino's recent book[64]) as demonstrating the force of contextual values in shaping the outcomes of scientific practice, and as a specific application of her epistemology.

The studies examined are in a very large literature concerned with differences between males and females in the prenatal synthesis of androgenic (male) steroids and the consequent effects on central nervous system development, and on the influence of the sex steroids (estrogens and androgens), at normal and pathologic levels, on the behavior of men and women. Longino's attacks on this work are not contained in the summary article, but are detailed in earlier papers and in her book. Here she merely states the conclusions in order to examine them in the light of her epistemology.

We must not proceed thereto, however, without noting our opinion that these criticisms of the literature are replete with special (highly articulate) pleading and are grossly unfair to the scientists (many of whom are women) who have contributed to it. Her standpoint is rigorously anti-essentialist; from it she claims that the studies examined "are vulnerable to criticism of their data and their observation methodologies." However, nowhere in the body of Longino's work do we find identified specific, recognizable flaws in the data and the methodologies. Indeed, the criticisms are not directed toward those at all. Instead, they are either banal (e.g., the argument that data from rodents should not be used to infer processes in people), or indictments of the investigators for making value judgments about departures from sex-stereotypical behaviors (of girls, for example, who have been androgenized in utero), and—almost—for homophobia. But, although *she* may detect such attitudes, we do not; nor do the investigators themselves, most of whom—men as well as women—are probably sympathetic to feminism. Led to expect serious criticism of data or methodologies, we find, not cooked data, uncontrolled experiments, or statistical gaffes, but implicit attitudes claimed to have been detected—by a hypersensitive anti-essentialist. By and large, the logic here is that, since the conclusions are unacceptable by feminist lights, the science must be flawed, and those flaws, in turn, are evidence for the posited influence of "contextual values."

Only one criticism is offered that addresses a possible concrete error—the matter of unrecognized intervening variables. This is an error to which scientists, true believers in the "standard assumptions," are very sensitive, because in complex systems it is indeed easy to overlook an internal or environmental process, or a feedback loop, that affects the measured endpoint. Still, good scientists make this mistake from time to time; and it is usually corrected by other scientists who repeat, often as done originally and then with modifications, the experiments. What unrecognized intervening variable does Longino claim to find? It is that fetally androgenized girls are or may be aware of their special condition, and that this awareness can lead to the patterning of behaviors (i.e., they become "tomboys"—a word that for Longino, although for nobody else, is derogatory). If that were true, then inferring a causal relationship between hormone levels during gestation and childhood behavior would indeed be a mistake: subject awareness of a gestational accident would be an uncontrolled, intervening variable. But this simply won't do. There is no reason to believe that any such awareness existed in general: many of the children studied were of preschool age. And if parental awareness were a factor, it would surely have influenced their fetally androgenized daughters in the opposite direction, away from the behavior of tomboys and toward that of "feminine" little girls.

So, the outcomes of this epistemological critique, applied to existing science, are neither stronger nor more convincing than those of the other feminist epistemologists. Longino's interest goes, however, beyond mere criticism of existing science. She assumes that there is enough of it in the literature already. Her project is more advanced. It is to substitute for what she calls the "linear" model implicit in the literature she has been criticizing—that model according to which the early production of testosterone in a Y-chromosome-bearing embryo makes a male, including male behavior—"by introducing one [a model] of greater putative complexity that includes physiological, environmental, historical and psychological elements."

As example and support for such an interactionist model, she adopts the neural selection theory of Gerald Edelman,[65] according to which the central nervous system is structured by a series of continuous interactions between groups of growing, synapse-forming, and synapse-breaking neurons and signals from outside the regions of active morphogenesis (the establishment of form during embryonic development). This is a "selectionist" theory, to do it minimum justice for present purposes, about the way local and environmental signals may affect brain structure and, inevitably, thought. While there are some doubters among neurophysiologists and artificial intelligence experts, Edelman's arguments are brilliantly presented, well-documented, and widely respected.[66] Unfortunately, they have nothing to do with Longino's argument about the "linearity" of hormones and behavior, nor is Edelman's theory in any significant sense "nonlinear." It would be perfectly consistent with Edelman's view for cascades of morphogenetic events, initiated by programmed molecular signals *such as hormones*, to be interwoven with and modulated by experience-driven cascades. We daresay that most of the investigators Longino has derogated are perfectly aware of that, and of the massive complexity of emergent behavior in any morphogenetic model.

What Longino is really after is a way of doing science that will negate *any* possibility of biological determination:

> Our preference for a neurobiological model that allows for agency, for the efficacy of intentionality, is partly a validation of our (and everyone's) subjective experience of thought, deliberation and choice. One of the tenets of feminist research is the valorization of subjective experience . . . When feminists talk of breaking out and do break out of socially prescribed sex-roles, when feminists criticize the institutions of domination, we are thereby insisting on the capacity of humans—male and female—to act on preconceptions of self and society and to act to bring about changes in self and society on the basis of those perceptions.[67]

Once again, the hard biology of embryogenesis is snowed-under by "valorization." Science as-it-is becomes, for such critics, an intolerable constraint, a terrible danger. To radical feminists as to dreamers of teleportation and transluminal space-travel, it represents abhorrent limits. Linear or not, it is liable at any moment to produce results that demolish one or another cherished preconception of ideology. Longino is at least honest about that. Her conclusion then follows: since standard science is likely to be affected by contextual values we don't like, let us, as feminists, drive our science by the contextual values we approve of. An answer is thus given to the original question: a feminist science *is* possible. Indeed, it exists! It *is* political. Its answers can well be different from answers obtained by investigators of the opposite persuasion. So be it.

Longino's essay ends on a sober note. Feminist science, when there is more of it, and since it will be much less concerned with quick results, less interested in manipulating nature than in understanding it in all its rich complexity, will have a hard time getting funded in this society, in which science is "harnessed to the making of money and the waging of war." Its practitioners, failing to satisfy the entrenched patriarchy as to the utility of its results, may have a hard time getting tenure. "So," she asks, "can there be a feminist science?" Her conclusion: "If this means: is it in principle possible to do science as a feminist? the answer must be: yes. If this means: can we in practice do science as feminists? the answer must be: not until we change present conditions."[68]

Hélas! What begins as an epistemological inquiry into science ends as familiar anti-science tricked out in the ambient clichés of the business—science "harnessed to the making of money and the waging of war"—the old moral one-ups*woman*ship, and the call to political action. It ends with the universal complaint of religious zealots, utopians, and totalizers generally. Science as-it-is untrustworthy. We can't bend it to our political will because, as a powerful institution of the present, compromised world, it is protected. It will not be bent until the enemy is weakened and the world is redeemed. (But then, once the world has been remade in the image of our ideals—then, we shall see.) We have heard this from ideologues, politicians, and thought police in various uniforms since Galileo's time. How sad it is that it should now emanate from the scholarly halls of universities and reverberate there among intellectuals who inherit the Enlightenment.

CHAPTER SIX

The Gates of Eden

If we are to construct an environmental movement powerful enough to enact needed reforms we must first relinquish our romantic fantasies. A meaningful environmentalism cannot be based on nostalgia, wishful thinking, and faith that the inherent goodness of humanity will manifest itself once civilization is dismantled.
MARTIN LEWIS, *GREEN DELUSIONS*

A few years ago, the novelist Ursula Le Guin, who is as much at home in ordinary fiction as in fantasy and science fiction, scored a considerable success with an unusual story, *Always Coming Home*, set in the far future. The scene is a North America peopled by tribes separated from each other politically and culturally, with hardly a memory of the days of a continent-spanning nation-state. The tale concentrates on one such group, the Kesh, whose ways and customs, although they are Le Guin's invention, strongly echo the folkways of pre-Columbian Native Americans.

The Kesh culture is elaborately described in the novel: it is the book's true hero. An appendix provides information about the language, poetry, music, and artwork of the Kesh, with specimens of each. At its publication, the book was accompanied by a series of ancillary products—recordings of Kesh songs and chants, replicas of Kesh paintings—all created by Le Guin in collaboration with artists and musicians fascinated by her vision of this possible future. It proved a popular line of goods at science-fiction bookstores and New Age specialty shops, and the full set—book and recordings—remains a popular item on university library reserve for students.

The strangest aspect of the Kesh culture is the degree to which it has rejected not only technology as such (the Kesh live close to nature, with minimal use of steam power and electricity, and every artifact they produce is handmade and imbued with the qualities of art), but also the entire set of attitudes, ambitions, and obsessions of what we tend to think of as civilization. They have no interest in abstract science. Their philosophy is embedded in their mythology. Nor do they concern themselves with history or social theory, or brood upon the destiny of man. The notion of knowledge for its own sake is alien to them. Their values are at once entirely in the present and timeless. But note: their ignorance does not come about because what we, in the present, call knowledge has disappeared from the face of the earth. In fact the human culture of the Kesh and the other tribes is paralleled by a "culture" of sentient computers, with which the Kesh are in occasional contact. The machines are willing to divulge to any curious Kesh such details of scientific theory or historical fact as might be sought. The point is that the Kesh simply aren't interested. Their world of myth, ritual, and song, and the slow turning of the seasons, satisfies them. Science and knowledge expand; but that expansion is the province of computers, which send probes to distant stars and seek the facts of history in the ruins of previous human civilizations. It is as if the Faustian impulse of humanity had been drawn off and perfused into the circuitry of machines, leaving humans in an Eden of contentment and forgetfulness, through whose gates they have at last returned.

Some readers are repelled by the somnolence of the Kesh and by their renunciation of ambition; but many are charmed and inspired (although Le Guin herself seems, at times, to be wryly ambivalent). The psychology and the ideas of those who admire the Kesh and extol the edenic virtues of such a society are what interest us here. Our interest derives not from any concern with minor literary fads—although the talented Le Guin has our best wishes—but from the fact that the admirers of the Kesh are so emblematic—so coextensive, in fact—with radical environmentalism, whose ideology has made itself widely felt in our time, but which is centered, in large measure, in the same congregation that includes the academic left.

Among them, support for environmental causes transcends doctrinal differences. The received version of environmental wisdom has an unmistakably radical—and apocalyptic—flavor. The finality of the conflict, the unflinching identification of one side with good and the other with evil, are diagnostic. David Day, whose view of Armageddon is entirely characteristic, introduces a practical workbook on ecological issues this way, modestly: "In the final analysis, the outcome of the 'ecology wars' will prove more critical than any ever fought by the human race. What is at stake is not the dominance of one nation over another, but the survival of life on the planet."[1] Tom Athanasiou, writing in *Socialist Review* and in a rational, analytic mood (for this is criticism of, among other things, ecological romanticism) is nevertheless certain that the devastation is imminent:

The coming devastation will breed a vast hatred. It may even be that the ideas of the green hard core . . . are poised for a breakout into larger domains. Fortunately, this is not the only possibility, and radical outrage is not likely to remain eternally constrained within the anti-communist frameworks of Cold War analysis . . . *capitalism—like the atmosphere— may soon cease to seem a part of the natural, eternal world.* (Emphasis added.)[2]

The academic left entertains millenarian hopes. Its vision of The Only Possible Future almost always includes an accommodation between human-kind and nature, a harmonious resolution of the predicated incompatibility between contemporary society and the sanctity of the natural world. Since it is the sprawling left we are speaking of, explanatory differences arise. To a feminist, the roots of environmental degradation lie in "the hegemony of patriarchal values." Marxists of a traditional stamp see the postulated ecological crisis as product of the more general "dialectical crisis of capitalism." Postmodernists might blame something like "a discursive practice that objectifies the natural and robs it of agency"; while anarchists, who are still to be found in odd corners, urban and bucolic, are likely to finger—of course—the *hierarchies* of the state. As usual, however, ideological syncretism is the prevalent note. All doctrinal variants are simultaneously endorsed to some degree; differences are submerged in a broad tide of indignation over environmental outrages, the list of which is continuously lengthened by selection of appropriate results from scientific journals (and by ignoring inconvenient ones). The typical dire warnings and portents conform perfectly to a model constructed more than twenty years ago by George Steiner: "We are told, in tones of punitive hysteria, either that our culture is doomed . . . or that it can be resuscitated only through a violent transfusion of those energies, of those styles of feeling, most representative of 'third-world' peoples."[3]

These days, it goes without saying, environmental concerns are widespread in the general population. Everyone who bundles old newspapers and separates recyclables is affected by them. Environmental piety, albeit in a diffuse and nonspecific form, has become an American civil religion. It is endorsed by multinational corporations and churches, as well as by politicians of all allegiances and at the highest levels. The *Wall Street Journal* reports with enthusiasm, on its Market pages, a meteoric rise in sales of cosmetics labeled "natural" (and the label of one risen brand—"Naturistics"—announces "Not Tested on Animals")[4]; the *National College Magazine*, distributed with thousands of college newspapers to a claimed million readers, offers an Environmental Issue devoted to "Earth Crisis," every third full page of which is a self-congratulatory advertisement from the Anheuser-Busch companies, whose people are to be presumed enlisted in the war against pollution.

The preferred form of environmental piety on the academic left, however, is laced with prophetic disclosures of doom. Under this view, adequate resolution of ecological problems is possible only through a revolutionary reconstruction of society, or, in the more favored language of theory, "dismantling the definitions of privilege and the diagrams of power that undergird industrial capitalist patriarchy." Moreover, it is almost always assumed that in the redeemed world industry and technology will have at most a minor role; that bureaucracy will have given way to a localized face-to-face democracy, in which trade and commerce are supplanted by a frugal self-sufficiency. Thus that Marxian will-o'-the-wisp, the "withering away of the state," has in our time taken on a hard, ecotopian specificity. It is now asserted as well, on the academic left, that "Baconian" science, an activity branded as toady and tool of imperialist expansion, is giving way to an holistic, nonexploitative way of knowing, a new form of science that expresses the spiritual pantheism of the new order. Of course it is supposed and proposed that reversion to decentralized tribalism will lead to the abolition of racism, of male supremacy, and to a stop to the rape of nature. It is easy to find this vision articulated in the theorizing of the most militant wing of the environmental movement. Dave Foreman, founder of Earth First! chants it: "We must break out of society's freeze on our passions, we must become animals again. We must feel the tug of the full moon, hear goose music overhead. We must love Earth and rage against her destroyers. We must open ourselves to relationships with one another, with the land; we must dare to love, to feel for something— some*one*—else."[5]

Clearly, Le Guin's Kesh, with their insistence that "person" may refer to a bear, a deer, a tree, or even a rock, are fictional models of the sort of human beings Foreman would have us become. Nor are he and his followers the only devotees of such a vision; dozens of academic radicals proclaim it in their work and it is increasingly reflected in the casual language (e.g., "environmentally friendly," "ecologically sane") and attitudes of students. Morris Berman, a onetime orthodox historian of science of leftist bent, is a typical convert. His vision of the future echoes Le Guin's fiction:

> Human culture will come to be seen more as a category of natural history, "a semi-permeable membrane between man and nature." Such a society will be pre-occupied with fitting into nature rather than attempting to master it . . . We will no longer depend on the technological fix, whether in medicine, agriculture or anything else . . . The economy,

finally, will be steady-state, a mixture of small-scale socialism, capitalism, and direct barter. This will be a "conserver" society with nothing wasted and with a great emphasis, to the extent that it is possible, on regional self-sufficiency.[6]

Berman's ideological purpose is to supplant, with an approach conditioned by "spirituality" and "ecstasy," the scientific vision that has reigned in Western intellectual life for three centuries.[7] He is a latter-day disciple of William Blake (and Carlos Castañeda!), eager to reject "single vision and Newton's sleep." Like many contemporary radical intellectuals who yearn for a recrudescence of irrationalism, Berman, in the tradition of Blake, focuses much of his scorn on Newton and, especially, on his philosophical precursor Francis Bacon: "The overall framework of scientific experimentation, the technological notion of the questioning of nature under duress, is the major Baconian legacy."[8]

Carolyn Merchant, a doyenne of the ecofeminist movement, is always eager to settle Bacon's hash. "But from the perspective of nature, women, and the lower orders of society emerges a less favorable image of Bacon and a critique of his program as ultimately benefiting the middle-class male entrepreneur."[9] Her larger proposition is that the emergence of the scientific worldview has somehow not only desacralized nature, leaving it prey to the rapacity of technologized capitalism, but has as well, in identifying "natural" with "female," undermined the dignity and autonomy of women and left them prey to a more absolute oppression than that which the prescientific world inflicted on them. Moreover this (in her view) has brought us to the brink of ecological catastrophe, which can now be averted only by a reversion to edenic simplicity (although in this new Eden, Eve will yield not an inch to Adam; and that old serpent, science, will have no power of persuasion).

> The pollution "of her [the earth's] purest steams" has been supported since the Scientific Revolution, by an ideology of "power over nature," an ontology of interchangeable atomic and human parts, and a methodology of "penetration" into her innermost secrets. The sick earth, "yea dead, yea putrefied" can probably in the long run be restored to health only by a reversal of mainstream values and a revolution in economic priorities. In this sense, the world must once again be turned upside down.[10]

Such sentiments are echoed ceaselessly in radical environmentalist literature, with or without feminist or New Age trappings (although both are present more often than not). The image of the earth as poisoned, as a deeply wounded victim, is central to the iconography. *Human* suffering, as such, while not neglected (especially when the victims are female or nonwhite) is notably secondary. Herewith is Earth's passion according to Jeremy Rifkin:

> The modern era has been characterized by a relentless assault on the earth's ecosystems. Dams, canals, railroad beds, and more recently high-ways have cut deeply into the surface of the earth, severing vital ecologi-cal arteries and rerouting nature's flora and fauna. Petrochemicals have poisoned the interior of nature, seeping into animals and plants, soaking the organs and tissues with the tar of the carboniferous era. The spent energy of the industrial revolution has choked the skies with layers of gases—carbon monoxide, carbon dioxide, sulfuric acid, chlorofluoro-carbons, nitrous oxide, methane, and the like—polluting the air, block-ing the heat from escaping the planet, and exposing the biota of the earth to increased doses of deadly ultraviolet radiation.[11]

Hysterical[12] as is Rifkin's prose to anyone having detailed knowledge of these "assaults," it is common currency in the environmental movement, and taken as unexceptionable wisdom by those whose environmentalism is linked to hopes for radical social transformation, whether along feminist, anti-capitalist, or racial lines. The particulars of the indictment seem to reflect the latest thinking of the environmental sciences, although the accuracy of that reflection, as we shall see, is very low. Similarly, the idea of a return to a more primal way of life, stripped of the arrogance and insolence of technology, may seem, especially to the young and historically naive, a newborn vision.

In fact the primitivist vision is a recurrent one, and is strictly in the Western tradition. The assumption that such thinking is a natural and unique concomitant of leftist (or "progressive") sentiments is utterly false. The idea that industrial society, with its dirt and noise, its depersonalization and anomie, is an affront to the natural order, and that the proper course for humanity is a return to a life bound up with the cycles of nature and tied to blood and soil, has arisen in the West with great regularity. Political philoso-phies with which it has been associated have run the gamut. Blame for man's alienation from natural virtue has been assigned to every possible malefactor. As Anna Bramwell points out in her remarkable history of ecologism:

> There are several different guilty parties in common usage. These are Christianity, the Enlightenment (with atheism, scepticism, rational-ism, and scientism following on), the scientific revolution (incorporat-ing capitalism and utilitarianism), Judaism (via either the Jewish ele-ment in Christianity or via capitalism), Men, the Nazis, the West, and various wrong spirits, such as greed, materialism, acquisitiveness, and

not knowing where to stop. The wrong spirit is a twentieth-century explanation, usually confined to the West and derived from the puritan element in Protestant and dissenting Christianity; therefore it is found mainly in Northern Europe and North America. According to this ethic, "bad" spirits are located in Western man, who is seen as the unsaved, expansive, nonecological dominator of nature. Only by rejecting the materialist heritage of the West will man be saved. [13]

Clearly, edenic ecologism, under one label or another, is an idea with a long pedigree in Western history. It has been conscripted for ideological use by radicals, conservatives, and reactionaries, by Communists and by Nazis, and by schismatic sects hard to place on that spectrum. At present, however, there is a particularly good fit between this view of the world and the overall perspective of the academic left. It strikes at most of their devils: industrial capitalism, white supremacy, imperialism, male supremacy, the various expressions of Western triumphalism. Simultaneously, it valorizes women (under the bizarre doctrine that women as a class are in a sympathetic resonance—that men cannot achieve—with nature) and nonwhite peoples, victims of European rapacity, who are assumed confidently to have been in an edenic state before their conquest.

Part of the package, in most formulations, is something like the Blakean view of Western science. It is assumed that science is crippled by its rejection of the subjective and the spiritual, by its reductive analytic methodology, by the artificiality of the conditions it creates for observation and experiment, by its insistence on a rigid distinction between thought and feeling, subject and object, and, finally, by its "Baconian" subservience to power and to the interests of the dominant social class. [14]

To display this set of *attitudes* in its most elevated form (in contrast to what will follow below), we quote a few lines from the conclusion to an article by Edward Goldsmith, published in the *Ecologist*. The point of this piece is to show that the scientific view of the facts and mechanisms of evolution is false because the reigning idea of science is fundamentally and comprehensively wrong; *and*, moreover, that we won't stop destroying the living world until we change our view of what science is and ought to be. Needless to say, this article provides no evidence whatsoever for said flaws in science, except by the application of standard (holist) epithets: compartmentalization, empiricism, induction; causality, and the like. Still, it is the *attitude* we are after here, and so:

> In order to understand evolution one must reject the Neo-Darwinist thesis and indeed the Paradigm of Science itself that this thesis so faithfully reflects.

We must also see evolution and life processes in general *as displaying precisely the opposite features to those that they are held to display by Neo-Darwinists and mainstream scientists in general.*

Rather than being atomistic they are highly organized and hierarchical; rather than being mechanistic and hence passive and non-creative, they are *living, dynamic,* and *creative*; rather than being random they are *ordered* and *highly purposive.*[15]

Such charges, and indeed such moralizing language, have cropped up among Lamarckians from the beginning of evolutionary theory, from time to time with striking results (as in the Lysenko era in the late Soviet Union); but nobody has so far been able to turn them to recognizable use either as explanation or for prediction.

Of course we, and we hope most thoughtful persons encountering edenic ecologism in explicit rather than poetic form, reject this tendentious and ignorant view of science. So deeply, however, are its tenets embedded in contemporary intellectual output, and so large is that output that, again, an entire volume would be needed to refute it. Here we can do little more than recall, following Bramwell, how much it has in common with other antirationalist posturings, including those of twentieth-century totalitarian movements and of religious zealots throughout history.

In any case we believe that here there is a more urgent problem than philosophical error and ideology masquerading as analysis. The threat of ecotopian enthusiasms is that they will, in fact and in the long run, weaken or eliminate the possibility of ecologically sound social policy, under whatever ideological banner that may materialize. We believe that such an effect must follow from the fervent antiscientism now embraced by radical environmentalists, an antiscientism that, if broadly influential, cannot fail to reduce the chances of success in answering questions and solving problems *that are quintessentially scientific.* As Michael Fumento puts it in his splendidly documented exposé of environmental alarmism:

Alarmists and people subjugating science to political ends don't want you to consider relative risk . . . Indeed, many of them haven't the slightest idea of what relative risk is . . . Understanding of how odds work is the last thing they want you to have. They want to be able to present you with a simple model that says that since this or that has been alleged to be harmful, it must be banned or at least heavily regulated.[16]

Environmental Realism

Let us be perfectly clear: we have no quarrel with environment-consciousness. As successor to "conservationism," it is based upon a sound conviction. The good life—now, and more so in the future—for our species as for all others, requires the clearest possible understanding of our interactions with nature. It requires avoidance of interactions that do or seem highly likely, on the basis of competent risk assessment, to deplete or damage nature. There is no reason to be concerned about an environmentalism so based: quite the contrary. To a large extent we share its basic fears. Most of the problems environmentalists point to have a real component. Our concern is, rather, with a revived, apocalyptic naturism that has, in several versions, caught the fancy of young people generally and engages a rapidly increasing number of well-meaning adults.

Exponential growth of the human population, for instance, has consequences, some of which are harmful and irreversible. Those consequences are inherent in the simple relationships among population size, ecosystem structure, and environmental carrying capacity. The last of these has limits; that is an established scientific principle. Technology cannot forever transcend carrying capacity. Once the limits are reached, the human consequences—not to mention those for other species—are certain to be horrible. Arguments to the contrary, whether from religious dogma or the Micawberish doctrine that "something will turn up" are bootless. The problem is that we do not as yet have a genuinely reliable or meaningful estimate of Earth's carrying capacity, in the sense of a constant for the equations of population growth. The prophets of doom, some scientists among them (who have a losing record of predictions to date), insist that it is a small number; religious fundamentalists and ecomillenarians *know* that God's (or Goddess's) world would work just fine, and forever, if we gave up civilization.

The current concern with the possibility of an enhanced and deleterious "greenhouse effect" is not just superstitious alarmism. It is only the public language and political style emerging with the concern that are dangerous. There *is* an atmospheric greenhouse. It has been there throughout the history of life on this planet, much as that may come as a surprise to some. Its temperature set-point is determined by physical and chemical variables, exactly as the temperature held by a thermostated heating plant is determined by the output of the furnace, the external temperature, and the rate of cooling of the building. Without a greenhouse atmosphere, the planet would be, like its neighbors in the solar system, sterile. Earth's atmospheric greenhouse is

the regulated heating plant that has turned its thin, wet skin into an incuba-
tor for life.[17]

It is distinctly possible that human activities, intensified as population
grows and fuel-hungry technology becomes ubiquitous, could change the set
point. Natural geological and astronomical phenomena, operative now as in
the past, *have*, certainly, done that—during successive ice ages, for example,
and perhaps as a consequence of the cyclic repositioning of Earth relative to
our star and the sun's to the center of the galaxy. Of course they will continue
to do so in the future. There *will* be global warming and cooling, whether we
are here or not. Volcanoes alone will see to it.[18] The question—and we
emphasize that it remains a *question*—of the effect upon this set point of
increasing emissions of carbon dioxide, methane, and other "greenhouse"
gases, byproducts of technology and agriculture, is of the highest importance.
It deserves the most comprehensive and scrupulous investigation.

So does, however, the question of the *climatic consequences*, a quite different and even more difficult question, about which there remain deep disagreements among atmospheric scientists. The depth and seriousness of these
disagreements is visible to every reader of such general professional journals as
Nature and *Science*.[19] The proposals of some (but certainly not all) atmospheric physicists that action to reduce greenhouse emissions should not be
deferred pending the outcome of this investigation have to be weighed seriously. There is only one Earth, and nobody in his right mind wants to use it as
a crash-test dummy. But recommendations *have to be weighed*. Such proposals
do not justify panic; nor do they call for anything like an immediate restruc-
turing of society, along lines sketched by somebody's derivative, post-
Marxist, poststructuralist utopianism.

Much the same point might be made with regard to the "ozone hole." The
hypothesis that the seasonal decline of ionospheric ozone over Antarctica is
caused in large part by active forms of chlorine (such as chlorine monoxide
radical, ClO) formed in the breakup of chlorofluorocarbons in the upper
atmosphere is an attractive one.[20] It is gratifying to the scientific ego, since
the observed ozone hole seems to confirm certain theoretical predictions
made as long ago as the early 1970s. Nonetheless, in all these considerations,
the meaning of the word *hypothesis* must be kept seriously in mind. It is not
only possible that the hypotheses are wrong; it is also quite possible that even
if they are right, measures taken in haste to prevent the hypothetical consequences may do more harm than good. There are countless examples of this.[21]

This is *not* an argument, please note, for the continued production of
chlorofluorocarbons. They are clearly significant reactants in the chain of

processes causing the seasonal hole in polar stratospheric ozone.[22] We trust that alternative refrigerants that will do no harm of their own will be used in air conditioners. We are not partial to styrofoam cups, or to the spray-can "art" that disfigures our cities. Exactness of scientific thought, however, and an honest comprehensiveness in the cost/benefit analysis that should be done before *any* solution to *any* global problem is undertaken, are of incalculable importance. Apocalyptic movements don't do honest and comprehensive cost/benefit analyses. They don't want to and they don't know how (again, see Fumento, *Science under Siege*). To the extent that science—the only reliable source of numbers for environmental cost/benefit analysis—is battered in the course of a primarily ideological crusade, so much greater will be the chance of making disastrous errors of policy.

Problems of human overpopulation and problems of drastic climatic change both bear on another threat that has to be taken seriously: the possibility of a further, even a catastrophic decline in biotic diversity. The earth's multitudes of species preserve in their DNA sequences not merely potential riches for commerce and medicine, but—perhaps—an as-yet only partially understood contribution to the large-scale stability of ecosystems; species house a history of life that we might translate one day, and—perhaps—may hold some unexpected clues to its meaning. The mass destruction of species solely for transient convenience, for profits, or simply from carelessness is stupid. The case for treating biodiversity as a problem of the first importance has been made with great skill and appropriate conviction by E. O. Wilson: we endorse it unreservedly.[23]

Given that we believe all this—and hope that every thoughtful person will come to do so as well—why are we so dismayed by radical environmentalism? Our concern is with the dogmas that inform the academic left's position on ecological matters, dogmas advocated and defended by means of a rhetoric that derives from theoretical positions having nothing to do with science and the facts of the case. It seems to us that this particular environmentalism of the intelligentsia leads, in the long run, to futility and disillusion. If intellectual and academic history are any guide, many of its votaries are likely to become apostates, not only from radical environmentalism, but from *all* engagement with environmental concerns. Some of its most passionately held positions seem ripe, after all, for refutation; its most emphatic predictions will, in some and perhaps many cases, crash headlong into reality, as did, for example, the memorable announcement of an ozone hole over Kennebunkport.[24] Other kinds of cataclysm than environmental will be predicted and will grow large in the oppositionist imagination. The need for a sober, scientific, environmentalism will, however, grow rather than decline with time. It *must* not become hostage to fashion.

Recall, in this regard, the injunction that adorned so many car bumpers ten or fifteen years ago: "Split Wood, Not Atoms." Those who know better regard such self-advertisements with deserved contempt; but those who stuck them on, innocent of any cost/benefit analysis of wood splitting (i.e., wood burning) versus atom splitting, have since mounted other, trendier stickers. The market for wood-burning stoves has collapsed with the last of the communes. Predictably—fortunately—the vast majority of bumper-sticker people never gave up on central heating; and even among the heavy users of bumper stickers, there are now many who have learned about fossil-fuel burning and its effects on health and the atmosphere. Out of simple, selfish concern for respiratory health, we would much rather live next door to a nuclear power plant (except possibly in the former Soviet Union), if that were the only alternative, than in a community of any substantial size that heats with wood—or coal. In fact: given the catastrophic effect on forests of large-scale, historic dependence on firewood, and the CO_2 produced by fossil fuels of any variety, there is something to be said for the proposition that in terms of environmental soundness, the self-congratulatory slogan got things exactly backwards. Without other choices, good sense would have called, and would still call, for splitting atoms, not wood.[25]

Attitudes

As we see it, radical environmental wrongheadedness is rooted in three interlinked attitudes. First of all, it is intensely *moralistic*. Among ecoradicals, there is a tendency for surmises to take on the character of articles of faith. Thus in any discussion of the greenhouse effect, alternatives to CO_2-emitting energy technologies such as nuclear power or hydroelectric projects are immediately ruled out of court because these have previously been assigned a place by the environmental left in its fixed demonology. Unchanging casts of devils completely exclude the careful, unemotional weighing of costs and benefits, of relative risks and relative certainties that is a necessary part of making pragmatic judgments. This is wryly illustrated by the fact that radical advocacy of "solar" power never puts hydroelectric into that category, in spite of the unarguable fact that hydroelectric power is civilization's oldest and most practicable application of captured solar energy. To a true believer, however, it is far easier to abandon a saint than a demon, and hydroelectric power is rife with demons, for example, men, machines, power lines, utility stocks and bonds, electromagnetic fields, and artificial lakes full of power boats.

A related radical instinct is to reject any form of amelioration. The radical mentality is, almost by definition, emotionally committed to change that is sweeping and wholesale, change that rewrites the terms under which we live. It is committed to punishing the wicked and rewarding the pure. It follows, then, that ecoradicals are never satisfied with gradual adjustment. The idea that our culture should go on much as usual while incremental changes are made to insulate the environment from damage is an unpalatable one for them, even if it could be proven that in the end the cumulative effect of such changes would be to create a permanent barrier to ecological degradation. The ecoradical has a tendency to insist that the *psychic* conversion he or she has undergone must be experienced by all others as well. The ecoradical insists upon redemption; and the language with which one calls for redemption —given the indifference and preoccupation of ordinary people with their daily lives—must be strong enough to scare them into attention.

Finally, we note the reflex dismissal of any but bad news by radical environmentalists (and, to a large extent, the media). [26] To the ecoradical, "No news is good news!"—provided we construe that old saw to mean, "It is impossible, under the current regime, for matters to improve, or even for it to turn out that worrisome threats have been overstated." As we shall see, the chances of overstatement are in fact pretty good. There is a supposition, unacknowledged but patently held, that worse means better. The worse things get, the closer the moment of rupture and overthrow, when the world's population will realize the folly of industrialism, of the comforts of The Way Things Are, and the sooner a return to the wisdom of its paleolithic (or neolithic, or hunter-gatherer, or smallholder) ancestors.

These psychological symptoms of radical environmentalism—moralism, disdain for the meliorative, a thirst for the decisive cataclysm—are of course characteristic of radical movements in general and, as Bramwell's book makes clear, they have found a home within ideologies of all sorts at different historical moments. There seem always to have been penitents among us, marching in their grim processions, lashing themselves for our sins against God or nature, crying out for the rest of us to join in the fun. At present the radicalism of the academic left is the most convenient symbiote for such fervor.

An Intellectual Embargo

In her latest work, *Radical Ecology*, Carolyn Merchant produces a taxonomy of the radical wing of the environmental movement. She examines the theories and dogmas of Deep Ecology, ecofeminism, social ecology, Marxist ecology, and a number of other variants, taking account of the similarities and inconsistencies that link and distinguish them. Predictably, her tone is that of cheerleader and evangelist. In Merchant's view, radical environmentalism is preferable not only to indifference to environmental questions but, as well, to *any* version of environmentalism that leaves intact the authority of scientists and technocrats on any question of substance. She endorses without reservation the cultural constructivist doctrine. Under the heading "Contributions of Radical Theorists" she asserts that "science is not a process of discovering the ultimate truths of nature, but a social construction that changes over time. The assumptions accepted by its practitioners are value-laden and reflect their places in both history and society, as well as the research priorities and funding sources of those in power."[27]

The motivation for Merchant's equation of science entirely with its social process (as opposed to its results), for her embrace of the strong constructivist view, is obvious. She is simply unprepared to accept the judgments of professional science as valid if they contradict her hopes or deflate the prognostications of doom that radical environmentalism needs in order to win converts. The relativism of cultural constructivist doctrine is the perfect tool for discounting science as biased or corrupt if and when it inconveniences one's political program.

Nonetheless, like all other environmentalists, radical or otherwise, Merchant stands in an unwilling and dependent relation to science as it is. The "hot" issues in environmentalism—the possibility of major global warming, the ozone "hole," species impoverishment, overpopulation and its consequences—are issues that would be unknown and unknowable but for the accomplishments of professional science. *They exist only because of, and specifically as products of, science.* Comparing CO_2 levels in "fossil" atmospheres with those of the present time is not the sort of activity the back-to-nature layman goes in for, nor is the construction of general circulation models of weather for running on supercomputers. Questions about species impoverishment arise from the dedicated work of biologists, not because a mass of worried citizens have daily experiences of species decline. *In short, environmentalism in its modern form, including the radical wing of it, is a reaction, occasionally appropriate, to specific discoveries of orthodox science.* The problem with *radical* environmentalism is, therefore, that its relations with science, upon which it must be based, have become so ridiculously acidulous and so dishonest.

There is a heavy traffic in ideas from science to the environmental movement—but, as far as ecoradicals are concerned, there is also a severe embargo. Radicals are willing, even eager, to accept work with alarmist or apocalyptic implications. They are unwilling or at least bitterly reluctant to accept scientific work that confutes or modifies alarmist theories. Examples of this sort of thing are legion. For years, it was an article of faith that dioxin is "the most toxic substance known to man," and that the use of Agent Orange in Vietnam has caused grave and exotic diseases in many or all war veterans and birth defects among their children. The initial estimates turned out to be overstated or simply wrong. Dioxin is nothing to fool around with, but it clearly is not *that* toxic. Actuarial and epidemiological evidence does not support Agent Orange effects *of the magnitude claimed* by activist and advocacy groups. Still, the issue remains mired in politics for the reason—among others—that vast sums (as compensation, for example) are at stake. It will surely *not* be settled by the recent, massive report from the Institute of Medicine, which associates exposure to Agent Orange with some health risks.

A more recent example: the Alar scare was just that, a scare. It did heavy damage, to no good purpose, to apple growers across the country. This is not to say that the initial guesses were necessarily dishonest or incompetent, only that there is an important difference, in science, between a hypothesis and a justifiable conclusion; and there is a further difference between a justifiable conclusion and a public policy. In science, moreover, all conclusions, especially about complex systems, are temporary. Therein lies the urgent importance of the fullest honesty, the most scrupulous avoidance of hyperbole, in discourses of policy.

The *Federal Register* for July 17, 1992, carries the extraordinary announcement that large numbers of substances, classified heretofore by the National Toxicology Program as carcinogens, are to be removed from that list. This announcement codifies understandings that have been growing slowly—and against bitter political opposition—in the scientific community, to the effect that the rodent-to-human extrapolations used in animal screening programs are invalid.[28] A distinguished editor of *Science* magazine had summarized neatly two years earlier the main arguments against standard methods of carcinogen testing with rodents (arguments ventilated in successive earlier issues of the journal):

Diets rich in fruits and vegetables tend to reduce human cancer. The rodent MTD test that labels plant chemicals as cancer-causing in humans is misleading. *The test is likewise of limited value for synthetic chemicals. The standard carcinogen tests that use rodents are an obsolescent relic of the ignorance of past decades.* At that time, extreme caution made sense. But now tremendous improvements of analytical and other procedures make possible a new toxicology and far more realistic evaluation of the dose levels at which pathological effects occur. (Emphasis added.)[29]

In science, this happens all the time. Hypotheses generated by a small amount of preliminary evidence frequently collapse when the studies become larger, more systematic, and better controlled. In such cases, however, the initial finding is usually welcomed by ecoradicals for its doom-laden implications, while the later and better finding is systematically ignored. To such people, dioxin will always be a particularly horrific demon; Vietnam vets who are afflicted by disease will always be victims of Agent Orange; apples will remain suspect unless grown by naturists having no truck with chemicals: anything but "organic" produce will be suspect at best. Science can indict, but it is never allowed to dismiss the charges.

Naively, one would think that environmentalists should welcome news that the planet is, in some respects, not in quite as much danger as we might have thought. This neglects, however, to allow for the psychodynamics of an apocalyptic vision. Radical environmentalists scan the news pages and the popular science journals eagerly for every scrap of evidence that the ecological End of Days is upon us. As with biblical millenarians, objections of logic and fact have little effect. They are already convinced that the world is in calamitous decline. They believe, and seem to *enjoy* believing, that nature is being violated—blasphemously—by their neighbors, and that ultimate retribution is on the way. Evidence to the contrary is viewed as a terrible letdown, not as a reprieve.

Similarly, inasmuch as ecoradicalism is a movement with a worshipful view of the primal, it takes little heed of sober findings that, in many cases, uncorrupted, nonwhite, primitive peoples have been just as contemptuous of what *we* call environmental values as are greedy Euro-American industrialists. As the geographer Martin Lewis notes in *Green Delusions*, "a large proportion of eco-radicals fervently believes that human social and ecological problems could be solved if only we would return to a primal way of life. Ultimately, this proves to be an article of faith that receives little support from the historical and anthropological record."[30]

The Indians of the Americas, for instance, are regularly depicted as paragons of ecological wisdom, at one with nature and the land. This not only collapses a vast and diverse array of cultures into a single "Native American" way of life, but, as well, neglects the fact that long before the hated Europeans made their way across the Atlantic, the earlier settlers, whose forebears came across the Arctic land bridge, wrought enormous—in some cases horrifying—changes to the biological landscape of the primeval Americas. For example, it is likely that most large North American mammals died out at the end of the last ice age because they were hunted to extinction by human newcomers: and there was as yet no population explosion! The Anasazi people of the Southwest turned their homeland into a treeless, eroded waste by their heedless use of timber. Mayan civilization may have collapsed because warfare, urbanization, and overpopulation depleted the fertility of its agricultural system. Slash-and-burn agriculture turned much of the Midwest from forest to grassland, effecting what was one of the most widespread impoverishments of an ecosystem in biological history.

Examples could be multiplied. It is as erroneous to view "primitive" peoples as walking hand-in-hand, philosophically, with a benign "nature" as it is to think of them as unoffending pacifists without knowledge of scalping, slave-taking, and human sacrifice—or to identify the latter three as inventions of Francis Bacon. The irony is that such misconceptions have long been embedded in the imaginative mythology of Western culture, certainly from Montaigne, Diderot, and Rousseau onward.

Ecoradicalism is as opaque to such insights as it is to scientific findings that might plausibly temper some aspects of environmental alarm. Science, indeed rationality itself, appears, to the radical mindset, as a Janus-headed beast, to be used when it warns, but dismissed with contempt when it attempts to reassure. Ecological science, in particular, is "a socially constructed science whose basic assumptions and conclusions change in accordance with social priorities and socially accepted metaphors."[31] In practical terms, this leaves the radical theorist free to accept what flatters his worldview and to reject what does not. As Martin Lewis points out, this leads to the ironic contradiction that radical environmentalism becomes objectively *antienvironmental* in some of its actual effects. The "natural" fibers and fabrics of which environmental radicals are so fond cost far more, in environmental terms, than synthetic substitutes. The high-density urban life from which "back to the land" environmentalists flee offers the best hope for decreasing human pressure upon wild areas and endangered biota; it is the homesteaders who add disproportionately to the strain. Unfortunately, common sense is slow to make headway among those fired up by a vision; rationality, scientific and otherwise, is, as we have seen, in bad odor among such people.

Crying a Fraction of a Wolf

It would be all too easy to place all the blame for the excesses of ecological radicalism on the philosophical fancies and whirligigs of the academic left— but that would also be facile and incorrect. Some of the responsibility must be borne by competent, in some cases distinguished, scientists who have taken up environmental causes. Much of the apocalyptic rhetoric of ecoradicals derives, albeit by a process of systematic vulgarization and overstatement, from the public pronouncements, books, and articles of such scientists. It represents the last stage of a process of simplification and elision, whereby speculation, hypothesis, and conclusions of the most provisional sort are transmuted into certainties. We do not claim that the process is a matter of wrongheadedness or error on the part of the scientists involved. Nor are we blaming them, nor suggesting that there is an easy remedy to be recommended. We are simply suggesting that the political tactics of some environmentally concerned scientists have a spectrum of effects; that some of these effects are of dubious value; and that among the latter are (1) the reinforcement of apocalyptic hyperbole among radical environmentalists and (2) the suppression of arguments and evidence that suggest caution as to derived social policy.

The underlying question is this: What is a responsible, socially aware scientist to do when his researches and legitimate speculations lead him to suspect that some aspect of modern technological society might, in the long run, have horrendous environmental consequences? We emphasize the hypothetical nature of such predictions simply because, in their early stages, novel scientific ideas are usually first approximations—even when they are not (as sometimes happens) false alarms. Moreover, it is often the case that the direst possibility appears not as a simple "prediction," but as one possibility along a spectrum of possibilities, one end of which is neutral or even good. This is clearly the case as regards the effects on global climate of a (nearly certain) doubling, in the next century, of atmospheric CO_2.[32]

In the course of routine science—where conjectures, hypotheses, and arrays of different alternatives are unrelated to any doomsday scenario—the matter of sorting out possibilities, segregating sound speculations from unsound, choosing the most parsimonious and, in terms of explanatory power, the strongest theory is carried through by the ordinary process of scientific publication and debate. It is the central business of empirical science, and rarely does it make headlines. But science with potentially deep environmental implications is not, cannot be, routine. Even the remote possibility of catastrophe sets off alarm bells so long as it is seen to have some scientific plausibility. This is as it should be. There is only one planet and one humanity.

We have already touched upon such hypotheses—the projected enhancement of the greenhouse effect, the seasonal attrition of the ozone layer, the predictions of catastrophic decline in biotic diversity. To dismiss these problems or treat them lightly is, in our present state of knowledge, irresponsible —as irresponsible as it would be to ignore evidence of an asteroid calculated to be on a collision course with Earth (which will happen eventually). On the other hand, in the social and political realm they illustrate the vicissitudes of debating and formulating policy in the presence of enduring uncertainty.

Many scientists argue that, in the face of social and political reality, the caution and tentativeness that, in the corridors of professional science, accompany the initial discussion of new ideas are ineffectual. They are inappropriate when one is trying to sound a warning about a grim environmental prospect, however yet uncertain. They reason that this is a society saturated by exaggeration and gross oversimplification. Consequently, ideas presented cautiously and hedged with warnings of provisionality have no chance of making a dent in public consciousness, or of influencing nonscientist politicians. A scientist who announces that preliminary estimates reveal a 2 percent chance that the sky will fall is unlikely to win much attention. But even a one in fifty possibility of having the sky collapse is appalling. Why not, then, stress the worst-case analysis? Why not proclaim in banner headlines, "THE SKY IS FALLING"? That way, at least, the potential falling sky problem will get public attention and the needed additional research will be paid for.

In some degree, this is the tactic adopted in relation to the possibilities cited above. Some scientists concerned with the anthropogenic contribution to atmospheric greenhouse gases have framed their public discourse so as to stress, as much as possible, the plausibility or even the imminence of disastrous global warming, and have dwelt at length on the worst possible ecological and agricultural consequences of that warming. The same applies to the ozone-depletion hypothesis, to the "nuclear winter" scenario, to the story of synthetic pesticides, and to a number of other questions.

In fairness, this does not amount to their throwing scientific caution to the winds. Rather it is a matter of emphasis and tone, of rhetoric and persuasive technique, of creating a public voice for oneself that differs quite consciously from one's "scientific" voice, of doing one's honest hedging very much *sotto voce*. A good example is the remarkably candid statement of the influential atmospheric scientist Stephen Schneider, of the National Center for Atmospheric Research:

To do that [reduce the risk of potentially disastrous climatic change] we have to get some broad-based support, to capture the public's imagination. That, of course, entails getting loads of media coverage. So we have to offer up scary scenarios, make simplified, dramatic statements, and make little mention of any doubts we might have. This "double ethical bind" that we frequently find ourselves in cannot be solved by any formula. Each of us has to decide what the right balance is between being effective and being honest.[33]

It is worth pointing out, for contrast with the radical environmentalist depiction of "science" as tool of repression and multinational polluters, that this same Stephen Schneider was honored with a major prize in 1991. This was the AAAS (American Association for the Advancement of Science) Award for the Public Understanding of Science. AAAS is the largest organization of professional scientists in the United States. While Schneider's approach embodies perfectly the elitism of hierarchs and syndics so hated by leftists, we are not eager to condemn it out of hand. It would be unfair to characterize it as hypocritical: even scientists must act politically when their convictions, whether or not based in science, call for political action. There is no reason to reproach the scientists who practice it openly, so long as their legitimate concern does not harden into dogmatism and they remain open (as most undoubtedly do) to the deflation of their original guesses by subsequent research. The idea that in addressing remote but terrible possibilities we should give more weight to the terror than to the remoteness is humanly understandable and legitimate, so long as it is understood that remoteness must count for something.

Nevertheless, there is a serious downside to the strategy of talking apparent science while actually doing politics. The most obvious danger, and, potentially, the most harmful, is that a long sequence of unrealized predictions of disaster—not so unlikely a prospect, perhaps—will desensitize the public and the policymakers to similar predictions coming later, somewhere down the line, no matter how firmly supported by evidence the latter turn out to be. Environmental alarmism is, to some extent, a fad, a cultural whimsy. Should it turn out that the alarms are false (as some of them have been—for example, the rather recent predictions of global *cooling*—and will be), the likely effect is that environmental concerns will diminish and that, even worse, the scientific community will have acquired a reputation as wolf cryers. That community has had many kinds of unfavorable reputation in the past,[34] but this one will be new and immensely more dangerous. We are going to have real environmental problems for a very long time, probably for as long as we are here on Earth in any numbers. The worst thing that could happen would be for environmentalism to go the way of the hula hoop. It will do no particular good for scientist-politicians to respond that they have cried only 2 percent of a wolf. Scientists, having deliberately blurred such niceties in raising the alarm, will be hard pressed to call them to their defense if a particular alarm turns out false.

Received Opinion

Important as this question is, however, it diverts us from our current concern, the effect of the rhetorical style of environmentally concerned scientists on the received opinions and clichés that circulate within the community of nonscientific intellectuals. Clearly, the millenarian longings of ecoradicals, and of academic radicals in general, have fastened, with a kind of perverse glee, on the gloomiest prognostications of atmospheric physicists and tropical biologists. If, shall we say, incontrovertible evidence were to appear tomorrow that current and projected levels of atmospheric CO_2 pose no danger of disastrous climate change,[35] thousands of ecoradicals would ignore it or, if unable to do so, be devastated. They would undoubtedly find a hundred specious reasons for rejecting the good news. They have stripped the last layers of caution and qualification from the warnings of scientists and converted them into articles of faith, upon which mere empiricism has never had any purchase.

To that extent, their capacity for understanding, let alone helpfully influencing, the public debate on environmental questions has been damaged. Furthermore, they have become so much more inclined to champion the wisdom of gurus and shamans—Deep Ecologists, Earth Firsters, James Lovelock, Rupert Sheldrake, and the like—whose views are open to serious question. Part—though only part—of the responsibility lies with the rhetorical style of the "public" science we have been discussing. There has been an unhealthy resonance between the propensity of some scientists to emphasize the worst, for what they genuinely believe to be good reasons, and the eagerness of radicals to *believe* the worst, and to build it into their theories of the world.

Having said this, we take leave of this issue in some perplexity. It is not clear to us that the emphatic style of some scientists on environmental issues—Paul Ehrlich on population growth, Carl Sagan on nuclear winter, Stephen Schneider on the greenhouse effect—is not for the best, all things considered. Nor can we recommend with confidence, in this volume, an alternative approach. The question of how scientists should address such issues in public forums is difficult. What we can say is that science, mostly of recent vintage, has established the existence of mechanisms by which the geophysical state of Earth *might* rather easily be upset by an ever-expanding human population. It has established as well mechanisms that *have done* and *will again do* the same thing independently of human activity. Environmentalists don't much relate to the Milankovitch theory, for example. In it the great, periodic climate changes on Earth that lead to ice ages and their reversal are driven by changes in the solar energy flux; and *those* are caused by the periodic changes of Earth's orbit. The Milankovitch theory has recently had a boost from a rigorous reexamination of some of the critical evidence.[36]

This is as devastating a set of discoveries for the human self-image, and for its constitutive myths of eternity, as was Copernicus's heliocentric solar system. To play with them upon the consciousness of a sometimes somnolent, often credulous public, to whose members—even the well-educated—the fundamental meaning of "probability" is impenetrable, is a terrible crap game. Every scientist-activist should be asking himself, "What if I'm wrong? What are the likely long-term effects of the social upheavals I am trying to put in train, and what if they are unnecessary?" Every scientist-participant should hold before himself the negative example of Jeremy Rifkin.

The Gospel according to Rifkin

The career and influence of the activist Jeremy Rifkin provide an instructive case study of the propensity of the academic left for persuasion by the worst kind of pseudoscientific alarmism. The case tells us how trustworthy are their judgments of scientific questions. Rifkin has long been a militant crusader against the supposed dangers of science and technology. In addition to producing his well-marketed books and articles, he continually organizes protests and, in many cases, lawsuits against what he professes to regard as urgent environmental threats, particularly those supposed to arise from biotechnology and genetic engineering. His latest crusade, for instance, is to organize the chefs of America's premier restaurants to boycott new varieties of vegetables produced by genetic modification in the laboratory.[37]

There is no doubt that Rifkin has achieved a kind of guruhood among most ecoradicals and vast numbers of their sympathizers on the nonscientific academic left. Here, for instance, is Steven Best's encomium:

> One of the best recent examples of this postmodern consciousness is Jeremy Rifkin's *Entropy*. Rifkin's book is a powerful and systematic analysis of the problems inherent in the modern worldview and the current

social and ecological crisis facing our planet. Rifkin identifies the modern worldview as the source of our current crisis and provides an illuminating account and critique of the modern will to knowledge and power. He demonstrates clearly how a repressive logic of development informs all aspects of society, from agriculture to transportation; he argues in favor of a radically different worldview and system of social organization that he explicitly terms postmodern.[38]

Stanley Aronowitz seems also to be a fan of Rifkin's postmodern consciousness. Recently, he denounced sarcastically a conference of biologists who *question* the wisdom and safety of recombinant DNA research (which makes them a rapidly diminishing minority within the scientific community) for refusing to invite Rifkin, and for rejecting the precept of his book, *Algeny*, that biotechnology must be utterly abolished.[39] We note simply at this point that Rifkin's *Entropy* is a book that defies its own title by being resolutely ignorant of fine (and not so fine) points of thermodynamics. *Algeny* is correspondingly vulgar when it comes to genetics and evolutionary theory, upon which it is necessarily based. Here, commenting upon it, is Stephen J. Gould:

> Rifkin ignores the most elementary procedures of fair scholarship. His book [*Algeny*], touted as a major conceptual statement about the nature of science and the history of biology, displays painful ignorance of its subject. His quotations are primarily from old and discredited secondary sources (including some creationist propaganda tracts). I see no evidence that he has ever read Darwin in the original. He obviously knows nothing about (or chooses not to mention) all the major works of Darwinian scholarship written by modern historians. His endless misquotes and half quotes are, for the most part, taken directly from excerpts in hostile secondary sources.[40]

Needless to say, Rifkin's ideological enthusiasms, which Best, Aronowitz, and a host of other would-be critics of modern science heartily second, are cut from the same anti-Enlightenment cloth as those of Merchant, Berman, Keller, and the like. He professes a hearty detestation of "Baconian" science as an expression of the Western urge to dominate, tyrannize, and torture; and he decries the "dualism" that severs us from nature. He hews closely to cultural constructivist dogma, deriving all the proposed sins and errors of science from the inhumane values of capitalism. (These folk seem all to have forgotten, or perhaps never to have noticed, the vast quantity of science, indistinguishable from other science, done over the past seventy-five years in aggressively *anticapitalist* societies such as that of the former Soviet Union.) In short, he has absorbed into his positions most of the clichés of the post-modern left.

The esteem for Rifkin of staunch postmodernists such as Best and Aronowitz highlights a noteworthy—and amusing—inconsistency of their thought. His use of postmodern slang notwithstanding, Rifkin, is, at bottom, an ardent believer in a timeless, changeless nature, whose realities are moral as well as physical. For him, nature is sacrosanct; it embodies absolute virtue. Thus, to him, genetic engineering—manipulating the genomes that have arisen during life's history—is blasphemous, not just dangerous. One would think that ardent postmodernists, who reject absolutes, especially moral absolutes, and claim that the notion of a natural reality antecedent to social convention is a bourgeois illusion, would steer clear!

Rifkin, clearly, is no fair-minded scholar, and his grasp of the technical aspects of science is spotty. His misunderstandings, moreover, are of the sort that stimulate rather than inhibit a desire to preach heady ecological sermons. His recent book *Beyond Beef* addresses what begins as a reasonably serious question: Does the use of beef as a staple food in Western society cost more in energy, land use, and externalities than is justifiable in ecological and economic terms? But the address is written in characteristic fashion, treating the issue with a blend of hyperbole, misinformation, and a flagrant selection of favorable sources. It provides us with an opportunity to examine why Rifkin is so popular a figure in academic radical circles. At the same time it demonstrates how appalling a fact that is. The following fragment is quite characteristic: "A number of environmental factors contributed to the agricultural crisis of the 1980's. They include 'eroding soils, shrinking forests, deteriorating rangelands, expanding deserts, acid rain, stratospheric ozone depletion, the buildup of greenhouse gases, air pollution, and the loss of biological diversity."[41]

It is a useful exercise to dissect these two sentences, phrase by phrase. First of all, was there, in the 1980s, an agricultural crisis, that is to say, a general agricultural crisis of global proportions? Certainly there were local agricultural "crises," as there have been ever since the neolithic invention of this alternative to hunting and gathering. The local disturbances, however, have little to do with Rifkin's general thesis. The well-publicized American farm crisis of the mid eighties, for instance, was the result of banking and credit policy, high interest rates, a decline of world prices for a number of foodstuffs, and so forth. Accelerating agricultural problems in Eastern Europe and the Soviet Union during the period were largely the consequence of an inept and demoralized central control in these command economies. The grim agricultural failures in the swath of Africa that lies south of the Sahara Desert

produced massive famine, and had an important environmental component
—drought, and more generally, desertification; but, in many instances, they
had an equally important political component as well—the long, gruesome
civil war in Ethiopia, the Arab *Kulturkampf* against the non-Islamic black
population of southern Sudan, the collapse into anarchy of Somalia. In
general there was no world-wide agricultural crisis as such; per capita food
production increased, although the situation is hardly one to be complacent
about.

What about Rifkin's list of causes? Might they, at least, have had important
local effects? Rifkin has quoted these intact from writing of Lester R. Brown,
who produced the passage under the auspices of the Worldwatch Institute—
publisher of the annual Worldwatch chronicle of impending environmental
catastrophe and, one should note, "the only book foretelling the end of the
world that routinely advertises next year's edition."[42] Agricultural soil ero-
sion is of course as old as tillage itself. It is sometimes a problem in industrialized countries that practice large-scale, mechanized farming. In terms of
human suffering, however, it is a far worse problem in the developing world,
where farming methods commonly remain primitive and where there are few
capitalists.

The shrinking of forests is, as we concede, a serious problem. However, it
is, in major part at least, the result of *increased* agricultural activity, not a
cause of *diminished* agricultural activity! Deforestation may be deplorable, but
in the short run—in the 1980s—it did not *cause* any agricultural crisis. In any
case, it is far from clear that the eighties witnessed a net *loss* of forest on a
global basis. And, as regards net decline of particular species in mature
forests, when it is in fact demonstrable, there is uncertainty about causation.
In his text on environmental ecology, Bill Freedman admits that "such decline also appears to take place in areas where pollution is unimportant, and
in these cases it has been hypothesized that dieback may be a natural process
that could variously involve climatic change, the synchronous senescence of
a cohort of overmature trees, or the effect of pathogens."[43]

Deteriorating rangelands were, indeed, a problem—most tragically, as we
have mentioned, in the Sahel, where the decline is associated with desertification. There is, however, no reason to ascribe this to any man-made cause; it
has been going on for centuries as part of the general shift of climate since the
end of the last ice age.

Acid rain, again, is indeed a problem—but not, for the most part, an
agricultural problem. The threat (and it is recognized that its magnitude was
much exaggerated in the 1980s)[44] is to certain forest communities and to the
biota of some lakes and streams. Cultivars are not particularly affected. This is

not to say that we advocate a return to high-sulfur coal, although one serious hypothesis offered in explanation of the failure of increased atmospheric CO_2 to warm up the Northern Hemisphere as predicted by the models is that the acidifying particulates emitted in fossil-fuel combustion protect, because they nucleate cloud-formation![45]

A significant or disastrous anthropogenic contribution to stratospheric ozone depletion, to state it once more, is still a conjectural matter. There may be an effect due to the sudden increase, after World War II, of chloro-fluorocarbon refrigerants and propellants, and it may become serious. To date, however, and in any case, there has been no solidly demonstrated biological effect (although there are plenty of speculations) due to ozone depletion. That is no surprise, since it occurs over the South Pole, where there is not much biology to be affected. In temperate and equatorial lati-tudes, there is no solid evidence as yet of a thinning ozone layer, let alone of increased amounts of ultraviolet light reaching the surface of the earth. In fact the evidence is that increased cloud cover has reduced the penetration of ultraviolet radiation to the surface.[46] There has thus been no discernible effect, adverse or otherwise, of ozone depletion on *agriculture*—which is the point at issue. It may yet happen, but that is another story: it is not the one being told, presumably to thousands of eager readers, by Rifkin.

Greenhouse gases are indeed increasing. The effect is still uncertain, al-though ultimately they will have an effect. Indeed they may be having an effect now: a benign effect in counteracting a slow, general cooling of the planet, brought about by geochemical or astronomical processes, or by the enhancement of agricultural production.[47] However, even the keenest augurs of global warming (among scientists, that is) admit that no such trend has been unambiguously observed, that we do not yet have the predicted "signal": in short, that the warming that should have been observed by now, according to the (general circulation) models and the unquestionable increase of green-house gases since the industrial revolution, hasn't happened yet.[48] All this may change: significant, early, greenhouse-induced global warming may be-come a reality. But however desperate things get, they will never become bad enough to produce *retroactively* a global agricultural crisis in the period 1980–90.

To round things off, we observe that air pollution is still chiefly an urban problem, not one that affects (or is caused by) food production. As for loss of biodiversity, this is a *result* of increased agricultural activity, but it is not yet the *cause* of any agricultural crisis!

We crave the reader's indulgence for worrying the logic, at such length, of so small a scrap of text. We do it (1) out of a sense of rueful admiration, (2) because it is entirely characteristic of its two authors, and (3) because we cannot do it for the entire literature. It takes a certain virtuosity to cram such a range of disinformation into two sentences. The nearest parallel we can think of, on the other end of the political spectrum, would be a televangelist contriving to blame the American family crisis on day care centers, gay pride, gun control, rock'n'roll, the abandonment of the gold standard, the theory of evolution, and the graduated income tax.

Why, then, should professional intellectuals, who have been trained in analytical thinking of some sort and presumably practice it for a living, be so eager to promote Rifkin as a major thinker on ecological issues? The sad truth seems to be that professorial tenure does not immunize people against spin doctors, political or scientific. Rifkin, as the quoted passage should demonstrate, hits all the hot buttons, deploys all the buzzwords at once. This tactic, together with a willingness to stoke all the postmodern leftist's prejudices against Western methodology, ontology, and epistemolgy, is enough to put to flight any lingering impulse to make careful distinctions, especially those that require scientific knowledge.

In Dubious Battle

There is something decidedly curious about this commitment of the postmodern academic left to environmental issues. It seizes eagerly on the pronouncements of scientists, judicious or otherwise, that hold out the promise of crisis or catastrophe. It does this despite a firm rejection of the notion that there is any special truth-value in science. The result is a farrago of scientific fragments—some of them sound in themselves, but taken out of context—myths, fancies, resentments and dreams both vindictive and utopian.

So what's new? Hasn't intellectual argument always been like that to some degree, the conceit of intellectuals notwithstanding? The answer is yes and no. Shoddy logic, emotion-laden exposition of what are offered as the hard facts of the human condition, special pleading, wishful thinking, posturing—all have long been with us as part of the public life of the mind and of political action. They appear even in the purest of natural sciences, although much more rarely (we believe) than in other disciplines. Among Western academic intellectuals, however, particularly since the Enlightenment, such argumentation has usually received, sooner or later, a sharp, public correction. That has been the game, particularly in empirical science.

The postmodern left, however, has introduced a new rhetorical wrinkle that shields it from such rebukes. They are ready, at need, to scorn the canons of logic, evidence, objectivity, and coherence on which most of them used to depend for a living. For them, the life of the mind is a dance on thin air. When dealing with academic arcana, this is merely obnoxious. One could well ignore the whole game and survive happily, except perhaps in certain departments. But it is a terribly dangerous attitude to take toward environmental questions, toward any questions, for that matter, of significant public import. Environmental issues are important precisely because they involve urgent questions of *fact*, whose investigation must be carried out, like it or not, by the methods and epistemic strategies of orthodox science. Epistemological hubris is a sin into which most scientists probably fall, from time to time. But it is an unremitting flaw of radical environmentalism, despite its pose of abandoning human arrogance and humbly seeking the counsel of nature. This pose is carried along on an undercurrent of unwavering self-righteousness. Ecoradicalism (like so much of contemporary academic radicalism generally) is really a movement of personal salvation. Consequently, in ecoradical lingo facts frequently devolve into mere tropes, and flat assertion is elevated to the status of evidence. Subjectivity is not only not suspect: it is demanded. Objectivity, on the other hand, is dismissed, curtly, as the delusion of a Western consciousness obsessed by domination, exploitation, and profit. Objectivity is Satan's wile.

"What is involved here is a reconceptualization of the human side of the human/nature dualism to free it from the legacy of rationalism," proclaims the feminist philosopher Val Plumwood, in an ecofeminist version of the credo.[49] With minor variations, the theme recurs, insistently, throughout the entire range of ecoradical literature. It is a staple in Jeremy Rifkin's books. And: it is a recipe for disaster in ecological matters—and in human affairs generally! It is the substitution of moonbeams and fairy-dust for thought, a frequent human practice, but one that has taught grim lessons in the course of history. Janet Biehl, in her devastating analysis of ecofeminism, has a few well-chosen words for it:

> notions focused overwhelmingly on women's allegedly quasi-biological traits and a mystical relationship that they presumably have with nature—a "nature" conceived as an all-nurturing and domestic Great Mother. This highly disparate body of hazy, poorly formulated notions, metaphors, and irrational analogies invites women to take a step backward to an era whose consciousness was dominated by myths and by mystifications of reality. It does not bode well for women—especially those who regard themselves as more than creatures of their sexuality—to follow this regressive path.[50]

Of course, the trick in avoiding hazy notions is to learn history; and that is difficult if one believes that, having been written before the discoveries of postmodernists, "history" is a delusion, not merely Henry Ford's "bunk."

An example of the ecoradical's gift for prognostication may be seen in Morris Berman's account of the social and historical mechanism by which the villainies of industrial society will wither away so that we may ascend to eco-paradise. "On the political level, decay will probably take the form of the breakup of the nation-state in favor of small regional units. This trend, sometimes called political separatism, devolution, or balkanization is by now quite widespread in all industrial societies."[51] Indeed, this trend is widespread, though hardly in all industrial societies (consider, per contra, the nascent Western European superstate). And it is called balkanization; and we may currently see what that entails, in the bloody, fratricidal stupidity that torments the Balkans! If Berman's prognosis is correct, the future is nothing to look forward to, and nothing the theorists can be proud of having recommended.

William Blake once wrote with pride that he was mad, yet retained his health: "mad as a refuge from unbelief—from Bacon, Newton, and Locke."[52] That kind of madness is forgivable, admirable, perhaps, in a poet of genius. Prophetic visions are indispensable to a culture. But we must remember how double-edged they can be, and how pregnant with unintended irony. As a guiding principle for a mass environmental movement, Blake's madness portends disaster. According to Martin Lewis, "Radical environmentalism enjoys substantial, and growing, intellectual clout. If its concerns merge with those of the broader academic left, a trend visible in the rise of both eco-marxism and of a self-proclaimed subversive postmodernism, we may well see intellectual hardening of uncompromisingly radical doctrines of social and ecological salvation."[53]

Notwithstanding Lewis's concern—which is nevertheless a point well taken—we must guess that the dangers he points to relate mostly to the contamination of discourse in intellectual circles. The variants of edenic environmentalism that Carolyn Merchant surveys in Radical Ecology are, as social movements, self-limiting sects. Taken all together, they hardly constitute a significant political fringe, except on the many campuses where the academic left already enjoys a high level of influence. In her more realistic moments, Merchant admits as much, though she rather wistfully hopes that ecoradical insights will percolate into the consciousness of the larger political community. Frankly, we think it is just as well that ecoradicalism remains stunted, even in relation to the mainstream environmental movement. It is far from clear to us that a true mass movement of fervent ecoradicals, hearts brimming with the precepts of Jeremy Rifkin or Dave Foreman, would not, in the end, turn out to be as pestilential as that of the followers of the Ayatollah Khomeini.

To us, it is self-evident that a 1 percent improvement in the efficiency of photo-voltaic cells, say, is, in environmental terms, worth substantially more than all the utopian eco-babble ever published. In this sense, we are unabashed technocrats, unashamed of the instrumentalism behind such assertions. An accomplishment of this kind will almost certainly not come from the ranks of the ecoradicals, most of whom would, no doubt, denounce it with scorn as a "techno-fix." Yet technology and the scientific thinking that stands behind it are, for all their vexed history, indispensable tools for providing humankind with a stable environment in which it can live on honorable terms with itself and with nature. The attempt to replace them with phantom visions of global consciousness change or cultural paradigm shifts are wrongheaded and, even worse, wronghearted.

Our experience with ecoradicals, as with the antiscientism of academic leftists generally, brings us round once more to poor old Francis Bacon, whose once restful shade has been recalled, in the last ten or fifteen years, to be harrowed endlessly by radical critics of Western science and culture. It is a bum rap. Let us raise a glass to Bacon! He wasn't much of a scientist or mathematician, but he made some shrewd guesses as to how our species might crawl out of the rut of ignorance. And here's to Baconian science—if that misattribution is to persist in our universities—Baconian in the sense of a rigorous adherence to the empirical, and a faith that what we learn that way can improve the prospects for human life. The more Baconian science we get, the easier it will be to believe that we have a fighting chance, if no more than that, on this lovely planet that spins its way through an unimaginably violent—and indifferent—space.

CHAPTER SEVEN

The Schools of Indictment

Meanwhile concentration on "how the meaning is generated" keeps theorists busy; especially proponents of Marxism, Feminism, Minority Discourse, Cultural Studies, Deconstruction, New Historicism, and other schools of indictment.
DENIS DONOGHUE, "BEWITCHED, BOTHERED, AND BEWILDERED"

Thus I think a good case can also be made that the AIDS pandemic is the fault of the heterosexual white majority.
LARRY KRAMER, *REPORTS FROM THE HOLOCAUST*

Evil is the oldest and most intractable of all enigmas. Alone among the species, *we* know that misfortune is inevitable. And yet, if this knowledge is universal, so is the propensity to see a malign agent behind our misfortunes. "But in each event—in the living act, the undoubted deed—there, some unknown but still reasoning thing puts forth the moulding of its features from behind the unreasoning mask."[1] So says Ahab, and so say we all in our most anguished moments. Most likely, this is a cognitive adaptation of our species, allowing us to function and continue sane in the face of actual or expected calamity. Out of it, however, there have emerged most of history's bloody and wasteful conflicts.

In this chapter we consider certain social and political responses to evils, real or perceived. In contrast to the arid disputes of the academy, these are matters that actively and continually roil our civil existence, and devastate the lives of thousands. Chiefly, we shall be concerned with the abiding problem of racial justice, and with the AIDS epidemic that has reawakened slumbering fears of plague and fatal contagion. Science, obviously, has a great deal to do with the latter problem. In fact it is our only defense, and our only source of hope. As regards racial injustice, however, it is not at all clear that science, even in its most forthright technological guise, has much to do with the situation. Yet its iconic status as the emblem of intellectual authority and material power in this era makes it one of the foci of the fantasies and dreams of the dispossessed.

These problems are *not* primarily located in the academy, and, indeed, where they intersect—in the blighted neighborhoods where AIDS is becoming omnipresent but a white face is rarely to be seen—the academy is but a rumor. Yet there is a peculiar and ultimately unhealthy traffic between the world of rarefied postmodern theory and the communities in which activism, though stemming from real and terrible problems, overflows into paranoia and fantasy. Many of the blind alleys down which activists charge have first been mapped out by academics bedazzled by contemporary theories of discourse. Thus, when questions of knowledge and authority arise in connection with scientific and medical matters, or in a general context, the strange combination of skepticism and credulity that characterizes the postmodern stance is strongly echoed. By a process of tacit reciprocity, the concerns of activists—their tactics and rhetoric as well—find defenders and advocates among the left-wing scholars and cultural critics to whom "resistance"—to the social norms of late capitalism—no matter how incoherent in its suppositions and doubtful in its practices, is always a welcome development.

We also consider here a social stirring—the so-called animal rights movement —that is, relatively speaking, small and eccentric, and that arises from concerns that we see as far less compelling than those that grip AIDS activists and crusaders for racial justice. This, too, is an arena where the language and attitudes of the academic left are deployed and, again, the cause is one that a growing number of academics have come to rationalize, albeit the response is on a far smaller scale than is the case with race and gender oppression.

AIDS

The story of AIDS, more so perhaps than any other in contemporary science, refutes, simply as it is recounted, the notion that science is not so much a matter of expanding knowledge as it is of competing, culturally constructed paradigms. To be sure, paradigms compete; and they contribute at any moment to the formulation of questions and the choice of "puzzles" (Thomas Kuhn's word); but the succession of paradigms does not involve starting each time from scratch. Theory choice is not just a matter of politics and style, as Kuhn himself insisted in defending his work against its critics.[2]

The ongoing story of AIDS is already the subject of a huge literature, a public literature as well as a new technical and professional one. We could not begin to do it justice here, even in summary; but some of its highlights illuminate the issues of our argument. If the outbreak of AIDS had occurred a decade earlier than it did, most of those ten years, it is safe to say, would have passed without any real understanding of its etiology, and hence with far less hope for its management. This is so because the disease results from an attack on the body's defenses against non-self—its immune system—by a particular and peculiar virus. The keys to understanding that attack and to identifying the pathogenic agent are bodies of very new knowledge in cellular immunology, molecular biology, and virology, knowledge that was spotty or did not exist until the late 1970s.[3]

The first papers describing a peculiar immunodeficiency syndrome appearing in homosexual men—soon to be known as Acquired Immune Deficiency Syndrome, AIDS—were published in 1981. By 1982–83, it was clear that a catastrophic epidemic had erupted in the gay communities of New York and San Francisco. The name given the disease stands for its underlying symptoms: a breakdown of normal immunity and the consequent, devastating appearance of opportunistic infections and unusual cancers that the body is unable to resist. The superficial symptoms were (as always) not informative, taken by themselves: diarrhea, fever, rapid weight loss, muscle wasting, weakness. Because, however, there were newly available and effective means of assaying immune function, the deadly origin of those secondary symptoms was soon understandable. Because indices of immune function were available—including those dependent on monoclonal antibodies—the primary target of the disease process was pinpointed: it is the particular subpopulation of T-cells known as $CD4^+$. We note that among the early reports was an important one from a group headed by Dr. Anthony Fauci, who will reappear below.[4]

There was no doubt, when the epidemic pattern became known, that this was a communicable, an *infectious* disease. It was to be understood by discovering what it is that attacks, and what then happens to, the susceptible cells of the immune system. The search for the pathogen was intense and worldwide (or, at least, Western worldwide). In 1984 an RNA virus (retrovirus) of the lentivirus ("slow virus") subfamily was identified as the causative agent by Luc Montagnier and his group at the Pasteur Institute in Paris, and independently by Robert Gallo in the United States.[5] This was the outcome of a worldwide effort unprecedented in the history of medicine: unprecedented in scale, scope, urgency, and—most importantly—in the extent to which applicable resources of basic science were brought to bear to create a new science ad hoc. The time interval between recognition of the epidemic and identification of this new and unusual infectious agent, which came to be known as HIV, human immunodeficiency virus, is also unprecedented in its brevity.

Had there been no biotechnology and molecular immunology based upon genetic engineering methods—methods so abhorrent to the radical critics of science, methods that biotechnology opponent Jeremy Rifkin and his "postmodern science" admirers consider to be not only superfluous but a kind of blasphemy against nature—there would have been no such developments. Nor would the screening methods by which the national blood supply was eventually protected have been created. Even as it was, the time needed for such work was long enough to allow a third major category of victims, after homosexual men and drug abusers who share hypodermic needles, to be infected—hemophiliacs and others receiving transfusions of blood and blood products.[6]

The epidemic has followed its predictable course within the at-risk populations. The rate of infection *seems* to be beginning only now, as we write, to level off, in the industrialized countries.[7] At least in part that local leveling off results from recognition among the original (American and European) victim population, practicing homosexual men, of the risks of casual sex with multiple partners. That recognition is a great credit to activists (including Larry Kramer, of whom more below) and more generally to the large numbers of well-educated, middle-class homosexual men who threw themselves into the effort of education about risky sex and behavior modification. It appears, however, that, in the main, potential victims educated one another. Neither their efforts nor the much larger government and media efforts of education and propaganda seem to have had much effect on the underclass subpopulation of victims, nor—sadly—on younger, less well-educated homosexual men.[8] The epidemic goes on, and despite the explosion of knowledge about the molecular biology of infection and response, neither cure nor vaccine is as yet to hand. In the meantime, as hard science moves ever closer to a description of the cellular and molecular basis of the disease,[9] the awful complexity of its pathogenesis becomes more evident. Hope for an early "cure" recedes rapidly, and the problems of designing therapeutic and vaccination strategies tax the powers of some of the world's most advanced laboratories and biomedical minds.[10] Mitigation of AIDS there is, then, in a growing medicine chest of new drugs and in evolving clinical strategies of their use; but as yet there is no cure and no prevention, *except* by avoidance of those acts that transfer body fluids from the infected to the uninfected.

Spreading the Fear

Almost from the start, activists have suffered a terrible ambivalence about the form to be taken by their education—and propaganda—efforts. There was and is a strong reason, on the one hand, to argue that *everybody* is under the gun. In the first place, this might have been true under certain conditions, for example, if transmission by ordinary heterosexual contact were as easy as it is by less common sexual practices. As we have found out, however, this clearly seems *not* to be true, at least in America and Europe.[11] Nevertheless, especially in the early phase of the epidemic, sober researchers and public health officials were understandably determined to err, if at all, on the side of caution, and to assume a "worst case" scenario, in which the general population was assumed to be gravely threatened. Secondly, the particular interests—and fears—of the at-risk groups came into play. If the whole population were believed to be at equal risk, the focus might be expected to shift away from unusual sexual behaviors, intravenous drug use, and the outlaw status of the groups practicing them. It was felt—correctly—that the perception of a universal threat would accelerate the appropriation of government money for treatment and research.[12]

On the other hand, the fact remains that the overwhelming majority of victims are members of classes originally tagged as "high risk": homosexuals and the inner-city poor, among whom drug abuse and "deviant" sexual behavior are commonplace. This invites a different and contradictory style of activist rhetoric. Here, AIDS is seen simply as the amplification and continuation of oppressive practices, long endemic in the society, that victimize gay men and the nonwhite underclass. The special moral claims of the victim groups, antecedent to AIDS as they undoubtedly were, can be invoked, at least to sympathetic audiences, as another argument for addressing AIDS with particular urgency.

In the conflict between these two strategies of education-propaganda, the first—AIDS is for everyone—has won hands down: it is now the generally accepted position among the public. Its iconic figure is Magic Johnson. The weird notion is that Magic Johnson is like everyone else, but perhaps nicer (the latter part of which may well be true). The second point of view—AIDS as a form of racist and capitalist oppression—has retreated to the ends of the spectrum: highbrow (academic left) and lowbrow (the conspiracy theories of militant Afrocentrists,[13] for example).

The AIDS-is-for-everyone thesis is now regnant in the media.[14] It is strongly reinforced by the image of thousands of celebrities of every description prominently displaying red ribbons during public appearances—perhaps a praiseworthy act in itself, but nonetheless firmly linked to the doctrine that AIDS is as much a threat to drug-avoiding and monogamous heterosexuals as to drug-using gay prostitutes. A wry comment on this situation appears in a recent *New Republic* editorial:

Without exception, the press has presented teen AIDS as a growing problem, which means you have to ignore the fact that the tiny number of such cases actually *fell* from 1990 to 1991 . . . The media invariably make teen AIDS a problem of white straights of both sexes, when most cases are minority male, and gay . . . Most teenagers with AIDS didn't get the disease from sex, but from hemophilia clotting factor, with gay male teens as the second biggest category.[15]

The *New Republic* prudently relied on the judgment of one of its correspondents, Michael Fumento, whose book, *The Myth of Heterosexual AIDS*, gives a detailed but accessible account of the epidemiological facts for those disposed to pay attention. (The hostile treatment of Fumento's book by AIDS activists, public health officials, and various groups having a stake in the perpetuation of the myth Fumento criticizes is one of the most depressing episodes to date in the history of the public perception of AIDS.)

Honesty must view this situation with an uneasy ambivalence. On the one hand, the imagery of stellar athletes (e.g., the late, justly famous Arthur Ashe), Hollywood figures, and well-scrubbed white teenagers succumbing to AIDS has probably enlisted support for AIDS care and research that pictures of suffering Africans and afflicted denizens of the gay ghettos of New York and San Francisco would never have elicited. On the other, there is the matter of fidelity to scientific fact—to the truth; and that is never a lighthearted issue. Myths have consequences.

L'Affaire Duesberg

From the first, there have been legitimate questions *within* science about the simple communicable-disease hypothesis, and these have been followed by illegitimate arguments (mostly from the extreme right, where "homophobia" is a reality) to the effect that there is nothing much for ordinary people to worry about, and that there would be nothing to worry about at all if certain people would stop doing unnatural acts.

Best known among *scientific* critics of the HIV-AIDS hypothesis has been Peter H. Duesberg, a distinguished Berkeley molecular biologist. Reduced to its essence, Duesberg's argument is that the full, classical proofs of causation that link a biological agent to a disease process remain, in this instance, somewhat lacking; and that if we are to be left with mere correlation, rather than proof, then the correlation between AIDS and certain predisposing "lifestyles" (including heavy drug use and the accompanying malnutrition) is just as strong as, or stronger than, the one between AIDS and HIV seropositivity.[16]

Nor is Duesberg alone. Although his arguments have been negated, successively, over time, as new science has appeared, the undeniable specificity of the at-risk population persists, and certain "lifestyle" characteristics of that population—of course—persist as well. Serious epidemiological work and examined clinical experience inevitably produce correlations: to a large extent, that is what they are *about*. Thus it appears that people who actually come down with AIDS are usually subject to immunosuppressive factors. Among them are: semen-induced autoimmunity following from anal intercourse; blood transfusions; multiple infections; chronic use of recreational and addictive drugs; and so on. This list is from arguments assembled by Robert Root-Bernstein, a physiologist at Michigan State University. It is a part of his argument to the effect that, whatever the role of HIV, it is the *predisposing* factors, which are themselves products of particular behaviors (some of them not choices but necessities, as in the case of hemophiliacs, who need the clotting factors in order to live), that determine whether AIDS develops. Put bluntly, as does one newspaper in which this argument has been featured, "healthy, drug-free people do not get AIDS."[17]

That line of argument would hardly deserve mention in a book concerned with the thinking of the academic *left*, since the lifestyle argument, in its most magniloquent form, is mainly a preoccupation of the right.[18] But it is in fact also used by the left,[19] when convenient, to demonstrate the uncertainties, the indeterminacy, of standard science, and in some extreme cases, to show that AIDS is a *social* malady first and foremost, whereas the scientists are merely playing their professional and careerist games of elegant research, heedless of the apocalypse. We shall see samples of this below.

Granting, however, that predisposition to a full-blown syndrome of any kind increases the likelihood of its appearance—which is, after all, a tautology—nothing adduced by way of evidence so far has shaken the infectious hypothesis of AIDS: quite the reverse.[20] In short, AIDS is a communicable—but not an especially contagious—disease. Its communication is limited to particular kinds of risky behavior and much more rarely to unprotected heterosexual[21] contact with infected persons or with their body fluids. The likelihood of infection in the first instance, or of progression afterward, *may* be affected by other factors of health status; but such modifications of the infection-disease sequence are very weakly determinative. The disease is caused by a virus; and despite brilliant, ceaselessly energetic work aimed at finding effective therapies, there is as yet no treatment that stops the disease in its tracks, let alone cures it. The most recent epidemiological data on AZT, the most powerful anti-AIDS drug, show that it has little or no effect on progression of the disease in seropositive, but still symptom-free, people.[22]

This is, sadly, no surprise: we haven't done much better with other viral diseases. Unfortunately, this virus, HIV, is a very smart, successful, and brutal one: it hangs around long enough, silently enough, in its victim to assure transmission to others, and only thereafter—long after—does it kill.

AIDS and the Academic Left

The relation between the scientific study of AIDS and public perceptions is thus vexed and equivocal. Of course, AIDS paranoia of the rankest sort endures and is in fact likely to increase. It continues to insist, in the face of all evidence, that AIDS is highly contagious and that the mere presence of an infected person, the lightest touch of the hand, carries a threat. Such superstition—there is no other word for it—is often concurrent with outright hostility to the at-risk groups—gay men[23] and intravenous drug users. As we have seen, however, a number of other factors—including a lack of forthrightness concerning risks to heterosexuals and unlikely but very well publicized alternative theories of AIDS causation—have bred misinformation and confusion among well-intentioned people. Among those segments of the population most deeply concerned—and this obviously includes at-risk groups and their supporters—uncertainty and ignorance have led to other kinds of paranoia. The nastiest version is endemic in poor (and sometimes not-so-poor)[24] black communities, where it is widely held that AIDS was deliberately developed and introduced by a white racist government in order to perpetrate genocide against the darker-skinned. This is a horrible, deeply saddening development in itself; but it falls outside our purview (although, inevitably, a few academics give this demented tale some sort of credence). Even among the highly educated and intellectually sophisticated, however, bizarre theories have taken hold. Often these are contemptuous of scientific opinion about AIDS and of science in general.

It is easy to find, in any bookstore and on most college campuses, versions of "AIDS activism" in which the hatred of scientific thought is indelibly manifest. It is characterized by antagonism toward controlled experiments of rational design, toward the effort to isolate variables in complex situations, toward the painstaking quantitative analysis that is essential to getting at the truth. Larry Kramer perhaps provides the best-known example, arguing as he does that "the heterosexual white majority" is to blame for the catastrophic AIDS epidemic. His book, *Reports from the Holocaust: The Making of an AIDS Activist*, is a collection of polemical pieces written in the decade that began in 1978. It concludes with a long and incoherent personal essay bearing the book's title. Personal or not, its tone is a scream of rage and fear. Of course, one cannot help sympathizing with that fear, with the sense of urgency, of someone who is a potential AIDS victim. Kramer is himself HIV positive, and has been living for years under a perceived death sentence. Like many other gay men, he has seen hundreds of friends and acquaintances cut down by the disease in what should have been the prime of life. If it were possible to view his writings as acts of personal catharsis, they might be received with unreserved sympathy. However, Kramer fully intends them to have an effect on public discourse and on public policy; and to an extent his expectations have been fulfilled. This puts his opinions in a much more ambiguous light.

A founder of such visible—and from some points of view, effective— groups as the Gay Men's Health Crisis and ACT-UP, Kramer has been, for the fifteen years of the AIDS era, a conscience of AIDS activism and one of its shrillest accusers. We do not mean to suggest that he argues, as do some black militants, for a literal conspiracy of infection, namely, that white heterosexuals have conspired in biological warfare against homosexuals. Kramer's indictment is different and more subtle: it has elements of paranoia, but it contains an argument. He doesn't deny that sexual promiscuity, promiscuous anal intercourse, to be specific, played a key role in the early and devastating spread of AIDS in America. That promiscuity, however, he contrives to blame on the heterosexual majority, who, by depriving gays of their ordinary, human identity, are supposed to have forced homosexuals to adopt an alternative one—an "identity" whose affirmation entailed a way of life in which sexual promiscuity is a necessary part, an affirmation of self.

This aspect of Kramer's thinking inadvertently reveals a theme that is present, but unacknowledged, in the pronouncements of most other gay AIDS activists. Many of the individuals who are most energetic and vocal in trying to cope with the epidemic were, some fifteen years ago, fervent advocates of an ethic of unrestrained sexual self-expression for gay men, a philosophy that played out, in practice, as a lifestyle that involved dozens, even hundreds, of anonymous sexual contacts per year. At the time, such frenetic sexuality was widely held to constitute a *political* act, one that held the key to the personal liberation of homosexual men and to the group solidarity necessary for effective political action against reigning heterosexual norms of society. The enthusiastic adoption of this point of view by thousands of "politicized" gays was, tragically, one of the chief factors in the explosive spread of AIDS through the homosexual community. This is not a matter of scapegoating—it is simply a fact. Yet it leaves a bitter residue of guilt among gay militants. It is a guilt that is hard to acknowledge directly, for fear of confirming the hostility toward homosexuality that strongly persists in the general community. Such guilt thus finds its outlet, not in explicit remorse, but rather in petulance and a redoubled militancy, in which the accusatory spirit is hugely amplified. Larry Kramer's polemics provide a prominent example.

Kramer drives his argument to further extremities, in which violent antipathy toward the majority culture in general and science in particular is plainly evident. He holds that once the epidemic was in full career, the heterosexual majority (and that includes the scientist members of it) were quite content with the situation; that, with utter equanimity, they *allowed* AIDS to devastate a part of the population: "I personally think that genocide is going on. The administration's determination—which has persisted for a very long seven years—not to do anything sufficient to fight the AIDS epidemic can only be construed as an attempt to see that minority populations they do not favor will die."[25]

Almost nobody escapes Kramer's denunciation, not even some gay men who happen to have succeeded him in positions of leadership of various activist groups. The then-mayor of New York, his entire administration, Ronald Reagan (of course) and his administration, the entire National Institutes of Health (NIH) and its Dr. Anthony Fauci, Dr. Frank Young (former director) and the entire FDA, the government–pharmaceutical industry complex—everyone, in fact, who is not a PWA (person with AIDS), and even some of them, come under furious attack, if not for active collaboration with the disease, then for failing to take appropriate action against it. All are denounced as irremediably stupid, or cowardly, or homophobic.

> In this country, our enemies include our President, our Department of Health and Human Services, the Hitlerian Centers for Disease Control, the U.S. Food and Drug Administration, the Public Health Service, the self-satisfied, iron-fisted, controlling, scientific frankenstein monsters who are in charge of research at the National Institutes of Health and who, with their stranglehold grip of death, prevent any research or thinking that does not coincide with the games their narrow minds are playing.[26]

It is painful to call attention to the absurdities of people whose suffering is undeniable. Yet these citations point to a widespread mythology well entrenched in our era, one whose grip is by no means confined to the unlettered. A deep ambiguity inflames Kramer's thought and echoes through the pronouncements of hundreds, perhaps thousands, of other activists. On the one hand, there is an almost superstitious awe of science and scientific medicine, a presumption that *if only they really cared*, researchers and clinicians could come up with a solution to the AIDS epidemic in short order. Kramer's polemic, for instance, not only denounces the failure of the medical establishment to take measures that *he* is certain should have been obvious early on, but also accuses it of withholding its knowledge and expertise, of perversely refusing to produce the miracles that, Kramer is sure, must have been there all along. There is something deeply atavistic about this. Scientists are seen as shamans possessed of limitless powers, and if these powers are withheld, the only possible explanation is the wonder-worker's malice.

On the other hand there is a countervailing sense that medicine and science are in fact impotent and even fraudulent, that they exist chiefly for the benefit of careerist researchers, that they are no better roads to the truth about health and disease than a thousand other "cultural practices." In ordinary times, these suspicions would be fleeting. But, even by its own singular standards, the gay community can hardly regard the last dozen years as "ordinary times." Physicians and biological scientists have failed at what they were thought, by a naive public, to be able to do flawlessly and on demand—cure infectious disease. There is a sense that an implicit pact has been broken, and the bitterness is enormous. Such feelings are endlessly recurrent in the rhetoric of AIDS activists, especially when that activism is conjoined to a mood of general cultural radicalism. As one gay rights activist puts it, "gay white men who previously considered themselves relatively privileged began to question the medical and scientific establishment, realizing as Blacks and women have long known that science is not always friendly."[27]

It would be disingenuous to dismiss feelings like this as an entirely baseless phobia. There is a long history of nasty behavior toward homosexuals by some spokesmen of the scientific community. Homosexuality has until recently been stigmatized in the judgments of scientific medicine and psychiatry as dysfunctionality. The conversion of mainstream medicine to a more tolerant view is a new development, and in some respects not entirely secure. Even more grave is the genuine conflict between AIDS victims, who look with desperate hope to any possibility of cure, no matter how farfetched, and AIDS researchers, whose professional standards—and moral conscience—impel them to conduct their studies so as to maximize the reliability of conclusions. The question of how to justify placebo studies in a population with an invariably fatal illness is one whose ethical dilemmas are always intractable. We have known for years oncologists whose spend sleepless nights over the implications of the chemotherapeutic trials in which their cancer patients participate. Muddying research protocols in order to allow everyone a fair shot at a minimally promising treatment may seem the only ethical course if you, or someone close to you, is ill with AIDS. Yet, from a wider perspective, it seems clear that the price of a few months of equivocal hope for a few hundred people may be genuine delay in the development of effective measures, delay that, somewhere down the road, will cost the lives of thousands of Africans. Nevertheless, those under the most immediate threat can hardly be blamed if they reject a long-term view that offers them little hope or comfort, and insist upon interpreting it as an excuse for the vanity of academic scientists for whom a "statistically clean" study seems to outweigh the value of a human life.

In the end, however, such emphatic dislike of science, no matter how comprehensible its emotional roots, worsens a wretched situation and darkens a prospect that has never been bright. Sooner or later, a price will be paid for such unearned antagonism, fed daily with hyperbolic activist statements, a price that will be reckoned in pain and death that could have been avoided. Even now, we begin to trace the cost in the numbers of AIDS sufferers who have forsworn orthodox methods in search of "alternative" therapies of one sort or another. For the moment, it is still possible to argue that such choices incur small cost—orthodox medicine is confessedly pretty helpless against AIDS—and may at least have the beneficial effect of keeping hope alive for a few precious months. Sooner or later, however, the balance will shift. Means to delay the consequences of HIV infection, if not to escape them utterly, will be found. When that time comes, those who are mired in the illusions and outright quackeries of alternative medicine will be the major losers, just as were the thousands of cancer patients who flocked to clinics dispensing Laetrile, or healing rays from black boxes, and died even sooner than necessary.

To the extent that the thought and culture of the academic left abets and amplifies the antagonisms between activists and researchers, between patients and physicians, they may be justly accused of working actual harm. We submit that this phenomenon is real and that concern about it is amply justified. Furthermore, it takes a form that ironically illustrates the seductiveness, as well as the silliness, of what passes for theory among campus radicals. Here is the view of Daniel Harris, in a widely read *Lingua Franca* article:

> From the rhetoric of critics like Cindy Patton, who says that ACT-UP "provides an interesting example of emerging postmodern political praxis using deconstructionist analysis and tactics," to the methodology described by ACT-UP spokesman Douglas Crimp, who maintains that the organization's graphic art addresses "questions of identity, authorship, and audience—and the ways in which all three are constructed through representation," the fight against AIDS is tainted with the faddish argot of postmodernism.[28]

Harris has leveled his guns at a very curious phenomenon indeed. The spell of postmodern theory has lured its acolytes into a bizarre philosophical cul-de-sac, where "reality" is effaced as a meaningful term and where representation, rhetoric, and discourse are the only allowable phenomenological categories. Confronted by an epidemic that is all too grimly real, these postulants are driven full circle into a giddy doctrine asserting that control over representation and rhetoric, over language and imagery, will, of itself, dispel the menace of AIDS. This, beneath its ostensibly up-to-date skeptical veneer, is purely magical thinking. It recurs to the ancient confusion between names and things, between mention and use. In a nearly literal sense, it encodes a faith in charms and magic words. It is, moreover, an approach that offers immediate satisfactions beyond the imponderable hope that AIDS will eventually yield to it. As Harris reminds us:

> One of the most seductive and overlooked attractions of the AIDS epidemic for postmodern theorists is that it uniquely engages an academic anxiety that has undermined the self-esteem of liberal arts faculties for decades—namely, their belittling awareness of the greater prestige of their scientific colleagues. The utter inability of the latter to find a cure, a vaccine, or even an effective treatment for the disease has created a kind of power vacuum in the university, a temporary eclipse of authority that affords a perfect opportunity for non-scientists to rush forward into an arena from which they have been previously excluded.[29]

Examples of what Harris has in mind abound in the discourse of the postmodernist clan. Hardly an issue of a trendy journal, nary a specimen of a theory-laden compendium exists without its disquisition on AIDS. Invariably, the disease is seen as a semiological construct, a phantom animated by the illusions of a reactionary culture, a creature of disordered discourse, a mere symptom of the tissue of social prejudices that surrounds us. We understand that this is, for those unacquainted with the genre, hard to credit; here is a specimen from Cindy Patton, one of Harris's exemplars of postmodernist thinkers on the subject of AIDS:

> This [orthodox] view of science not only obscures the power relations between science and public policy; it is fatal to people in danger of HIV infection and catastrophic for the communities and nations in the developing world which are currently and inextricably the objects of scientific research on AIDS. It masks the way in which medical research reconstructs colonial relationships under the dual guise of scientific objectivity and efforts for the "good of mankind." It obscures the ways in

which pressure to adopt the organizational scheme of science as representative of lived experience reinscribes hierarchies of social difference. And finally, it reads as progress the destruction of vernaculars and the adoption of scientific language.[30]

Patton's jeremiad not only echoes all the standard postmodernist clichés but takes as given the critique of science being cranked out in radical-feminist redoubts, by left-wing social theorists, and by the remaining deconstructionists.

Here is another example from no less a source than Derrida himself:

Given both time and space, the structure of delays and relays, no human being is sheltered from AIDS. This possibility is thus installed at the heart of the social bond as intersubjectivity, the mortal and indestructible trace of a third. Not the third as the condition of the symbolic and of the law, but as the destructuring structuration of the social bond.[31]

Here we have the Master in a characteristic mode: banality with an honor guard of double-talk. Closer to home, from a cultural critic and would-be AIDS activist:

Yet, in the context of the new global order, our society (US AIDS) is still able to construct a political epidemiology in which its own internal Third World of blacks and Hispanics are "objectively" identified as the principal threat to America's immune system . . . Moreover, this same "map" is deployed to trace the "African" origins of HIV with the intention of sexualizing the transmission of diseases, which historically has followed the trade routes of commerce and war . . .

. . . Thus, I prefer to speak not of "persons with AIDS" (PWA's) but rather of a "society with AIDS" (SWA). The issue here is how such a society is to respond to itself, having discovered that the autoimmunity it believed it enjoyed as an advanced medicalized society is a fiction.[32]

And another:

My first thesis is that a psychoanalytic perspective on AIDS must begin by acknowledging that *each of us is living with AIDS*: we are all PWAs (Persons With AIDS) insofar as AIDS is structured, radically and precisely, as the unconscious real of the social field of contemporary America.

. . . The analogy of social psychosis enables us to understand AIDS as a condition of the *body politic*, an index of the socialized body of the American subject caught in a network of signifiers that renders it vulnerable to AIDS precisely because, by refusing a signifier for AIDS, it faces the prospect that what is foreclosed in the symbolic will return in the real. By persistently representing itself as having a "general population" that remains largely immune to the incidence of AIDS, America pushes AIDS . . . to the outside of its psychic and social economies . . .[33]

These examples—and we have selected some of the least bewildering rhetoric from each of them—are quite typical. Certain themes are recurrent: The basic problem is the *language* and the social typology with which America thinks about AIDS. Get this under the control of the right-minded, and a remedy will appear. Most especially, it is imperative to get Americans to stop thinking about "high-risk groups"; this is the error from which all others flow. Failure to heed this insight will unleash a terrible retribution on the smug majority.

Pedestrian virology, immunology, and epidemiology are, needless to say, banished from such analyses. The moralizing impulse is in full career, albeit with Jacques Lacan's rhetoric rather than Jerry Falwell's. Such hallucinatory language, which would not for an instant be accepted in connection with lupus, myasthenia gravis, or hepatitis B, is somehow justified because of the deep sense of victimhood (preceding AIDS and likely to outlast it), the gnawing feeling that AIDS represents but the latest and worst of a series of persecutions that beset the afflicted communities, the gay community in particular. Scientific or merely rational argument is irrelevant. Scientific standards are what is being castigated. Articles of faith (e.g., that the unrepentant white heterosexual majority will ultimately be visited with the full force of the plague) are being declared. The language is religious, that of Exodus. Moses and Aaron are warning Pharaoh.

The postmodern version of "philosophy of science" is often cited in jeremiads of this sort. Cultural constructivist dogma is invoked with depressing frequency to describe, or rather to deride, scientific work done in connection with the AIDS crisis. As Daniel Harris points out, "In fact, academic AIDS theorists malign the presumed 'objectivity' of science every step of the way as Donna Haraway does in her essay 'The Biopolitics of Postmodern Bodies,' in which she jeers at our society's veneration for science, for the 'univocal language' of empiricism that, in fact, conceals "a barely contained and inharmonious heterogeneity.' "[34]

The need to formulate one's thinking in language that endorses the postmodern critique of standard scientific epistemology distorts even thoughtful and sensible evaluations of the AIDS epidemic. Steven Epstein, a sociologist writing in *Socialist Review* on the "democratization" of AIDS research by activist groups comprised largely of lay persons, is impelled to use terms like "contested construction of knowledge" and to brood at length over Foucauldian apothegms concerning knowledge as power.[35] Though he wisely rejects the most radical epistemological proposals—that AIDS is merely an artifact of the way in which outcast groups are "represented," that supposed knowledge about AIDS is merely an artifact of linguistic convention and the contest for social power—his preoccupation with these questions derails any serious investigation of methodological and statistical matters, topics that would have been indispensable were he to prove his case for the value and importance of the "research" undertaken by the activists he so admires. One is left wondering what evidence there is to substantiate Epstein's claim that "the AIDS movement has turned science into politics, but also turned politics into science; and the combined effect is to carve out a large space of scientific inquiry within which grassroots participation comes to be seen as useful, desirable, and even necessary."[36] Ruminations on postmodern discursive communities do nothing to authenticate such claims, and put them in fact on the slippery slope to pure bombast.

Likewise, David Kirp, an eminent academic expert on social policy aspects of AIDS, writes in *The Nation*[37] to tell us that scientific and medical practice is being remade by the gay activists who defy the wisdom of medical orthodoxy to conduct their own field trials of new and innovative therapies. At least that is what the tone of his piece initially suggests. Closer examination reveals a far more modest claim, as well as considerable misgiving concerning this putative paradigm shift. After an initial flourish, *maestoso*, Kirp merely points out that some credential-bearing researchers have been sloppy and unethical, while some individuals without official credentials have learned enough in the way of standard methodology and experimental design to perform useful studies. This is interesting, but it hardly constitutes evidence that science has been remade, or that its fundamental assumptions are crumbling. In the end, and to his credit, Kirp reserves most of his energy for an attack on those outsiders who scorn scientific protocols, ignore controls and statistical measures of validity, and who, in many cases, end up migrating to the camp of one or another "alternative" healer. It is deeply disquieting to note the way in which even sober writing on the subject of AIDS displays the need to genuflect to the idols of the postmodern pantheon. Kirp's piece, is, in fact, a plea on behalf of orthodox science; yet it is garbed misleadingly in the language of those who decry that orthodoxy.

What finally comes of this, we suspect, is a slow but inexorable degradation of understanding among activists, members of high-risk groups, and the public in general. It is not misinformation about AIDS, per se, that is the threat, so much as it is misinformation about what standards of judgment have to be brought to bear in order to decide whether theories, studies, compilations of data are to be trusted. Grumbling about the arrogance and authoritarianism of standard science has its place. These are real enough at times and in places. They can harm as well as annoy. But far more substantial harm is done when all that is proposed in place of "scientific objectivity" is the overpriced vaporware of postmodern skepticism, conjoined with the understandable, but insidious, delusion that victimhood puts one in direct contact with the wellsprings of truth.

The activism of cutting-edge theorists has a definite and concrete downside. Political pressure from such activists has led on occasion to loosening and changing some of the rules for conducting clinical trials, particularly in the direction of eliminating placebos, depending more on the statements of participants (who lie, more often than used to be expected, in order to get treatment), and broadening participation to include all elements of the affected and potentially affected population. Unfortunately but unsurprisingly, no instance has yet been reported of a significant "grassroots contribution" to the cellular immunology, to the molecular biology, to the virology, to the effective clinical management of AIDS, or to relevant epidemiological theory. Since it is *knowledge* we are concerned with, we must judge that even the authentic claims of commentators like Kirp and Epstein are more important for morale among AIDS sufferers than for any implications for biomedical knowledge.

In the meantime, the modification of clinical trial designs in response to political pressure has upon occasion introduced not only design absurdities but the potential for horrifically increased expense of the simplest of testing assignments. Benjamin and Janet Wittes have contributed a short but exact analysis of some of these new rules.[38] As Wittes and Wittes point out, the really dangerous implication of such politicized experimental design is *not* the inclusiveness itself, but the requirement, if the most elementary statistical honesty is to be maintained, "that all subgroups must be included in sufficient numbers to demonstrate possible treatment differences between them and 'other subjects,' presumably white men."[39] Their critique is not gaseous theorizing: it reflects the experience of nearly a hundred years of epidemiology and analysis. In the face of the realities of funding and of chronic personnel shortages, emotional arguments for "inclusiveness" and "compassion" in medical research come close to irrelevance (notwithstanding political statements to the contrary from scientist-administrators, who must answer to the Congress and the media, and who must avoid unpleasant confrontations with activist groups). This is not to deny that medical research and clinical testing designs are sometimes very cold-hearted. But that is not the question; the question is: What will contribute to the growth of vital knowledge about AIDS? Issues of compassion and supposed fairness in the immediate instance, of giving everyone a shot at every treatment that shows the least early promise, of giving ear to every voice from the most afflicted communities, may tug at the heart; but true compassion is not necessarily a short-term matter, and attempts to view it as such run the risk of turning it into mere sentimentality.

We are not here arguing for science as the only way of getting answers to questions about how the body—or the universe—works. We don't think we need, for present purposes, to defend scientific objectivism as the single admissible philosophical standpoint. We simply observe that science is, as all the world's experience clearly tells us, overwhelmingly the best trick we so far know for getting the upper hand against disease. And we *know* that the politicized, overtheorized "criticism" that is our subject offers nothing at all in that direction. Its main effect has been to reassure aspiring cultural critics that they can play a significant role in combating AIDS without having to do anything so tiresome as, for instance, abandoning the joys of lit-crit for careers in medicine or molecular biology. In the case of the AIDS tragedy, such "criticism" may have the merit of allowing some desperate people to let off steam—and God knows they need that. But it has not in any degree hastened the arrival of those desperately needed insights into chemotherapy, biochemistry, molecular biology, and immunology without which we cannot much help those millions throughout the world who have the AIDS virus, or those additional tens of millions who are in danger of being infected by it.

Animal Rights: Doctor Doolittle Meets Professor Foucault

"Either Dr. Moossa stops the course or I will shoot him in the head." That was the anonymous telephone message from an animal rights activist to the teacher of a postgraduate surgery course in which, necessarily, animals were to be used. [40] Dr. Moossa, who has a wife and children, and an employer—the University of California at San Diego—not even lukewarm about backing him up, stopped the course. And so his students went without instruction or practice. Had Moossa not stopped, he might have suffered the same fate as did the obstetrician recently killed by a crazed activist of the antiabortion movement. We make this comparison advisedly. Dr. Moossa's travails are but one item on a long list of harassments, threats, and sabotage directed in recent years against individuals and institutions that do biomedical research employing animal subjects. Actions include verbal abuse and invective, picket lines, raids on laboratories to wreck equipment and "liberate" laboratory animals, destruction of research notes and records, and, in a few instances, reasonably well-demonstrated plots to inflict actual violence on research workers.

These actions represent a surprising resurgence of the antivivisectionist sentiment that prevailed in the United States and much of the rest of the Western world in the nineteenth century. As it happens, much of the renewed fervor can be traced to the foundational work of one individual, the Australian philosopher Peter Singer. Singer's arguments first appeared in an article in the *New York Review of Books*, later expanded into an immensely popular book.[41] Singer's theories on the rights of animals, at least those whose neurological organization is complex enough so that we may deem them "sentient entities" capable of emotional states, derive, in the main, from his own extension of nineteenth-century Utilitarianism. These arguments are neither mystical nor antirational; Singer is an authentic descendant of the Enlightenment. Although his politics are, roughly speaking, left of center, we could not with honesty affiliate him with the academic left, in the sense in which we have been using that term.

Others, however, have picked up and amplified Singer's ideas. The philosopher Tom Regan, in particular, has probably been the leading American advocate of this school of thought.[42] The influence of such ideas has grown exponentially. Many campuses, including our own,[43] now have organizations dedicated to the vindication of animal rights, a point of view that calls not only for universal vegetarianism and the rejection of animal products like fur and leather but also for the abandonment of animal research by scientists and physicians.

The relation of the animal rights movement to the academic left is a question of some complexity. Clearly there *are* connections. There is evidence in the popular stereotypes, jokes, and catchphrases that encapsulate the common understanding of what the academic left, the "political correctness" crowd, is all about. The standard caricature of the full-blown PC personage limns an individual who is not only deeply mindful of the sensibilities of nonwhite peoples and non-Western cultures, committed to the necessity of introducing nonsexist neologisms into every nook and cranny of English diction, militantly devoted to the abolition of every last vestige of gender stereotype, and deeply sympathetic toward homosexuality as a form of Otherness. She is, as well, exquisitely alert to the status of nonhuman animals as an oppressed and exploited class. In the prevalent folklore, she is careful to call pets "animal companions"; intent on converting all her friends to meatless (or at least eggs-only) diets; and inclined to heap vituperation on fur-bearing humans wherever she encounters them.

In this instance, the antennae of popular humorists have, as they often do, detected something significant about our times and foibles. Stereotypes aside, one ought to be quite careful in any serious discussion of the issue. "Animal rights," in the direct sense, and as opposed to concern for animal welfare, is a position that commands the support of only a small fraction of the academic left. It is quite different from a question like "multicultural education," which is without question a defining issue for all postmodern radicals. There are thousands of campus types who have systematically replaced "black" by "African-American" in their lexicon, who ascribe all the world's ills to the white, capitalist patriarchy, who would drive three hundred miles out of their way to avoid heterosexist Colorado; but who nonetheless eat steak with relish and wear, for appropriate occasions, leather bomber jackets or deerskin moccasins. Even among committed ecoradicals, animal rights is at best a lukewarm issue: it is a tricky thing to champion hunter-gatherer cultures as paragons of ecological wisdom without allowing that hunting may be a justifiable activity. Likewise, feminists are, for the most part, nervous about endorsing an ideology whose rhetoric and emotional appeal so closely parallel those of the "pro-life" movement. AIDS activists are equally edgy about aligning themselves with efforts that, if successful, will bring contemporary medical research to a grinding halt.

On the other hand, there are positions within the spectrum of the academic left that passionately embrace animal rights doctrine in its most unmitigated form. One chapter of Morris Berman's recent book is an asseverative paean to animal rights sentiment.

It should be clear, then, that how any culture relates to animals says much about how people in that culture feel about their bodies. This in turn reveals the essential structure of the Self/Other relationship and does much to explain the particular history of that culture, the body politic . . . And it is in technological societies that we find the greatest terror of the organic, in fact the deepest hatred and fear of life, that this planet has ever known. [44]

Similarly, there is a strong current of support for animal rights dogma within the ecofeminist community, particularly that wing which embraces "holist" or "organicist" points of view. A recent issue of Hypatia, the leading journal of feminist-inspired philosophy, was devoted to ecofeminist issues. [45] Most of the articles endorsed (albeit with occasional quibbles) the central beliefs of the animal rights movement. Here is an example, from Carol J. Adams, a militant advocate of both feminist and animal rights causes:

To eat animals is to make of them instruments; this proclaims dominance and power-over. The subordination of animals is not a given but a decision resulting from an ideology that participates in the very dualisms that ecofeminism seeks to eliminate. We achieve autonomy by acting independently of such an ideology.[46]

And, in a triumphalist vein, Deborah Slicer:

And while the [animal rights] movement continues to take its undeserved share of ridicule, it has, for the most part, advanced beyond that first stage [ridicule] and into the second, discussion. There is even some encouraging evidence that its recommendations are being adopted by a significant number of people who are becoming vegetarians; buying "cruelty-free" toiletries, household products and cosmetics; refusing to dissect pithed animals in biology classes or to practice surgery on dogs in medical school "dog labs"; and rethinking the status of fur.[47]

Most of these instances of outright support for animal rights on the academic left are to be found in the region where nominally "left-wing" politics begins to merge with the intellectual junk food of the "New Age" movement. Left theorists of more abstract and cerebral bent tend, with some exceptions, to avoid such questions. Nonetheless, there are crosscurrents that convey attitudes from one realm to the other. The spirit of postmodern critical theory nourishes contemporary animal rights doctrine; it coexists, logical inconsistencies notwithstanding, with old-fashioned pantheism and unvarnished sentimentality. The key point is the perspectivism and relativism central to the postmodern stance. On the one hand, it endows a mythical community, the supposedly "sentient" animals, with a status parallel to other communities of oppressed, exploited, voiceless beings. The eagerness of academic leftists to run headlong from their ostensibly privileged positions as members of the hegemonic white European technoculture overflows, in the case of animal rights sympathizers, into the impulse to denounce "species-ism," and the assumption that human needs and interests must always come first with us humans. "Humanism"—already a word in bad odor among postmodernists—now takes on a doubly evil connotation. The indulgent relativism that declares all cultures, all narratives to be equally valid, is stretched to accommodate entities capable of neither culture nor narrative.

At the same time, the hyperskeptical aspect of postmodern thought comes into play. With science reduced, on this view, to a "truth-game" played by a narrow interpretive community under self-referential rules, it becomes easy to dismiss, without close argument or factual investigation, the claims of science to produce results essential to human well-being. Theoretically, of course, an animal rights purist would be quite willing to ban animal subject research even in the face of evidence of its enormous usefulness in extending human life and relieving human suffering. Some people take that precise position.[48] But others, out of all-too-human motives, hedge the issue by declaring, on their own authority of course, that animal research leads to useless science and wrongheaded medical practice. No doubt this forestalls an inevitable crisis of conscience.[49] Also, it makes it possible for the activist to represent his position to the unconverted as something other than a doctrine enjoining extreme and painful sacrifice on individuals and communities.

If animal research is misleading anyway, if it exists only to make profits or to gratify the ambitions of scientists trapped in a perverted system of rewards and incentives, why not ban it and spare the poor animals all that suffering? Postmodernism, with its insistence that science is "just another discourse," provides this impulse with a highly efficient lubricant. That is the nub of the problem. The postmodern position, incorporated into the sloganeering of animal rights militancy, is not only academic nonsense—it is concretely dangerous nonsense. Animal subject research is, without any question, enormously important to efficient medical practice, and its abandonment would entail incalculable human costs.

Let us consider the question of necessity. In 1985 the National Research Council issued, following a long series of studies, meetings, and workshops, a summary volume on "Models for Biomedical Research." The volume includes a long table that lists all the Nobel Prize winners in physiology or medicine from 1901 through 1984, a few words about the particular contribution of each, and the experimental system used for the research leading to the prize committee's recognition. One hundred and thirty-nine Nobelists are listed. The coded descriptions of their work are a bird's-eye view of the progress made by science—and against human (*and* animal) disease—in the first eighty-four years of this century.

In just the first four decades, for example, prizes were awarded to discoverers of the significant mechanisms of: digestion, signal transmission in the nervous system, immunity, hormone function and hormone action, vision, hearing, heredity, respiration, blood clotting, trace factors (such as vitamins) in nutrition, the electrocardiogram and the cardiac cycle, and a dozen others. The diseases (again of animals as well as of humans) about which significant, direct understanding and vastly improved treatments resulted from such knowledge *in those four decades alone* included, *inter alia*, malaria, tuberculosis, parasitic infestation, diabetes, heart disease, cancer, typhus, anemia, neurological disease, dwarfism, pathologies of sight and hearing, and diseases of immunity, such as anaphylaxis.

The experimental systems employed by the 139 Nobel laureates included all levels of biological organization, from isolated cells in culture, through bacteria, yeasts, and molds, to higher plants and invertebrate animals, to humans and other mammals. Many of these scientists were physicians, and it is therefore no surprise that twenty-five of them—18 percent—did research on humans in all or a part of their work. Most of those, however, studied other systems as well. Fourteen percent of the investigators used microorganisms or plants; but 86 percent of the investigators employed animals as experimental systems, usually as *models*.

We dwell on these simple numbers in order to focus on the key arguments, other than those in moral philosophy, that surround animal research today and establish the general atmosphere that encourages activism. Was the research tabulated in the "models" volume necessary? That, obviously, depends on one's definition of necessity. But, given an honest and informed choice, would rational and humane people have voted at any point to stop it? We think not, not under our definition of what it means to be humane. In fact, had the research not been done, there would have been far fewer rational people around to vote, and the world would clearly have had a burden of suffering, and of early death, far more diverse, far more terrifying, than today's burden, however great that is.

Let us ask a somewhat different and more technical question: could all, or most, of that life-saving and life-enhancing science have been done without the use of animals? The answer is certainly no. To understand why, one must understand the purpose of animal models in experimental biology and medicine. Even the simplest organism is an almost unimaginably complex system, whose fundamental chemical and physical processes reflect a heritage of several billion years of trial, error, and modification. Yet a disease can, although certainly does not always, arise from a single, simple molecular difference buried deep within the continuous, labyrinthine reaction system that is the chemistry of life. The accidental substitution of a single nucleotide "letter" out of tens of millions in an individual genome, within the gene for just one of the several kinds of hemoglobin used over the course of a life to make red blood cells, can produce a life-destroying anemia. Equally simple mistakes account for a large array of other diseases. How are the errors to be found, so that they can be dealt with?

In the whole patient, they are not even the proverbial needle in a haystack: they are, far worse, an errant sand grain on a thousand miles of Atlantic beach. The *process* to be examined must be isolated and disassembled in order to determine its elementary components: it must at some point be studied in the simplest possible system, whose variables are under control. And once that has been done, once there is a mental construct of its components and operation, it must be isolated and observed again intact, under conditions of minimum intrusion by unexpected, uncontrolled environmental variables. That is the first step of modeling. But it is only the first step if what is ultimately at stake is a disease.

There are unities, astonishing unities, of biochemistry and physiology among all the living things of Earth; but there are also divergences. It is the latter that give us biological diversity. Thus, while the basic, ionic mechanism by which a message is sent from one cell to another seems to be the same in sponges and in mollusks and in people, the brain of a child is not the same thing as the sensory cells of a sponge, or even as the giant axon of a squid. While research on that marvelously large and hardy squid cell does tell us how information is processed, at the most fundamental level, in all nervous systems, we cannot use squid axons—much less computer simulations of squid axons—to determine in detail what has gone wrong in a human neurological disease, or how to treat it. For that, the modeling has to proceed *upward*, step by step, to more complex systems, closer to the ultimate problem, but not yet as complex or as subject to the operation of confounding variables.

The advance of technology, in cell culture, computing, enzymology, drug design, does not remove the necessity for appropriate modeling and experimentation. In fact, the exhilarating growth of such technologies produces hypotheses about process, and candidate drug structures, at a rate unimagined even a decade ago. Every one of those hypotheses or synthetic drugs has to be studied under controlled conditions, in a biological system of appropriate complexity, if it is to be of any use at all as knowledge or as therapy. And the choice is clear: the final models, at the highest level of complexity, must be either human beings or other mammals, specifically those whose relevant systems mimic the particular human ones as closely as possible. Cell cultures alone, computer graphics alone, do not and cannot decide, in the absence of biochemical and physiological information derived from studies on animals, how a drug works, or whether it is likely to work as is predicted from theory. In fact, the better we get at cell culture and invertebrate-animal molecular biology and computer graphics, the more good candidate drugs are proposed, the more need for higher-animal modeling there is, that is, for experiments in which animals are the subjects.

We have, therefore, a range of possible choices. First, give the whole thing up, bravely, as a sacrifice, on moral grounds. Accept life on Earth as it was before science. Decide that humans should be no more "privileged" than the bacteria, the yeasts, the trees—or the viruses. Second, test and model, if we must, but do it on *people*, not on defenseless animals. If our species wants to fiddle with nature (i.e., with disease), then the fiddling should be done with human subjects, not with innocent mice, rats, dogs, cats, monkeys. Third, model, and use animals as necessary, but with every care to minimize their discomfort or suffering.

Most sensible persons, confronted with these choices, and recognizing that there really are no significant additional alternatives, opt for (3). As it happens, (3) is, and has been for a very long time, the general position of biomedical science and scientists. Of course there are lapses from time to time, as there are in all organized (i.e., hierarchical) activities, although they are far, far rarer than one is led to believe by animal rights enthusiasts. The facts and arguments of the case are available for any fair-minded inquirer to assess.[50]

Science and Afrocentrism

As we write, we are confronted with the spectacle of a revived ethnic tribalism in Europe, where Serbs, Croats, and Bosnian Moslems rape and murder one another in the charnel house of the former Yugoslavia; we may soon see something similar between Balts and Russians, Russians and Romanians, Romanians and Magyars, or Magyars and Slovaks. The murderous hatreds that rend Northern Ireland no longer seem anomalous. Elsewhere, the racial and religious chauvinism that pits Sikh against Hindu against Moslem, Sinhalese against Tamil, Arab Sudanese against black Sudanese goes on unabated. We might expect the humanitarian conscience to be especially aware, in such a time, of the horrors lurking in tribalism.

Yet in the decidedly less lethal venue of academic life, we find that tribalism, in one form or another, is the most-favored project of leftist ideologues, who appear to have abandoned, for the moment, the universalism that once shone through even the dreariest left-wing cant. The "politics of identity" is now sanctified on the campus.[51] Increasingly, many groups are held to deserve their own separate and inviolable space. Nor has the crusade been in vain. Women's studies programs are ubiquitous; many of them now have departmental status. Latinos and Native Americans are similarly favored at a growing number of schools. Gay/lesbian students and faculty are organizing nationwide to demand the same kind of accommodation for their communities, as are the physically disabled ("differently abled"). Of course, in point of chronological priority and intensity of separatist feeling, the black community easily holds pride of place. Black studies departments, in one form or another, have been around longest, maintain the greatest distance from the rest of the scholarly community, and command the fiercest loyalty from their ostensible constituency.

The pros and cons of the currently fashionable separatism may be debated endlessly. It is not our intention to judge it, although we admit readily to serious reservations concerning at least some aspects of this "balkanization" of the academy. We are, however, chiefly concerned with how the politics of identity has affected the teaching and learning of science and, more generally, the *perception* of science among the various "communities" that are becoming privileged with special academic status. We have already considered the first-line, new critiques of science, whose constituencies consist of the major communities of academic feminists (which we insist upon distinguishing as a *subset* of women students and faculty), together with a relatively small number of male adherents, of cultural constructivists, and of radical environmentalists. Here, however, we address the relationship between Afrocentrism and the sciences, a phenomenon that has a rather different profile and that resonates emphatically beyond the walls of academe.

A great many scholars who are associated with black (or African-American) studies programs are decidedly on the left of the political spectrum, at least in the sense that they are profoundly unreconciled to current political arrangements and tend to identify with movements, in this country and abroad, that embrace some form of socialism. In the alignments of campus politics, its uproars and showdowns, they are usually affiliated with the academic left, as we have defined it. The legitimacy of ethnic studies is a favored leftist shibboleth; and there is reciprocity in the form of black support, albeit sometimes grudging, for the programs and slogans of the left.

Notwithstanding this alliance, the intellectual and scholarly style of black studies is decidedly different from that of the trendy left. The language and posture of postmodernism is almost wholly absent from black studies. The philosophers and social theorists who enthrall white leftists are not much in favor among militant blacks. The respective student constituencies are largely disjoint. The white left is, of course, eager to offer its own analyses as justification for the protected status of black studies within the university. Indeed, one reason that the left finds postmodern doctrines that denounce "universalism" and celebrate "difference" so attractive is that these notions are perfectly suited to the idea that different "communities," especially those arguably oppressed, are entitled to institutionally separate facilities (such as buildings and "studies" departments) insulated from the scrutiny of conventional scholarship. To share facilities with other departments, faculty, and students is considered demeaning.

The situation is not completely symmetric, however. Black studies specialists are largely indifferent to the fine-grained ideological concerns of contemporary left-wing theory. Their style, for better or worse, is very much their own. (In offering these generalizations, we are keenly aware that they disregard important distinctions. The black studies departments at Harvard and Princeton are markedly different, in style and substance, from those at, say, the University of Massachusetts or the City College of New York; frankly, it is the latter, better represented type that we have in mind.)

The argument most often put forth for teaching a "black" perspective on various subjects is that in the present climate young black people cannot be expected to make an ungrudging adjustment to the styles and assumptions of the traditional white university. They are said to need an approach that respects their singular cultural experience and that validates their sense of self-worth. Moreover, they need examples and role models that counter the presumed discouragement and disparagement that the white-dominated culture has inflicted on them. In short, they need a milieu in which white values are not "privileged."

Whatever the value of such an approach in general, it would seem to be highly questionable when applied to the teaching of natural science, especially at the college level. These reservations, however, are of little force in the face of the current mania for "Afrocentric" approaches to teaching everything in sight. A small, but growing, literature is emerging that intends to provide at least the beginnings of an Afrocentric science pedagogy, and, as it grows, the demand that the university accommodate such a curriculum is increasingly heard. So far as we are aware, no reputable college catalog now offers Afrocentric calculus or microbiology. We can certainly not discount the possibility, however, that such a catalog is now in the press: developments in elementary and secondary education point immediately and ominously in that direction.

At first glance, some of the key elements of the Afrocentric approach seem benign. What is wrong with the idea that talented black kids, who have met with little opportunity or encouragement in the sciences, should be familiarized with the lives and achievements of black scientists and the accomplishments of African cultures in the areas of technology and speculative science? At worst it seems a bit mechanistic and unnecessary—thousands of bright young people from other marginalized communities have made strong scientific careers without the benefit of emplaced "role models"—but there would seem to be little actual harm in it. Unfortunately, in the grim comedy of American education, lower and higher, things are not so simple.

If one examines the nascent literature of Afrocentric science, one is immediately struck by two things: the enormous amount of Afrocentrism, and the remarkable paucity of science. Even worse, however, is the flagrant falsification of science (and of history and ethnography as well) in the service of Afrocentric chauvinism. The notion that intelligent but naive students will first encounter "science" in this form is chilling indeed. There is absolutely nothing in it to encourage the legitimate hope that in the not-too-distant future blacks will occupy positions in science proportional to their numbers in the general population.

A good example of the substance and range of the "Afrocentric science" literature is to be found in the volume *Blacks in Science, Ancient and Modern*, edited by Ivan Van Sertima, a professor of Africana studies at Rutgers University. A few of the pieces are inoffensive and possibly, in some limited way, useful. They consist of brief biographies of black scientists, engineers, and inventors. One must take it on faith that they are sufficiently inspirational to do some good. Presumably, they are intended for children of ten or eleven, since they are written at that level of comprehension. The volume's idea of providing inspiration from purely African culture is illustrated by a piece by Claudia Zaslavsky that briefly describes the arithmetic terminology of the Yoruba and Benin peoples. This amounts, intellectually, to little more than a minor curiosity, although the cheerleading spirit in which it is presented may, for all we know, give it the desired morale-building effect. Some genuinely impressive facts are to be found here and there in the book, such as the two-thousand-year-old tradition of steel-making in Tanzania, and the eyewitness account of a caesarean section in nineteenth-century Uganda.[52] On the other hand, even the biographical pieces are usually guilty of exaggeration, as in John Henrik Clarke's sketch of Thomas Edison's assistant Lewis Latimer, or the description of quotidian inventions as great scientific breakthroughs. Even worse is Van Sertima's own sketch of the chemist Lloyd Quarterman. This is so abominably written, so vague, and so afflicted with Van Sertima's scientific ignorance as to do a great disservice to a man who has had an honorable, possibly a distinguished, career in science.

These sins, however, are venial. Far worse is the bombast and the wretched logic that pervade most of the book. One finds here in abundance the antic confusion that is the unfortunate hallmark of so much Afrocentric literature. There is the refusal, repeated in article after article, to recognize that "Africa," a geographic term, is not synonymous with race or culture. Those terms are, in turn, conflated with language, religion, and political system. The great cultural variety of the African continent over the course of history is flattened into a simpleminded Africanness. The contrary sin is also present; cultural unities are artificially sundered when they straddle "Europe" and "Africa." Thus we have, for instance, a short section on Euclid, the great compiler of Greek mathematical knowledge, circa 300 B.C. Why Euclid? Because Euclid worked in Alexandria, which makes him an Egyptian, hence an African! Presumably, this makes him black as well (although this is not explicitly stated), since the reigning assumption of the book is that the Egyptians were of the same racial stock as sub-Saharan West Africans.[53] In short, we have one more depressing instance of the inane but by now ineradicable historical fallacy that anything south of the Mediterranean is "black."

Naturally, the claims on behalf of Egyptian science and mathematics are correspondingly exaggerated. We are told in no uncertain terms that the Egyptian contribution to geometry has been grossly undervalued by white scholars.[54] After all, the Egyptians approximated π by 3.16 and computed the volume of a truncated pyramid! Therefore, their achievement is comparable to that of the Greeks, if it does not, in fact, outshine it! Such statements conveniently overlook the fact that every mathematician with an interest in the history of the subject (and that means most of them) knows perfectly well that the Egyptians approximated π by 3.16[55] (not terribly good compared to the Greek approximation $22/7 = 3.1428 \ldots$ —even worse when one considers that the Greek mathematicians, unlike the Egyptians, had a well-developed and general methodology of successive approximations) and developed various formulae for geometric mensuration, so that what is asserted to have been suppressed is in fact widely known. What is truly irresponsible in this piece, especially from the pedagogical point of view, is the failure to come to grips with the enormous conceptual gap between the systematic synthetic geometry of the Greeks and the clever but ad hoc mensural geometry of Egypt (and other civilizations). To set the one up as the equal of the other, for the sake of racial pride (overlooking, once more, the non sequitur of Egyptian and therefore black), is to deprive students of an indispensable mathematical insight, and, moreover, to prime them to react with hostility to any attempt to convey it to them.

Silly as this is, it pales by comparison to some of the other articles. Khalil Messiha tells us, for instance, that a small wooden figure of a bird, presumably made in Egypt during the Hellenistic period, is an example of "African experimental aeronautics." The evidence? If you build a copy of balsa wood (rather than the original sycamore), and then add a vertical stabilizer (not present in the original) to the tail, you get a so-so version of a toy glider!

Of course, the tale of the Dogon "discovery" of the dwarf companion of Sirius—the great celestial beacon of Canis Major—resurfaces in two articles by Hunter H. Adams (of whom more below). The "evidence," aside from unreferenced diagrams by the author, is of wretched quality. We are offered a picture of a carved gourd from Guinea and a Babylonian zodiacal diagram, which resemble each other in only the vaguest way. A Dogon doodle, said to represent Sirius, is compared to an astronomical photograph of the star. Indeed, they do resemble each other! The Dogon "Sirius" is a disk with four points coming out of it: the photograph shows Sirius as a disk with six points projecting. The author neglects to inform us that the "points" in the photograph are an *artifact* of the design of astronomical reflecting telescopes. (In this respect, it does no good to argue, as Adams does on the basis of the flimsiest evidence, that ancient "Africa" invented refracting telescopes; only *reflectors* produce this effect.) To make matters worse, Van Sertima's preface solemnly endorses and amplifies the Adams claims.

If, by some act of a mad jokester, a book of similar content, rigor, style, and tone, say, "Norwegians in Science, Ancient and Modern" were assembled, no reputable publisher would touch it. If it were published, there would be protest marches in the capitals of the Great Lakes states, at least, and a break in diplomatic relations with Norway would undoubtedly ensue. Yet in the current climate, academic presses are eager, with (presumably) the most honorable intentions, to publish such Afrocentric material.

Somehow, the condescending belief has taken hold that black children can be persuaded to take an interest in science only if they are fed an educational diet of fairy tales. The white academic left—to which some of the contributors to Van Sertima's book belong—is resolutely and astonishingly unembarrassed by such stuff. To apply rigorous scrutiny to it is, after all, to insist on the horrid Western paradigm of objectivity, and to deny blacks the right to form their own "interpretive community" and to create their own metaphors; thus to the postmodern left it is indelibly racist. The most awful aspect of the situation, however, is the way it testifies to the desperation and confusion of black people themselves. To resort to such tall tales is to reveal a deep and tragic insecurity and a willingness among black intellectuals—including scientists—to hold their tongues about nonsense damaging to their children.

Blacks in Science is but one venue for the egregious Hunter Adams. His notoriety has been much enlarged by the so called "Portland Baseline Essays,"[56] a series of texts designed to put "multicultural content" (for which read, typically, "Afrocentrism") into the public school curriculum. Adams is the author of the *Baseline Essay in Science*, which bears the innocuous title, "African and African-American Contributions to Science and Technology." A measure of the reliability of this work is Adams's use of the model bird already mentioned. He goes far beyond the claims previously cited, dubious as they may have been, and now asserts that the figurine proves that the Egyptians possessed full-sized, working gliders that were in common use. Even this pales, however, beside Adams's most bizarre pronouncements. "Early African writings," he claims, "indicate a possible understanding of quantum physics and gravitational theory."[57] Further, he credits the Egyptians with the full panoply of psychic powers—they were famous as masters of psi.[58] Thus the

Egyptian "science," which is supposed to inspire young people, turns out to be the worst kind of New Age boobery.

Here is the biographical note on Adams that appears in *Blacks in Science*:

Formerly on staff (1969–70) at the University of Chicago in the Chemistry Department, where he was in charge of operations of the mass spectrometer and also assisted graduate students in their research.

Since 1970 he has been at the Argonne National Laboratory at the ZGS Atomic Accelerator. There he has been advancing the state of the art of proton beam detection and diagnostic equipment, such as proportional wire counters.

He is currently researching the impact magnetic fields may have had on the rise of civilizations. He has written science-related articles for *Ebony Jr.*

Similar claims are made in the Baseline Essays. Behind this fustian lies the fact that Adams has completed no work toward a degree beyond a high school diploma and has no record of scientific publication. He was employed at the Argonne National Laboratory as an industrial hygiene technician; and he did no research there.[59] He is, however, associated with the secretive but increasingly controversial group known as the "melanin scholars," whose doctrine of an innate racial superiority of peoples with dark skin forms a bizarre echo to the Aryanism of J. A. Gobineau, H. S. Chamberlain, and Adolf Hitler. One can imagine what is permitted to "scholars" under this strange new ideological confection by noting that Adams now feels free to assert that the Dogon people's "knowledge" of the Sirius system need not have involved telescopes after all.[60] Apparently, he agrees with other "melanists"[61] that the melanin of deep-hued Central Africans enhanced their psychic powers to the point that such astronomical knowledge could be acquired by direct and unmediated insight! We leave it to the reader to draw the obvious inferences about the pedagogical competence of projects—and of entire school systems—that rely on such "materials" to instruct and inspire young people potentially interested in science.

There have been far worthier attempts to claim science as part of the black heritage—inspirational biographies of Benjamin Banneker, E. E. Just,[62] and so forth. There are plenty of honest stories of struggle and scientific achievement to be told. But the full-fledged Afrocentrists find this altogether too unfulfilling. The confabulations of Adams, Van Sertima, and their ilk are much more to their taste. Leaving aside "objectivity," simple sanity cries out on the dangers of this stuff to the prospects of students who are prompted to take it seriously. We can report from personal experience that there are many such students. At the very least, they are wasting their time with nonsense during a stage of their lives at which they should be learning basic science as rapidly and efficiently as possible. Furthermore, they are likely to face confusion and disillusionment as they acquire even a fragment of intellectual sophistication. Far from inculcating self-esteem, teaching science from the Portland Baseline Essays or from their proliferating equivalents elsewhere must do the gravest harm to the minds of minority students.

How is such foolery to be dealt with? Ideally, competent black scientists and mathematicians, whose numbers are not as large as we would like but certainly large enough for the purpose, would refute it in no uncertain terms, while at the same time stepping in to provide healthier role models than the invented ones of the Afrocentrists. In reality, however, competent black scientists and mathematicians are beleaguered with symbolic as well as practical responsibilities; and—in any case—why should they be obliged to take on a burden that—initially, at least—must make them the focus of opprobrious charges of racial disloyalty, if not worse? Still: in the current political climate of universities and especially of public education, it is hard to see how a predominantly white scientific establishment could hope to have much of an effect, even if it had the courage to speak up.

One might hope, at a minimum, that responsible universities and scholarly presses would find their own courage to put distance between themselves and the worst kind of Afrocentric fantasy mongering.[63] Competent referees are, after all, easy to find, and among them a few might be willing to be quoted. Unfortunately, the academic left, in its misguided loyalty to "multiculturalism" and the politics of identity, is not likely to play any useful role. Its ideology has, in fact, opened the doors of universities and schools of education to the nonsense we have described.

The Academic Left and Afrocentrism

It is not hard to see how poorly placed the hard-core academic left is to protest against the cited mischief. Postmodern relativism undercuts any possible protest grounded on the notion of objectivity. It entails a perspectivism that finds no basis for epistemological distinctions between science and fables. Feminists, indignant as they have been at the strictures of scientific orthodoxy, bogged down in their own school of indictment, are in no position to call Afrocentrists to account for sins more ostentatious, but, at root, hardly more reprehensible, than their own. The introduction to Hunter H. Adams's *Baseline Essay in Science* glancingly reflects the influence of postmodern academic attitudes and their usefulness as a first line of defense against the demands of scientific thought. Adams's quotation from the great physicist Louis de Broglie, for instance, is subtly amended to a misquotation, almost certainly on the basis of a reading—but hardly a misreading—of feminist science-criticism.[64]

Beyond reluctance to criticize Afrocentric pseudoscience, which seems to be general, we find among some prominent academic leftists a positive eagerness to endorse it. This is evident in a recent article by Bell Hooks (or, to render the name in her chosen orthography, bell hooks). Hooks, a black woman scholar at Oberlin College, is well known as a militantly feminist, postmodernist scholar of literature and culture. Her piece "Columbus: Gone but Not Forgotten" is, for the most part, yet another recitation of the stylish counter-myth: Columbus as founding father of all the iniquity and violence that followed upon the European intrusion into the Americas, and thus of the continuing injustice visited by the ruling elite in the United States upon Native Americans, blacks, women, homosexuals, and so forth. To the extent that white European settlement begat a swarm of evils—and, beyond denial, it did so—some indignation (tempered by a minimal understanding of history) is warranted. On the other hand, Hooks's implicit portrayal of the pre-Columbian America as a pacifist Elysium is a type-specimen of the fatuous cliché that has been repeated endlessly in left-wing diatribes prepared for the Columbus quincentennial. Thus it is not especially remarkable. What *is* remarkable is the thesis that serves Hooks as the springboard for her ruminations: again, we are confronted with the work of Ivan Van Sertima, this time in the form of his book, *They Came before Columbus.* Says Hooks:

> Thinking about the Columbus legacy and the foundations of white supremacy, I am drawn to Ivan van Sertima's groundbreaking book *They Came before Columbus.* Documenting the presence of Africans in this land before Columbus, his work calls upon us to recognize the existence in American history of a social reality where individuals met one another within the location of ethnic, national, cultural difference, who did not make of that difference a site of imperialist/cultural domination.[65]

Leaving aside the grave difficulty of making a site out of a difference, which must be akin to the problem of silk purses and sows' ears, we can see that Hooks accepts unquestioningly Van Sertima's thesis of ancient contact between seafaring West Africans and Meso-American cultures, together with the further assertion that this encounter was wholly peaceful and mutually enriching. We shall not examine Van Sertima's book in any detail, noting only that the assumption of transatlantic travel by ancient Africans is, in itself, unsupported by any evidence.[66] (We have already had a glimpse of Van

Sertima's evidentiary standards.) Conjoined with the further proposal that black explorers and Native Americans met without violence or exploitation, and that the ideas of Afro-Egyptian civilization provided the seed from which the South and Central American high cultures sprang, Van Sertima's hypothesis clearly belongs in the category of wishful thinking. Thus it is hard to characterize Hooks's embrace of Van Sertima's ideas as anything other than superstitious credulity.

Yet unlike Van Sertima, whose reputation does not extend much beyond fanatically Afrocentric circles, Hooks is a mainstay of the academic left and a ubiquitous presence at conferences and symposia devoted to questions of racial and gender justice. As even the brief excerpt quoted above reveals, she has mastered the postmodernist lexicon and the style that is de rigueur for fashionable campus radicals. She is, in short, a far more prestige-laden and influential figure than Van Sertima, ranking with highly respected black scholars such as Henry Louis Gates of Harvard and her sometime collaborator Cornel West of Princeton. Her susceptibility to Afrocentric fantasy and the pseudoscience that supports it is thus particularly ominous.

An even more startling example confronts us in recent work of Sandra Harding. As noted earlier, this feminist philosopher of science has openly called for a revolution against science, for replacing it with a multicultural, multiracial, ethnically diverse discipline, claiming that it will be more "strongly" objective than the existing version. One can garner some notion of what her enthusiasm really endorses by taking note of her recent book, *Whose Science? Whose Knowledge? Thinking from Women's Lives*. Several pages of that work are devoted to repeating slavishly the claims of *Blacks in Science*, without so much as a hint of skepticism or reservation.[67] In the gospel according to Harding, skepticism is to be reserved exclusively for scientific work done by white males and backed by the methodologies of scientific orthodoxy. "Strong objectivity" turns out to be another name for pathetic gullibility. Clearly, to the extent that Harding and her feminist admirers have any influence on the situation, they can only intensify the pedagogical damage being wrought by Van Sertima, Adams, and most of their collaborators.

To take another example, the sociologist Stanley Aronowitz has also championed the cause of multicultural science and asserts that it will revolutionize the *content* and the *conceptual foundations* of all the scientific disciplines, as well as the demographic picture of those who practice them. He is hardly well placed to object, even were he so inclined, when his summons to remake science along multicultural lines is answered by the purveyors of Afrocentric "science." Likewise, we have seen that Andrew Ross, Aronowitz's friend and ally among the "cultural critics," waxes indignant at any attempt of the scientifically literate to inculcate a widespread public understanding of what distinguishes authentic science from pseudoscience, New Ageism, and superstition in general. In Ross's view, such people are mere bullies, intent on preserving the unjust privileges of a scientific elite that works hand in glove with other purveyors of bourgeois mystification. He is not a very reliable ally in any attempt to undo the gross miseducation of young black people.

All this strongly suggests that even if the universities of this country eventually succeed in developing effective antidotes to the myths of fervent Afrocentrists (and failure to do so will leave many black students in an intellectual ghetto), they will do so without much help from most of the campus left; more than likely, they will have to proceed in the face of its indignant opposition. This may seem an unkind characterization, but the direct evidence, sadly and shamefully, supports it. The experience of Bernard Ortiz de Montellano provides an example. Ortiz de Montellano is an anthropologist and ethnohistorian originally trained in organic chemistry,[68] who has been active in trying to develop honest and legitimate "multicultural" approaches for encouraging minority youngsters to scientific careers. Yet his encounter with the Afrocentrists has sidetracked him into a new line of work, one that has produced a series of painstaking and impeccably documented refutations of Van Sertima, Adams, and their friends among the "melanin scholars."[69] His attempts to convey these findings have met, however, with severe frustrations from an unexpected quarter.

When Ortiz de Montellano and some of his colleagues—all of them nonwhite (he is himself Mexican-American) or female—attempted to present a critique of Afrocentric pseudoethnography at a recent meeting of the American Anthropological Association, their proposal was rejected. The tone and manner of this rejection suggest strongly the heavy hand of a new orthodoxy among cultural anthropologists, one that pretends to atone for the putative sins of ethnography during the era of Western imperialism and colonialism by abandoning the "Western" prerogative to judge the narratives of "non-Western" peoples in the light of objective knowledge and scientific methodology. All of this further confirms the rueful judgments of Robin Fox, cited in an earlier chapter, concerning the decadence of a subject that was, at its height, not only scientifically important but intellectually bold and morally brave.

In the face of all this, one is left with a melancholy question. Can the university, as the ultimate locus of scientific education, find the courage to stand up to Afrocentric (and related politicized) nonsense—to the degree necessary to sustain and enhance the possibility of scientific careers for young black Americans? We don't know the answer: it is certainly *not* clear that it will be "yes." Far more likely, in fact, is increased agitation for black-separatist "science" to be institutionalized alongside the standard variety, in a ghastly parody of affirmative action. This has already and notoriously occurred in a few places. And: on the basis of its record to date, we foresee that the wider academic left, with its relativist and perspectivist intellectual armament, its snide postmodern insistence that all narratives are equally valid and equally invalid, is likely to cheer the latter process on. Let us hope that all this is bad prophecy for the ultimate response of higher education (at least as regards science), to AIDS extremism, animal rights agitation, and Afrocentric fantasies. It must be a dim hope, however, tempered by fear for the consequences of nonfulfillment.

CHAPTER EIGHT

Why Do the People Imagine a Vain Thing?

Why do the Nations so furiously rage together? And why do the People imagine a Vain Thing?

PSALM 2 AND *MESSIAH* NO. 36

El sueño de la razon produce monstruos.

CAPTION TO PLATE 43 OF GOYA'S *CAPRICHOS*

Our sense of historical motion is no longer linear, but as of a spiral. We can now conceive of a technocratic, hygienic utopia functioning in a void of human possibilities.

GEORGE STEINER, *IN BLUEBEARD'S CASTLE*

How shall we read the Psalmist? Let us attempt it dialectically. The rage of nations—and of races, social classes, genders, creeds, and pariahs of all sorts—is the hammering pulse of history. That rage and the horrors that boil forth from it evoke all manner of imaginings. Some few of those may honor us as a species: they are the recurrent dreams of peace, justice, universal dignity, the power of human reason allied with kindness. The others are indeed vain: they are the flux of racism, of rabid nationalism, of misogyny, of religious fanaticism, and of superstition. These vanities feed in turn the rage that generates them, and inflame it further with malignity.

What of the sleep—or is it the dream?—of reason? Goya's epigram has always had a double meaning. When reason sleeps, the monsters of human pride, foolishness, malice, and cruelty emerge to do their worst. Thus it may be read: a maxim of the Enlightenment. Yet it is true that utopian fancies that flow from an unjustified esteem for the power of reason can also breed monsters of violence, vengefulness, and tyranny, monsters the equals of those overthrown in reason's name. The judicious historian will always have both interpretations at his elbow, for history has abundant examples of each. Yet this book is unashamedly an affirmation, in one particular context, of the first. Thankfully, we note that this is the one that seems to have Goya's explicit endorsement: "Imagination deserted by reason creates impossible, useless thoughts. United with reason, imagination is the mother of all art and the source of all its beauty." The same, we insist, holds true of science, or at least of that complex human activity deserving of the name. United with reason, imagination is indeed the mother of scientific insight.

We have been censorious, though not, we hope, unfairly so, toward critiques of modern science that have gained currency and popularity in contemporary scholarship, particularly among those thinkers and theorists who advocate sweeping, egalitarian changes in the economic, social, racial, sexual, and ecological mores of our culture. We emphasize again that the underlying grievances that ignite their anger are by no means wholly imaginary or capricious. Racial bigotry and the deification of greed have clogged cities with ruined men and women, and have come near to turning crime into a rational career choice for tens of thousands of young people. Casual and unremarked brutality against women is a continuing fact of our culture as, of course, of others. Easy as it is to mock the self-righteousness of scholarly bluestockings in their academic sinecures, we must keep in mind the real fear of violence that attends the life of any woman, no matter how privileged. The crime of rape remains a brutal expression of power, not only over women, but over other—vanquished—men.[1] Matters of sexual taste and private choice that a true civilization would cede ungrudgingly to individuals remain subject to an intolerance that is sometimes expressed as violence and is never less than humiliating. And when we contemplate the mess and the real, if invisible, dangers that are at times created by the frenzied processes of industry and technology, we are once again face to face with the mordant power of greed, with the shortsightedness that can offend the landscape in aid of a few more years of cheerful annual reports to stockholders.

Yet, as far as this book is concerned, we find ourselves in the position of scorning groups seemingly alert to such outrages and committed to doing something about them. This may seem a moral paradox. We justify it by insisting that the rage of nations, even that nation of persons pledged to constructing a just and ethical society, can and does beget its own vain imaginings. We have seen how easily a redemptive vision can slip free of reason (at least as regards science, one narrow but crucially important realm of thought) to produce monsters in the form of theories and conjectures as silly—and possibly as dangerous, too—as they are self-important.

These are so far, we admit, monsters of a lesser kind. They have tortured to date no more than the logic of discourse, and they skulk mainly in a secluded academic setting. We shall argue subsequently that matters of greater impor-tance are ultimately at stake; but even if that were not so, we believe that the health of a culture is measured in part by the vigor with which its immune system responds to nonsense. Such an immune response, although sometimes slow in the mounting, has been the richest heritage of the Enlightenment. We are vain enough to hope that our effort might form a small part of such a response.

We see it as an act of fairness (although, no doubt, our subjects will take it as evidence of deepest hostility) to inquire into the underlying sources of the antipathy toward science and its standards of validity that we have traced and attempted to refute. We are trying, in a sense, to turn the tables on the cultural constructivist theorizers whose ideas we find so unconvincing. What, we ask, are the social and ideological roots of their antiscientific theorizing? If their ideas were sustained by decent arguments and adequate evidence, this would be an unfair attack ad hominem. But we have been at pains to answer those ideas on their own ground; and we think we have shown that they are ill-founded and obtuse. What follows is meant then, candidly, not as addi-tional refutation, but as a sincere effort to comprehend why such shaky doctrines have been embraced so enthusiastically by individuals who are by no means stupid and who have often, as it happens, made penetrating analyses in other areas of social and political thought.

Our society has had an astonishingly general mood swing: the tonality of the new state goes far toward explaining the antagonism and suspicion toward science that we have undertaken to examine. The last of this chapter's epigraphs, that of George Steiner, comes close to epitomizing it. Steiner's darkly brooding essay *In Bluebeard's Castle* examines the fate of optimistic humanism in an age that has proved merciless toward hopeful illusions. Steiner notes the great paradox bedeviling our civilization and tormenting its most sensitive spirits. Humanism—post-Enlightenment Western humanism—has created, in the face of all the narrow particularism and dogmatic absolut-ism that has eternally plagued our species, an ethic of universal justice and universal tolerance. Moreover, history attests to the value of the humanistic view in tokens that go far beyond sentiment. The Western culture that grows from, extends, and intensifies the Enlightenment proves itself and displays its uniqueness most impressively by its ability to fathom nature and nature's regularities, to a depth unimaginable in prior civilizations. Western culture converts that knowledge into the instruments, conveniences, and perceived necessities of daily life with a swiftness that far outspeeds the traditional pace of historical process. Even for the gloomy Steiner, the impulse to celebrate the unique range, depth, and virtuosity of Western civilization and its antecedents runs so strongly that it leads to language that must enrage a dedicated multiculturalist: "But it remains a truism—or ought to—that the world of Plato is not that of the shamans, that Galilean and Newtonian physics have made a major portion of human reality articulate to the mind, that the inventions of Mozart reach beyond drum-taps and Javanese bells—moving, heavy with remembrance of other dreams as these are."[2]

The counterargument to this sentiment is, alas, all too obvious. Whatever may be said in praise of Western civilization and its most exalted visions of itself, it cannot be seriously maintained that the level of bloody-mindedness, selfishness, and cruelty found therein has fallen, overall, very much below that characteristic of the species in other times and places or embedded in other modes of social existence. One need only to open the daily papers for confirmation. In the meantime the instrumentalities available to our worst impulses have grown unimaginably lethal, and clearly we have no more power than any other culture to nullify those impulses. The stench of history is ever present in our nostrils; it rises from the killing fields on which Napoleon fought his fellow tyrants, from Gettysburg, from Sedan, Verdun, Stalingrad, Dien Bien Phu. Most especially it rises from Lidice and Dachau and Treblinka, from the Gulag, from Hiroshima and Nagasaki. The same science that Steiner celebrates for its articulation of human reality is the accomplice, sometimes the eager accomplice, of these atrocities.

Yet the mind framed by the ethic of the Enlightenment has no natural path of retreat. The very clarity of vision, the insistence on calling things by their true names, that defines us as heirs to the Enlightenment makes it impossible for us (that is to say, the honest among us) to disguise the rancid corners of our history under gaudy banners of nationalism, religion, progress, or justice. The very scope of the knowledge we have insisted upon rules out a comforting ignorance. We are bound to Enlightenment values—the universality of moral principles, the sanctity of individual volition, a detestation of wanton cruelty—and yet we have no choice but to indict the very civilization that begat those values as it goes careening through time leaving pain, death, bewilderment, the wreckage of aboriginal tribes and of rain forests in its wake.

But again, the terms of that indictment can be spelled out *only* in the language of those values. This, and not the mincing word games of the deconstructionists, is the true aporia. The criminal is also accuser and judge. Again, Steiner:

And it is true also that the very posture of self-indictment, of remorse in which much of educated Western sensibility now finds itself is again a culturally specific phenomenon. What other races have turned in penitence to those whom they once enslaved, what other civilizations have morally indicted the brilliance of their own past? The reflex of self-scrutiny in the name of ethical absolutes is, once more, a characteristically Western, post-Voltairian act.[3]

(Hardly post-Voltairian at that. Consider Montaigne.[4]) Today, of course, the cultural mood described by Steiner has crystallized in a politically strident form, under the name of multiculturalism or some such. It is the driving impulse of much of the intellectual self-abnegation that parades as theory in many university departments of literature. "A deconstructionist is not a parasite but a parricide. He is a bad son demolishing beyond hope of repair the machine of Western metaphysics."[5] It is clear that the father-victim of this ritual murder wears the vestments of the Enlightenment and that he has been tried by the light of his own stern, universalizing code.

When we examine how this mode of self-abnegation, sustained by the very code of values it deplores, deals with Western science we shall not be surprised that the leading sentiment is revulsion and retreat. This is not a new theme; the notion that science is poisoned knowledge, the fruit of a Faustian bargain, has been with us for a long time, and its cry has more often come from reactionaries than from progressives. The idea that we should back away from science, set its temptations and gratifications aside, is at least as old as *Frankenstein* or "Rappacini's Daughter." For good or ill, however, science and scientists have been largely deaf to these remonstrances. In ethical argument, they can give as good as they get. For every bomb, there is a vaccine; for every ICBM, a CAT-scanner. Yet those driven to throw off the guilt of Western man by the pricklings of Western conscience find the answer unsatisfactory. New evils pile on old—Chernobyl, Bhopal, the threat if not the assurance of an excessive greenhouse effect in the atmosphere or of a growing ozone hole, possibly bored by our air-conditioners and hair sprays. New epidemics add to the toll of old, familiar ones. Science seems potent to destroy, but appears ineffectual as a savior.

Science, as the term is now understood is, moreover, uniquely associated with Western culture. It arose only once, an invention that is unlikely to be repeated in detail no matter how many other cultures and peoples eventually come to produce fine scientists. In a hundred years, the greatest theoretical physicist in the world may well be Maori or Xhosa by descent; he—or she, as may well be the case—will nonetheless be a Westerner in the most important aspect of his or her intellectual temperament. The argument may be made, *pace* Steiner, that even Mozart is outdone by the polyrhythmic splendors of the Javanese gamelan or by the sonorities of Indian classical music. No such possibility exists for the multiculturalist challenging the intellectual hegemony of Western science, aside from pure falsification as is practiced, say, by the most deluded Afrocentrists or New Age mystics. No other civilization bears a like gem in its crown. Thus science becomes an irresistible target for those Western intellectuals whose sense of their own heritage has become an intolerable moral burden.

But science will not allow itself to be abandoned. It is too powerful and, when all is said and done, too interesting. Therefore it must be exorcised, castrated, at least at the symbolic level, if no material recourse is available. The natural view—that science gives power to those who understand and underwrite it precisely because it sees accurately into the workings of nature—is, of course correct; but it sorts ill with the temperament of the would-be exorcist. Thus the drive, fragmented and incoherent but energetic, to impeach science not merely as amoral handmaiden of the wickedly powerful but as flawed at its conceptual roots. The moralistic imagination always demands such an iconographic degradation of that from which it wishes to turn away. Science cannot be seen merely as dangerous; it must also be revealed as *false* in some essential way.

It is actually this moralism, rather than any solid philosophical commonality, that unites the various critiques we have examined. Moralism has the bad intellectual habit of excusing itself, on its own grounds, for weak and shoddy arguments. Moralism of this kind is, for instance, untroubled by the fact that its denunciations of Western scientific epistemology are composed on word processors whose very existence derives from a subtle understanding of the universe encoded in quantum mechanics, perhaps by a writer whose indispensable spectacles depend upon the light—via the science of optics—of Enlightenment. It lives very comfortably with all such contradictions.

At the level of broad cultural inclination and underlying symbology this analysis accounts, we think, for much of what we have been examining. Yet it would not be amiss to give a more specific and local view of the phenomenon, one centered in the particulars of American intellectual and political culture and paying some attention, without condescension, to the psychological generalities involved. We pass, in other words, to a less lofty and more concrete level of investigation than that which preoccupies George Steiner.

As we examine the process by which the current hostility to science within the academic left was incubated and nurtured, we find ourselves naturally turning to the 1960s. Commentators such as Roger Kimball have placed heavy emphasis on the sixties as the breeding time for all sorts of malfeasance. He sees the confrontational style of campus multiculturalists, feminists,

Marxists, and postmodernist advocates of nontraditional scholarship as having descended from the attitudes and tactics of the sixties student left, namely, Students for a Democratic Society, the Student Nonviolent Coordinating
Committee, and other activist groups that engaged in militant civil rights and
anti–Vietnam War politics. There is some justice to this charge. The rude
tactics of sixties campus militancy—picketing, sit-ins, a generalized rhetoric
of suspicion and scorn toward the nominal academic hierarchy—are recycled
in our day when questions of making the curriculum more "diverse" or initiating women's studies programs become hot issues. Many veterans of the sixties
are still on hand as leaders or advisors to radical undergraduates, and indeed,
to those who know these folk well, the trace of nostalgia is strong and
unmistakable.

On the other hand, at a higher level the styles of intellectual radicalism
have diverged from what was current in 1968. Racial separatism, which in
that era was reluctantly accepted by the white left as a temporary tactical
necessity, has long since hardened into a major component of the so-called
politics of identity. In this respect, the term *integration*, which was, after all,
the inspirational watchword of the sixties civil rights movement, is now
scorned in all politically fashionable quarters. Feminism has become institutionalized, and, while neither its emotional tonality nor its concrete agenda
has changed appreciably over two decades, it has evolved a large body of
densely esoteric doctrine informed, to a large degree, by developments in
literary theory and to a lesser extent by psychoanalysis. Marxism, as understood by the American left, has mutated from a revolutionary program driven
by a strong sense of economic forces, to a philosophical impulse that mixes
with other strains—feminist, deconstructionist, Foucauldian, Lacanian, ecological, and so forth—to create the eclectic brew of postmodern radicalism.
In the parlance of the academic left, "radical scholarship" formerly meant
historical research into the catastrophes of capitalism or the careful tracing of
the interlocking relations among economic, political, and institutional
elites. Nowadays it has devolved, typically, into a murky theoretical project, a
sort of abstract, unempirical sociology preoccupied by semiotics and bur-
dened by portentous overinterpretation.

Environmentalism itself, first mooted in the sixties as an essential political
concern, has become entangled in a body of mystical or quasi-religious impulses carrying it progressively further from scientific discourse and leaving it
vulnerable to such dubious dogmas as goddess worship, a "living" planet, and
animal rights. Many radicals schooled in the politics of jobs and wages, voter
registration, opposition to the military draft, reproductive rights, and health

care are put off by this turning away from practical, real-life concerns on the part of academic leftists, and ascribe it, with considerable scorn, to the "leisure of the theory class."

Despite such ideological shifts and revisions, there remains a strong resonance between the style of the contemporary academic left and the wider cultural mood of the sixties, the so-called "counterculture." This goes beyond sharply defined points of political doctrine. It has to be remembered that in the sixties the culture of rebellion and alienation was far more diffuse and generalized than can be captured in the manifestos and position papers of self-identified radical sects. The mood of the period was one of widespread disdain for all kinds of middle-class norms and expectations. "Sex, drugs, and rock 'n' roll" was as important a rallying cry as any antiwar slogan. This enthusiasm persists in our day, of course, but now it is chiefly found (among whites) in a completely apolitical and somewhat racist youth culture, quite segregated from the intelligentsia. Even so, as an index of the academic mores of the late sixties, it demonstrates why it has become possible for professors of English or art history to abandon—ostentatiously—all concern with the treasures of the high culture and to celebrate popular culture and its vulgarities quite without apology.

The sixties were also notable for a fascination with "non-Western" modes of thought. The drug culture of the period was tied to an eclectic mysticism into which all sorts of esoterica—authentic or bogus—from a variety of cultures were enthusiastically imported. Astrology flourished alongside Tantric Buddhism, and the fictitious shamans of Carlos Casteñeda echoed the vaporous posturings of Timothy Leary. Of course, on the literal level much of this was sternly rejected by the solemn ideologues of the "serious" left; but, inevitably, there was leakage from one subculture into the other—the legendary Woodstock Festival, for instance, incorporated both. Thousands of radical students, some of whom went on to become today's radical professors, toyed with these more colorful heterodoxies. What prevailed was a kind of generous latitudinarianism, a willingness to accept all sorts of oddball ideas at least provisionally, and if necessary, whimsically. Implicitly, what had taken shape was an informal *conspiracy of the heretical*, a community of defiance toward the conventional, a range of wacky doctrines linked mainly by the glee they all took in spitting in the face of received opinion. The progeny take the appropriate form, in our time, of slogans such as "in your face," and of the continuing fad for legible clothing whose legends are flamboyantly meaningless.

This inclination to regard heterodoxy, per se, as intrinsically valuable persists in the oppositional subculture of our own day, in its attitudes and in its

choice of philosophical godfathers. One can attribute the popularity of relativism, for instance, as much to the traces of omnivorous credulity that linger
from the sixties as to the dialectical skills of influential thinkers. Likewise, the
skepticism of deconstruction or Foucauldian determinism is but the obverse of
the generalized will-to-believe, the impudent *credo quia absurdum*, of the hip
radical (or radicalized hippie) counterculture.

We might also point out that, seen apart from its characteristic idiosyncrasies, the supposedly novel mood of the sixties was, in its scatterbrained
way, part of a long tradition of rebellious, Romantic individualism that goes
back at least as far as Rousseau. Some of the monumental figures in the history
of this sensibility—Blake, Wordsworth, Goethe, to name a few of the
giants—were notable for their rejection of the worldview suggested by the
orthodox science of their day. To them, science seemed to be both the servant
and the sponsor of a blinkered and incomplete vision of human possibility
and spiritual destiny. It suggested closure and limitation, where open-
endedness of vision was desired. It emphasized abstraction and schematism
while rigorously censuring the subjective. It hardly needs saying that this
perception, justified or not, was sufficient to create a lasting breach between
the Romantic impulse in our culture and the scientific temperament. Thus
the tatterdemalion romanticism of the sixties, deficient as it may have been in
enduring poetry, was heir to a long-standing tradition in which science and
the modes of thought sustaining science are seen to epitomize a small spirit
and a stunted imagination. They had their personification in the figure of the
"nerd," now known as the "techie." By contrast, pretensions to cosmic insight and mystic wisdom tend to be given every benefit of the doubt. Were
Madame Blavatsky still with us, she would be a winner on the student-
sponsored lecture circuit of our colleges.

Obviously, this generalized susceptibility to unorthodoxy—which may, of
course, manifest itself as "radical skepticism"—does not coexist happily with
a conventional respect for the competence and authority of the sciences.
Indeed, even the most playful mock-belief in astrology, the Tarot deck, or the
I-Ching usually betokens a certain restiveness under the doctrine that science,
as typically practiced, is the most reliable way of learning about the universe.
While it would not do to overstate this point, we can detect in modern-day
challenges to science, whether from humanists or sociologists, a strong echo
of the whimsies of the sixties. The current phenomenon is, in our view, rather
more obnoxious because it has entirely lost its playfulness and now parades as
serious scholarship and solemn theory.

This thesis is strengthened by the fact that many of the academics who are
most actively hostile toward standard science are affiliated, formally or infor-

mally, with areas of study that first arose during the sixties—women's studies, ethnic studies, environmental studies, and so forth. In their very origins, these subjects were linked as much to the oppositional culture of the sixties as to the formal traditions of research and scholarship. It is not surprising that a whiff of the sixties mentality, of LSD mysticism, shamanistic revelation, and ecstatic nonsense, still clings to them in some places. All this ties into a wider observation about the psychological underpinnings of the rebellious or oppositional intellectual stance. We put it forward as a general rule that the indulgence of one kind of heterodoxy betokens a further susceptibility to eccentric or highly speculative ideas. In the eighteenth century, for instance, the nascent political radicalism of groups such as the Freemasons was often associated with esoteric magical doctrine; historical figures like Cagliostro and the Rosicrucians attest to this. Earlier in this century, radical intellectuals, their patrons and sympathizers, mingled with theosophists, spiritualists, Jungians, and the like. A predilection for the unconventional almost always reigns among rebellious spirits. Science, nowadays the most stable and unassailable convention of them all, presents an irresistible challenge to such contrary and defiant natures. It stands as a metaphor for the smug self-assurance of the ruling culture and the stability of its institutions; therefore it must be brought low.

This analysis, please note, is not intended as an indiscriminate sneer to be employed against all social radicalism. We are merely taking note of what seems to be a general fact of social psychology, a fact that ought to enter into our understanding of the strengths and weaknesses of radical thinking.

We must consider the sadly ironic role that valid criticism of science and technology has played in fostering a far more dubious philosophical antiscientism. Much that has been said and written against the dangerous and foolhardy abuse of technology and the distortion of scientific knowledge is true and wise. (The reader who has remained with us can easily create his or her own catalog of such vices, needing no help from us on this point.) Nor is an emphasis on the social roots of the perversion of science pointless or particularly misleading. One needn't be a mystic or a radical antiempiricist to associate technologically generated disaster with the acquisitive propensities of industrialism. The industrial world's catalysis of population growth and its consequent need for resources attacks the forest and pollutes the ocean; its drive for profits adds toxins to the air long after it has become clear that this is in the end a stupid thing to do. It is disheartening to note, however, that such well-founded criticism often leads further to bizarre and shoddy theorizing about the epistemic status of science, as if allegiance to such doctrines could somehow magically generate the power to halt the technological misuse of scientific knowledge. Even more discouraging is the fact that among the thinkers who most clearly and accurately emphasize the dangers into which technocratic society falls, there are quite a few who lose perspective and begin to expound dubious positions concerning modern science, excoriating it for nonexistent philosophical flaws. Of course, the ultimate result is likely to be that they will cruelly subvert their own hopes.

In the current situation, however, the strategy of thumbing one's nose at science has few immediate costs. This, after all, is a society that remains saturated with superstitions of the grossest kind, and debased by a ubiquitous credulity. Any trip to a supermarket checkout-counter tabloid display will confirm this. Neither an occupant of the White House nor a celebrated literary scholar runs much risk of damage to professional standing simply because of a belief in astrology.[6] It would be inadmissibly cynical to suggest that the antiscientism of the academic left is just a ploy, playing to the ignorant in the fashion of New Age hucksters. On the other hand, the pervasive superstition of the culture does not stop short at the gates of the university. Many college newspapers, after all, run astrology columns without the slightest irony. It is quite possible that the criticisms of science we find unwarranted and, indeed, perverse are instigated by the mere propinquity of so much silliness in the general cultural atmosphere.

Furthermore this high-handed attitude toward science and toward rationality itself has its instrumental uses within the narrow world of left-wing campus politics. It is far easier to put up with Afrocentric nonsense, for instance, if one's immediate tactical reasons for doing so are seconded by a philosophical relativism that scorns the supposed canons of objective knowledge and shrugs indifferently at the abandonment of logic and evidence. If one believes that science is nothing but metaphor, then one need have no qualms about substituting for it other metaphors. These last points, however, are relatively minor. There are more telling factors deriving from the psychology of intense political commitment. We have in mind the propensity of ideological systems, however noble their ultimate aims, to induce a totalizing mentality in their adherents.

Totalism, as we would define it, is the impulse to bring the entire range of human phenomena within the rubric of a favored doctrinal system. It erects ideological categories which are viewed as primary, privileged, and comprehensive. Totalism of this kind has been the historic tendency of organized religion as we have known it since the end of classical paganism. Religion, however, has no monopoly in this respect. Marxism, to take one inevitable example, is just such a totalizing system, especially as amplified and interpreted by a monomaniacal thinker such as Lenin. Hard-core Marxists assume that all history and culture, all social phenomena in fact, are to be explained by the so-called laws of historical development set forth by Marx and elaborated by certain privileged successors. Marxists assert that the economic order of society, as reflected in class relations, underlies everything and that all other aspects of social life—morals, law, art, political principles, and all the rest—are derivative, mere epiphenomena.

The exact sciences too are subject to this scheme. Indeed, they are regarded as lesser branches of a master science unknown to all but Marxists, the science of dialectical materialism. Dialectical materialism, on this view, has the power to oversee and indeed to rectify all the other sciences. Supposedly, it is uniquely situated to escape the illusions in which ordinary science, because of its historical and social situation, may become entrapped. Marx himself took these ideas quite seriously and was eager to sit in judgment on scientific questions despite his modest and sometimes absent competence to do so.[7] This sorry tradition was carried on by Lenin in a number of notorious pamphlets;[8] under Stalin's regime in the Soviet Union it was at least a factor in the wretched Lysenko affair, which crippled Soviet biology for decades.

This is not to accuse all modern scholars who describe themselves as Marxists of such simplistic and reductive attitudes. Nowadays, Marxism is a rather flexible conviction, and many versions avoid the occasional crudities and intellectual barbarities of Marx himself. Nonetheless, old habits die hard, and the notion that science is a fallible part of the cultural "superstructure" of bourgeois society is reborn as "social constructivism" among Marxist, quasi-Marxist, and purportedly non-Marxist intellectuals. Here too, however, behind this piece of doctrinal refurbishment, there stands the habit of totalizing thought. Sociologists qua sociologists are hardly immune to it. Who would not want to see his own particular discipline regarded as the "master science"?

Leaving aside the vanity of sociologists, which, on the scale of human folly, is hardly worse than that of physicists, the totalizing instinct is found to be firmly embedded in some current modes of radical thought which in other respects have put themselves at a distance from traditional Marxism. Feminism, at its most aggressive and confrontational, is obviously one such system of thought. As totalizers, radical feminist theorists are easily a match for the most rigid Marxist. The lurking idea behind this presumptuousness seems to be that the situation of women in society cannot be viewed merely as a "problem," susceptible to pragmatic amelioration. Rather, it must be seen as *foundational*, as having cosmic dimensions, and thus it can be redeemed only by a wholesale reconstitution of the entire social fabric. Radical feminism in this vein not only makes a claim on received notions of equity and justice but appropriates the whole notion of justice to itself. It demands to be recognized as morally omnicompetent. It follows that no institution of the existing order may be viewed as free of the original sin of sexism, for to exempt anything from the surveillance of the feminist ethic is to deny the absolute priority of feminist values.

Since science in particular is incontestably central to contemporary society, it too must be made to run the gauntlet of feminist censoriousness. The historical fact is that science, as a social institution, has perpetrated systematic injustices against women, most notably the unforgivable injustice of excluding them until recent decades. This is a genuine grievance which even the most down-to-earth feminist could not overlook. The radical feminist critique of science, however, goes far beyond a demand for redress on this point. It insists, as we hope we have demonstrated, on a reconstitution of science at its most basic conceptual level, so that it may be incorporated, as no more than a subsidiary element, in the moral and ideological universe of militant feminism. We should not, in theory, expect this attitude to produce a dispassionate and well-reasoned evaluation of science and, as we have seen, the most widely known attempts at feminist analysis of science—those of, say, Harding, Keller, and Haraway—are all undone intellectually by the moral ferocity that motivates them in the first place.

The radical ecology movement is another site of the totalizing instinct. Radical ecology, as embraced by the extreme wing of the European Green parties and such American groups as Earth First!, is hard to understand if we think of it as a mere political doctrine. It is more an embryonic religion, combining syncretically a genuine fear of environmental degradation with all sorts of sentimental, mystical, and ecstatic attitudes toward an idealized nature. As time goes on, it is increasingly impatient with utilitarian arguments giving priority to the safety, health, and well-being of human beings. Rather, it deifies the "natural" in a way that condemns implicitly most of what humanity has been up to since the end of the Neolithic period. Humanity is seen as worthy only to the extent that it takes its place in a static, unchanging world as one species among millions, undistinguished by any special moral worth. Characteristically, ecological radicals call for a drastic reduction of the earth's human population, not so much to make life healthier and more fulfilling for the people remaining as to create new space for all those species shoved aside by the heretofore exponential growth of humanity.

Such a dogma, outwardly gentle but horribly fierce in its inward essence, is impatient with the analytical and empirical style of science. Deep Ecology, as this tendency is often called, insists on facing nature with an intellectual passivity wholly inconsistent with the scientific attitude. Such a creed is by its very content a totalizing one. After all, its devotees cannot head off into an isolated wilderness all their own; its goals can be met only through the universal acquiescence of mankind. It is entirely inconsistent with pluralism. Therefore it comes to see the tradition of scientific thought as an impiety, a stiff-necked refusal to merge with nature on nature's own terms. Couched in such language, these doctrines are not open to rational refutation, for the entire tradition of rationality is under the deepest suspicion. On the other hand, rank superstition is entirely acceptable—if it comes garbed in the vestments of nature worship.

Deep Ecology as such may not have many adherents. Its doctrines, however, have emotional appeal to a range of people who are disgusted by the evils of contemporary industrial civilization, by the filth and environmental destruction it creates. The formulations and catchphrases of ecological radicalism have infiltrated with remarkable efficiency the language of more moderate environmental groups and of oppositional politics generally. Distrust of science is naturally borne along with such terminology; as some of the slogans of Deep Ecology become platitudes, its antiscientism and antirationalism are quietly absorbed into the intellectual bloodstream of oppositional discourse. This tendency is reinforced by the degree to which the language of ecological radicalism tends to echo the New Age clichés of popular culture, and by its insistence on the futility of mere human reason, which chimes well with the radical skepticism of the trendy postmodernists whose own totalizing pretensions we have already explored.

To this point our argument has implicitly rested on a certain benign and, perhaps, slightly condescending assumption. We have tacitly assumed that there is in sober fact no real contradiction between science as such and the social and political goals (in their nonfanatic forms) of the various oppositional movements we have studied. It would therefore seem that in adopting a view of science as an enemy or as a corrupt institution in need of deep ideological reform, these movements have created a bogeyman visible only to themselves and have wasted time and intellectual energy. Moreover, in the rather parochial environment of the university at least, they have irritated and annoyed people who initially bore them no ill will; they have made skeptics of potential allies. There will be no large-scale conversion of scientists to the point of view of these critics. What is more probable is that the scientific community will increasingly come to feel that since the criticism is asinine, the moral claims of those who advance it must be negligible. The recent antipathy toward science of various social radicals is not only pointless but self-defeating—or so it would seem.

We must, however, qualify this view, even at risk of seeming professorial. It is true that on most of the larger ideological issues, science as such has little to say. Whether or not capitalism must always retain an exploitative aspect, whether or not it is possible to construct a viable and humane socialist society, are not, for example, questions on which scientific knowledge has much bearing. On the other hand, science has been in some crucial respects a positive source of support to oppositional movements. It has dispelled much of the nonsense that sustained sexual and racial discrimination, in spite of early attempts to ally it with the forces of repression. It has given powerful evidence of the ineluctable importance of environmental consciousness in the application of technology.

All the same, there remains a small residue of political issues on which the conclusions of science have some bearing without necessarily flattering the hopes of passionate reformers. These issues are another source of friction and they motivate, beyond the factors we have already considered, the hostility that exists. The examples are not numerous but they are important.

If we examine feminist doctrine, for instance, we find it split, for the most part, into two camps. On the one hand, there is what is usually called "essentialist" feminism, which hews to the idea that there are indeed innate differences between the sexes in emotive and cognitive style and in ethical predisposition. Of course, it is assumed that the "feminine" side of humanity, its *good* side, has been cruelly neglected and suppressed, and that the purpose of feminism is to restore it to its merited preeminence. On the other hand, "antiessentialist" feminism insists that there are no innate psychological differences of any importance, and that to posit their existence is not only chimerical but invites the continued repression and exclusion of women. The grounds for this fear are obvious; myths of essential difference have provided a host of societies with their justifications for mistreating women and cruelly circumscribing their lives.

Not unexpectedly, there have been attempts to synthesize these apparently conflicting views, most commonly by invoking a species of social construc- tivist doctrine in order to argue that while there is no congenital difference between men and women on the psychological and behavioral level, the strictures of a sexist society induce children to grow up thinking and behaving as though there were, whence women end up being more admirable in their ethical and philosophical outlook even as they are intolerably degraded. This putative reconciliation, an attempt to have it both ways, is precarious, unsta- ble, and vulnerable to its inherent and quite obvious contradictions. (Femi- nist philosopher Sandra Harding's work provides a cautionary example of the pitfalls.) Most feminists sooner or later fall to one side or the other. Unfor- tunately, both factions eventually run up against hard facts that are less than encouraging to them. Science, to the extent that it is the bearer of these bad tidings, becomes the focus of the resulting hostility.

To consider the essentialists first, we observe that they form the subculture from which goddess worshipers and believers in a supposed golden age of matriarchy are usually drawn. Of course, insofar as science is generally hostile to superstition, it grants little encouragement to devotees of the goddess and offers a worldview by and large antagonistic to its mystical whims. Matriarchalism, by contrast, need not be overtly superstitious or antirational. However, to the extent that it relies on historical precedent, it is doomed to be disappointed by orthodox historiography, anthropology, and archaeology. There is not much that the matriarchalists can say in answer short of a retreat, acknowledged or otherwise, into the misty uncertainties of wishful thinking.[9]

The case with antiessentialist feminism is more nuanced and ultimately more important. Antiessentialism is the common creed of most feminists involved in serious intellectual life in or out of the academy. The reasons for this should be fairly obvious. The doctrine of innate mental differences between the sexes holds obvious perils for women embarked on scholarly careers in a society that until recently barred them from such roles. It would be natural therefore to assume that the relevant branches of science— behavioral and cognitive psychology, neurophysiology, and so forth—are the allies and benefactors of the antiessentialists precisely because they *have* done so much to dispel the myths of female intellectual limitations. Because of their insights, one cannot, these days, deny the capacity of women for any kind of intellectual or creative activity without revealing oneself as an ignoramus.

Paradoxically, however, this kind of science figures high on the antiessentialist-feminist enemies list. The problem is the absolutism—the totalizing inclination we spoke of above—that afflicts even the highest intellectual circles of feminism. The fact is that the behavioral sciences have given an inordinate amount of time to the question of sex differences and their origins. By and large, the notion of hard-and-fast, rigid, categorical differences has been shown up as an absurdity, which ought to give feminists all the ammunition they need for political arguments in favor of equality of opportunity. On the other hand, there is evidence, strong to begin with and growing stronger over time, that a number of perceived behavioral differences between males and females, especially at an early age, are in fact innate and congenital. This inference is not easily impeached as the warped product of sexist science, since many of the researchers who came up with it were, in fact, women who would have been happy to reach the opposite conclusion. It is, moreover, a fact whose political implications are generally negligible. In no way does it imply the inadvisability of complete legal and social equality between the sexes. Nonetheless feminists who fear the lurking dangers of essentialism are outraged and frightened by this innocuous work.

Apparently, no evidence, however strong, is to be allowed to dent their conviction that all but the obvious anatomical differences between men and women are "socially constructed." This proposition amounts to a credo, a virtual article of faith, a test, in certain circles, of one's loyalty to feminist principles. As a result, an extensive and dogmatic feminist literature has grown up around the "genes and gender" question and the trick of finding farfetched arguments for evading these harmless but (to the absolutist mentality) displeasing truths has grown into a minor art form. The relations between the scientific community and militant feminism have been, to say the least, soured by this curious episode.

In the environmental movement we find another example where scientific standards are distrusted because of their capacity to bring ideologically unwelcome news. We are not speaking here of ecological extremism—Deep Ecology, spiritualized ecofeminism, animal rights absolutism, and the like—but rather of certain tendencies to be found even within mainstream environmentalism. Historically, of course, environmentalism arose precisely because of fervent and conscientious advocacy on the part of ecologists, climatologists, geophysicists, and other scientists who were the first to perceive the dangers into which unthinking abuse of the environment and spendthrift attitudes toward our finite resources were leading us. Distinguished scientists with unimpeachable credentials are still among the most prominent figures in the movement, and, moreover, the issues that most intensely excite environmentalist fervor—the greenhouse effect, for instance—would be invisible in the absence of acute scientific work. Still, a rift exists, if only embryonically, between science as a mode of thought and environmentalism as a political force.

The problem is that environmentalism, like any political movement, needs a credo, a stock of beliefs to form the core of its motivations. Those upon which environmentalism draws are reasonably obvious and certainly valid as a general rule. A dismissive attitude toward environmental threats is, after all, symptomatic of a psychotic break with reality. All the same, when we get down to cases we find that many rank-and-file environmentalists—the sort of people who flock to Earth Day celebrations—understand environmental issues in a way inconsistent with the scientific temperament.

To a scientist functioning in his professional capacity the questions that arise in connection with ecological issues are subject to the same methodological constraints as any others. Most of these questions cannot be answered with anything like perfect confidence. As scientists, we cannot say, for instance, that a runaway atmospheric greenhouse is the certain result, within a fixed number of years, of continued combustion of fossil fuels, and that it will inevitably bring about ecological catastrophe. We simply know too little, and what we know is imperfect and provisional. Complexity and its attendant uncertainty is part of the scientist's everyday intellectual environment and he is well aware of his predictive limitations. It may well turn out that the greenhouse scare was a false alarm (which does not excuse us from taking it very seriously and considering public policies that, for the sake of prudence, assume the worst).

Environmental fervor, on the other hand, requires a large degree of subjective certainty. It will impose a strict binary logic on the probabilistic estimates of cautious scientists and convert a theorist's tentative hypotheses to dire inevitabilities. Most environmental activists *know* that the greenhouse world is upon us; scientists know no such thing, although they may fear it. Perhaps the activists are right when they convert suspicions to certain conclusions, in the sense that to do otherwise would leave them dispirited and unmotivated. On the other hand, there inevitably will be false alarms every now and then; every once in a while apologists for industry or government or the military will turn out to have been right, from a scientific point of view. The most likely result, unfortunately, is not an intellectually flexible environmental movement that finds ways to live with uncertainty and is capable of revising its agenda as new scientific information emerges. Rather, given the emotional intensity of political commitment, it is probable that organized environmentalism will, by stages, come to reject the exasperating caution of scientific thinking and move closer and closer to the shallow gratifications of dogmatism, a position as silly as that of the know-nothings who are confident that no man-made activity can change the world or God's plan for it. In a sense, the prediction of an environmentalist rejection of science has already been borne out by the recent success of the most radical and doctrinaire form of ecological thought and the celebrity of its demagogic proponents. To those of us for whom environmental science is of desperate importance, this is a deeply melancholy conclusion.

We have a final thought to offer, unfortunately a pessimistic one. In a world of recalcitrant evils, the would-be reformer is most likely to share the fate of Sisyphus and to live with frustrated hopes and pitifully transient achievements. One of the saddest facts of life is that frustration rarely begets wisdom; but it frequently ignites irresponsible fantasy. In the final analysis, the curious and mostly regrettable phenomenon we have been talking about—the anti-scientism of intellectuals—may be yet another instance of this general principle.

CHAPTER NINE

Does It Matter?

There is today a broad, generational, postmodernist current of irrationalism with its roots deep in the seductively brilliant thought of Nietzsche and Heidegger, which is at its core elitist and antidemocratic, even though that thought is often absorbed only in the flattened, simplified version popular among today's students and intellectuals. This is the cloven hoof of earth mother communitarianism, the need for the organic, the authentic feeling and for passion as against the cool "patriarchal" logic of the broad Left. This trend includes the rejection of science as well as scientistic fetishism. And, of course, it is permeated by utter contempt for the warp and woof of genuine democracy, for discussion, give and take, compromise, and elected representative bodies.

BOGDAN DENITCH, *AFTER THE FLOOD*

Discounting the Critics

Science is, above all else, a reality-driven enterprise. Every active investigator is inescapably aware of this. It creates the pain as well as much of the delight of research. Reality is the overseer at one's shoulder, ready to rap one's knuckles or to spring the trap into which one has been led by overconfidence, or by a too-complacent reliance on mere surmise. Science succeeds precisely because it has accepted a bargain in which even the boldest imagination stands hostage to reality. Reality is the unrelenting angel with whom scientists have agreed to wrestle.

Yet those who insist that science is driven by culture and by politics, by economics, by aesthetics, even by a species of understated mysticism, are not for that reason alone to be dismissed as wrongheaded. On the contrary, these assertions, if "driven" is replaced by "influenced," come near to being truisms. Great difficulties arise, however, when such insights are wielded as ideological blunt instruments in the name of "demystification," or to nourish the political vanity of factions and the academic vanity of scholarship more notable for its novelty than for its profundity. A serious investigation of the interplay of cultural and social factors with the workings of scientific research in a given field is an enterprise that requires patience, subtlety, erudition, and a knowledge of human nature. Above all, however, it requires an intimate appreciation of the science in question, of its inner logic and of the store of data on which it relies, of its intellectual and experimental tools. In saying this, we are plainly aware that we are setting very high standards for the successful pursuit of such work. We are saying, in effect, that a scholar devoted to a project of this kind must be, *inter alia*, a scientist of professional competence, or nearly so.

This is not a dictum that sits at all happily in the minds of those thinkers whom this book has criticized. They read it as an ideological demand, emanating from the arrogance of a priesthood that denies the intellectual fitness of anyone who is not an initiate. Such resentment is quite understandable on an emotional level. But as logic, it is of little avail. The critiques of science and the political misuses of it that we have evaluated vary greatly in perspective and argumentative strategy. Yet common to almost all of them is a failure to grapple seriously with the detailed content of the scientific ideas they propose to contest, and with the scientific practices they wish to impeach.

This alone, aside from other defects of reasoning and evidence, aside from the routine intrusion of special pleading, almost invariably guarantees that the results, however aggressively proclaimed, will lack accuracy and specificity. Such work founders because it treats science as a token of the illegitimacy of the current social order, rather than as a coherent body of ideas requiring the most exacting attentiveness. Scientists—aside from a small cadre of ideologically motivated sympathizers—generally ignore these critiques, not out of blind defensiveness of their own turf, not out of snobbery, but because the critics simply sail so wide of the mark, and have so little to say about the actual ideas with which scientists contend every day of their working lives.

Science, from the most cloistered and abstruse to the most directly applicable, has taken little notice of recent critiques of its underlying conceptual basis. Instead, it has tended to be vaguely receptive to the political rightthinking to which the critics lay claim, but without examining the details. There has been no rethinking of fundamental ideas, or of how these are to be articulated to experimental and observational reality, in response to recent criticism by feminists, sociologists, and postmodern philosophers. Occasionally, the opinions of such thinkers are voiced in scientific journals of the more general, informal sort (as opposed to specialized research publications).[1] This illustrates an admirable intellectual hospitality, or—sometimes—a lazy and weak-minded one. But it should not be misread as general acceptance or influence on actual scientific practice. Science as such—molecular biology, solid-state physics, polymer chemistry, nonlinear differential equations, and the thousands of other specialties—would have taken the same course over the last couple of decades had no feminist philosopher or postmodern social critic ever addressed a line to scientific matters. To put it bluntly, the probability that science will sooner or later take these critiques sufficiently to heart to change its fundamental way of knowing is vanishingly small.

This is not to say that criticism of the social implications of scientific practice has been without effect. In a few areas of applied science, medicine, and technology, practices have been rethought and modified on the basis of specific critiques (for example, of workplace safety monitoring), a good many of them having a left-wing provenance. Some of the modifications have been useful; some, as we have seen, are less than justifiable on logical grounds. In any event this must be distinguished very clearly from revisions at the level of concepts and fundamental methodology. There, despite the hyperbole and the earnest attempts of ideologically charged enthusiasts, the effect is imperceptible.

Thus we come round once more to the question of why the critiques of science generated by the academic left should be taken seriously enough to require an honest rejoinder from science. If, as we believe, science will not, in any serious way, be influenced, deflected, restricted, or even inconvenienced by these critics and those they influence, why should their work draw more than passing comment from anyone outside that mutual admiration society? Why, in fact, have we troubled to worry about it? From the broadest point of view, we worry, and believe that scientists as well as lay persons well disposed toward intellectual progress should worry, because unanswered criticism *must* in due course have effects. We worry for the reason articulated by Arthur Potynen, for example, among many others who have begun to ask this question:

> Those attempting to ignore Post-modernism are many: for example, the natural sciences and business departments often hope that the affected, yet essentially harmless, humanities will remain isolated and irrelevant. But if power is the essence of all human endeavors, then can science escape being labeled willful and coercive? Can business be anything other than rapacious? *Can either science or business continue to function in a political culture that assumes them to be oppressive?* (Emphasis added.)[2]

Those who choose to do so may dismiss this remark as the carping of one to whom "business as usual" may be the most sacred of values. The epigraph of this chapter, a *cri de coeur* from Bogdan Denitch, a life-long socialist of unswerving faith, fully demonstrates that such misgivings do not necessarily go hand-in-hand with a benign attitude toward "business." A similar complaint is heard from the left-wing social theorist Alan Chalmers:

> I am by no means alone in viewing social trends in the contemporary world with dismay and alarm. The gulf between rich and poor and between developed and undeveloped countries widens, the environment is destroyed, and the threat of annihilation looms. The social and political problems facing us are urgent and vital. I do not think this cause is helped by construals of science as a capitalist male conspiracy or as indistinguishable from black magic or voodoo.[3]

Plainly, one needn't be in any sense a conservative to view the antiscientism of the academic left with deepest apprehension. One needs only to have paid some attention to it, to have understood its meanings, deep as well as superficial, to be concerned.

Academic Recognition and Fairness

The issue of academic recognition derives, we admit, from what some will descry as academic priggishness. Much of the work we have cited has been received with enthusiasm in certain academic quarters, where trendiness in the humanities and social sciences combines with "identity politics" and with the residue of Marxist intellectual totalism that persists among leftist thinkers. Paradoxically enough, in the United States—this bastion of free-market capitalism and reflexive hostility to socialist ideas—such enthusiasm is usually enough to guarantee success and even celebrity in the narrow world of the academy. The critics whose work so disturbs us will, in one sense, have the last laugh. For the most part, they have made it to the upper rungs of the academic ladder, from which no rejoinder, however well founded, will dislodge them in any likely future. Nor will the richest charitable foundations cease to support and honor them.

Most of the science-critics with whom we have dealt in this book, for instance, have high positions in such notable universities as Princeton, Berkeley, Brown, MIT, Rutgers, and the CUNY Graduate Center—and, if we allow some British examples, Oxford and the University of Edinburgh. They do not struggle in outer darkness: many of them are influential, politically as well as intellectually. We shall not illustrate this point with a list of the chairs, fellowships, and academic titles to which they lay claim at those institutions, but the facts are a matter of public record, and readers so inclined can easily satisfy themselves that most of our subjects, far from struggling as assistant professors, hanging on by their fingernails, are near the summit as regards academic prestige and seniority. Moreover, they enjoy that status principally by virtue of the work we find so unsatisfactory. Since we are hardly alone in such judgments, as we have tried to show, this fact raises serious questions about the presumed intellectual meritocracy of the academy.

It seems clear to us, in fact, that on the whole the academic left's critiques of science have enjoyed an astonishing free ride. Most evaluation of this work has come, so far, from scholars whose own ideological commitments are strong and whose scientific backgrounds tend to be deficient. It has been shielded, we suppose, from more skeptical treatment for a number of reasons. First of all, its egalitarian posture has won it the benefit of the doubt from prospective hostile critics whose own political sentiments run along the same lines. A critical response to particular feminist assertions, for instance, poses a problem for well-intentioned academics, many scientists among them, on whom the general moral claims of feminism exert a strong, positive influence. There is a reluctance to dissect them with the same rigor that might be applied to work that is not put forward as part of a wider struggle for justice. We have repeatedly run into this attitude on the part of left-wing humanists, in particular, who are willing to overlook the inconsistencies and evidentiary inadequacies of anti-science critiques in part because, as nonscientists, they are imperfectly aware of those flaws, but more importantly because such work strikes them as courageous and pathbreaking in its *political* ambitions. To these sympathizers, the very act of putting on the table questions about the competence and objectivity of science, about its complicity with the injustices of capitalism, racism, and patriarchy, is praiseworthy in and of itself; nagging doubts about the competence and fairness with which this is done are easily deferred.

This attitude is seconded by a frame of mind that derives from envies, rivalries, and resentments that have long beset academic life. Humanists and sociologists alike take a certain pleasure in the notion that the mighty principality of the exact sciences, with its arsenal of laboratories and observatories, its inexhaustible sources of funding, its imagined stentorian voice in public policy, its intimidating intellectual mystique, is now itself put on the bench for demystification. They are eager to find virtue in any analysis that claims to have accomplished this trick. Its novelty alone insulates it from severe questioning.[4] Moreover, as many of those questions would themselves require a reasonably deep knowledge of scientific particulars, it is far from clear that humanists and sociologists are even able to frame them. On the other hand, they are not immune to the flattery implicit in claims that the perspectives of postmodern literary theory or radical social theory have brought to light certain aspects of, shall we say, theoretical physics or evolutionary biology to which the scientists themselves were heretofore blind! They are correspondingly uneager to subject such claims to skeptical scrutiny, especially in light of the painful fact that to exercise it, they would have to acquire expertise in a field they have studiously avoided.

More than once we have been told by Y, the eminent professor of English and cultural critic, that X's critique of science shows her to be intimately familiar with some specific technical subject. On looking into X's work, however, we seem to find a great deal of psycho-talk, a good deal of literary language, and a glaring absence of knowledge of the supposed subject. In this case Y's wish has clearly been father to the thought. Likewise, we have been told over and over again that the keen methodologies of contemporary literary analysis leave up-to-date literary critics uniquely placed to analyze the rhetorical strategies and semiological underpinnings of scientific treatises. This claim—it amounts to an unquestioned truth in some circles—has, however, remained entirely unsupported. We have never seen it illustrated with reference, for example, to a paper on the instability of induced magnetic fields in high-temperature superconductors. We doubt we ever shall. Nonetheless, it remains an effective sop to the vanity of contemporary literati.

At the same time, we find that scholars who are eager to praise cultural and political critiques of science are reluctant to take into account the (admittedly rare) actual responses of scientists to this work. Having, as they see it, reduced the scientific community to an object of study, they are quite unwilling (despite their repertory-theater "feeling for the organism") to allow the specimen to wriggle free of its new restraints. Their logic is that any objections on the part of scientists are tainted, prima facie, by self-interest and special pleading. It is never taken into account that the same defects might afflict the critiques themselves. In any event, this kind of exclusion serves the further purpose of acquitting the humanist and social-scientist enthusiasts of recent science criticism of the tiresome chore of learning the specifics of the biology, chemistry, physics, or mathematics being criticized.

In sum, we are accusing a powerful faction in modern academic life of intellectual dereliction. This accusation has nothing to do with political correctness or "subversion"; it has to do, rather, with the craft of scholarship—a craft that has always had consequences, independently of the behaviors of individual scholars. We allege that eagerness to praise a certain spectrum of work has disarmed skepticism and careful critical attention. Political sympathy has combined with professional vanity to give undue weight, prestige, and influence to a decidedly slender body of work.

As we have observed, substantial careers have been made and salvaged on the basis of it, to the relative detriment of scholars whose own work is more competent but less ideologically intoxicating. As scientists, we observe this fact with some dismay. We know how hard it is to achieve even a modestly successful career in science or mathematics. In addition to native talent and appallingly hard work over many years, it requires unremitting self-scrutiny and attention to the possibility of error. There is the constant risk, unknown outside of the sciences, that the rewards of one's patient effort will vanish beyond recovery because a colleague announces the same result a few weeks before one's own work is ready for publication. Moreover, there is rarely any lasting payoff for mere self-promotion, or for the ability to talk a good game— although, as in life generally, such talents don't hurt.

To top it all, for the typical faculty scientist, teaching responsibilities, charged as they are with the necessity to cram huge amounts of hard technical information into a few dozen lectures every semester, afford little opportunity to play to the grandstand. The displays of classroom charisma to which many humanists are frankly given, and that make campus celebrities of the cleverest,[5] are simply out of place in most science courses. Science has its staginess and its celebrities, to be sure, but the style is different. Papers, for example, are never literally "read." Delivery is almost ostentatiously casual, even at the great international meetings. It is a gaffe not to credit one's graduate students and collaborators, in great detail. Emphasis is supposed to be upon the idea, the data, the structure, the equations: forceful or rhapsodic presentation is taken to be amateurish. To be sure, this is as much a style as is the page-turning and premeditated diction of the humanists; but in science it is the findings that are supposed to enchant, not the person offering them. The classroom offshoot of this is that even a young and inexperienced science teacher, following the example of his mentors, does not rhapsodize on the grandeur of, say, the Hertzsprung-Russell diagram, in which simple, single graph the life, death, and evolution of all the stars is displayed or implied: that grandeur is supposed to be grasped by the student who is prepared, by prior study, to grasp the implications of such a plot. Some, but not many, are.

In view of this, we can hardly be indifferent to the spectacle of major academic honors and emoluments accruing to work whose lack of substance and whose reliance on special pleading and appeals to political solidarity seem perfectly obvious. The situation is the more provoking in that, when all is said and done, the central appeal of such work is the pretext it provides to disparage the natural sciences—to dismiss their astounding achievements as so much legerdemain on the part of a ruling elite. Would that we could—but we cannot—recommend practicable remedies for this situation that are also simple and fair. The best we can do, for the moment, is to bring our point of view as honestly as possible, but also as forcefully, to the attention of the academic and scientific community.

A Schism in the Academy?

Emeritus professor of sociology Lewis S. Feuer has recently written:

> If multiculturalists succeed in acquiring control of the curriculum and if they then institute a kind of force-conditioning of students with the "literatures" and ideological apologia for backward peoples, the consequence for the universities will be quite other than they foresee. Science students, with their essential preparatory studies growing all the time, will finally rebel against the "humanities" requirements, and for all practical purposes, the colleges of science will secede from their traditional association with the liberal arts college . . . The free marketplace of students and professors will, unless politically intimidated, decide for those institutions loyal to scientific values.[6]

Even if one judges Feuer's vision to be unduly pessimistic, it must be admitted that he has his finger on something. The mood he detects is real, if unfocused. The National Association of Scholars, antiradical if not in any simple way "conservative," may intend to exploit it.[7] One of its principles seems to be that in campus disputes over "political correctness" and the like, among the things to be done is "get the scientists on your side."[8]

The immediate subjects of Feuer's ire—"multiculturalists"—have not, under that description, received much attention from us, aside from our remarks concerning "Afrocentric" science. But it would be disingenuous to pretend that our subjects and his don't have a large overlap. Left-wing critics of science such as Stanley Aronowitz, Sandra Harding, and Helen Longino salt their analyses heavily with appeals to the presumed superiority of "multicultural" epistemologies. If their views become part of the general intellectual baggage of the academic left, a process we believe is well advanced, then Feuer's scenario, or something resembling it, becomes very much more likely. Departments of chemistry or electrical engineering, other things being equal, have little to say, and will continue to have little to say, about the kind of "multiculturalism" that infuses the reading lists of courses on the modern novel with works by Third World women. If, however, the same academic factions are seen to be agitating for a similarly politicized view of science, one that draws heavily on the critiques we have analyzed above, academic life, already fractious, is likely to become considerably *more* belligerent. The chemists and engineers are quite likely to insist on a say about such agitation, once it is their own courses, rather than the vague metaphors of a discipline, that are being proposed for purification or diversification.

On the whole, it is regrettable that serious students of the exact sciences rarely encounter, in their training, courses in the history of their disciplines that pay close attention to social, cultural, and political factors. Such knowledge is usually acquired ad hoc by those scientists who take a particular interest in it. But, as Feuer points out, the burden of essential preparatory studies is enormous, and is continually growing. Time is precious to a young scientist, and the optimal career path leads to the frontier of the subject as quickly as possible, leaving little opportunity for historical rumination. Nevertheless, much as one might lament the rarity of historically oriented science courses (and we join in that lament), in our judgment their absence is, on the whole, preferable to a hypothetical curriculum that requires such courses but hands responsibility for them over to historians and sociologists whose views derive from the science critiques of the academic left. We should be very surprised if our opinion is not shared by the vast majority of our colleagues, including some whose political outlook is unequivocally egalitarian and feminist.

The humanities, as traditionally understood, are indispensable to our civilization and to the prospects of living a fulfilling life within it. The indispensability of professional academic humanists, on the other hand, is a less certain proposition. Academic scientists have acceded to it, and properly so, out of respect for their colleagues as well as a deep concern that the great traditions of Western humanism should not be buried under the shabby detritus of popular culture and philistinism. The current stir over the postmodern style of humanist scholarship invokes the possibility that this body of sentiment may erode. Scientists are not the only skeptics, or even the most important ones; nor is such skepticism, we insist, necessarily a badge of right-wing leanings. Many of our wholeheartedly left-liberal, humanist colleagues are increasingly embarrassed by the spectacle of flamboyant celebrities in their respective fields playing "such fantastic tricks before high heaven as make the angels weep." How far things would have to go before the disenchantment becomes severe enough to provoke a genuine crisis in university life is anyone's guess. It is, however, obvious that to the extent that the various misconceptions about science we have examined become part of the general stew of postmodern assertions, the irritation of scientists will grow increasingly acute.

This tendency will be amplified because of the *activism* inspired, to a great degree, by left-wing antiscientism. In this regard, we might mention the raids on laboratories by animal-rights extremists, the successful attempts to inhibit funding of innocuous studies on the relation between heredity and socio-pathic behavior and to close down scientific meetings on the subject, the assaults on evolutionary biologists for their advocacy of a not particularly doctrinaire sociobiological perspective, and the endless string of nuisance lawsuits brought by Jeremy Rifkin and his allies. All these actions have had the support, or at least the sympathy, of a substantial faction of the academic left. To most if not all scientists, however, they smack of the deepest anti-intellectualism. It would not be going too far to label them as superstitious outright. Yet they dovetail rather neatly with the emerging dogmas of the left concerning the innate fallibility of Western science.

Our speculations on the growing antipathy toward academic radicalism on the part of scientists are influenced by a certain sense of how the humanities and the arts enter into the actual lives of our scientific colleagues. On the whole, scientists are deeply cultured people, in the best and most honorable sense. The image of the scientific monomaniac, of science departments de-voted to a "naive scientism,"[9] is, to say the least, highly misleading. The range of knowledge of music, art, history, philosophy, and literature to be found in a random sample of scientists is, we know from long experience, extensive, and in some fortunate venues enormous. Most of this learning has been acquired, of necessity, at odd moments here and there—not through formal or systematic study. As humanists, therefore, scientists are auto-didacts. One obvious consequence of this fact is to undercut the argument that traditional humanities departments, in their role as educators, are indis-pensable bearers of the great treasures of our cultural heritage. There are other, albeit less efficient, routes to erudition.

Let us be blunt: having come so far, we have little left to lose. If, taking a fanciful hypothesis, the humanities department of MIT (a bastion, by the way, of left-wing rectitude) were to walk out in a huff, the scientific faculty could, at need and with enough released time, patch together a humanities curriculum, to be taught by the scientists themselves. It would have obvious gaps and rough spots, to be sure, and it might with some regularity prove inane; but on the whole it would be, we imagine, no worse than operative. What the opposite situation—a walkout by the scientists—would produce, as the humanities department tried to cope with the demand for science education, we leave to the reader's imagination. This little exercise in one-upmanship is, of course, utter fantasy. But it does point to something real. The notion that scientists and engineers will always accept as axiomatic the competence and indispensability for higher education of humanists and so-cial scientists is altogether too smug. Other sentiments are clearly astir. How these matters play out in American intellectual life will depend, to some degree, on the ability of the non-scientists to rein in the most grotesque tendencies in their respective fields.

We are not loath to propose that some, at least, of the "critiques of science" we have addressed must be counted among such grotesqueries. If they retain their current ascendancy among humanist radicals, it seems likely that the gap between C. P. Snow's "two cultures" will become a rigid barrier, maintained by mutual disesteem between scientists and their would-be critics. The sense of unity, of sharing in a common, though diverse, enterprise, will certainly diminish. The comity, the sense of mutual respect, that has been one of the graces of university life, compensating for its difficulties and penuries, will atrophy. At the same time, the tendency of the university to devolve into a cluster of rival satrapies, each eager to serve exclusively its own clientele, will be amplified. The thinning of the curriculum into a list of narrow, mutually incomprehensible specialties will be accelerated. Feuer's suggestion of a formal secession of the sciences from the radicalized world of no-longer-liberal arts may seem overblown. It would be premature, however, to rule it out.

The Debasement of Science Education

It is self-evident that active and interested citizenship in this country, the frame of mind that follows public affairs and stands ready to participate to some degree in ongoing debates, requires a usable knowledge of science and technology—at the very least, a seat-of-the-pants ability to track disputes concerning science and public policy. In a republic that counts Franklin and Jefferson among its founders, and whose culture heroes prominently include the participants in the Manhattan Project and the entrepreneurs who have endowed every desktop with its own computer, one might hope that such intellectual endowments would be commonplace, if not ubiquitous. All too obviously, this is not the case. Outside the community of professional scientists and engineers, understanding of even the most elementary science is thin and vague. Indeed, most of the population, including its iconic voices— the television entertainers who comment on and not infrequently distort the news—seems to take a perverse pride in the self-abnegation "I'm no rocket scientist."[10] The mass media have acquired a habit, deriving equally from fear and laziness, of presenting scientific matters in the most stripped-down terms; and they bail out in terror when any kind of nuance or subtlety threatens to intrude on the *story*. If scientists have acquired a quasi-sacerdotal status in the popular imagination, it is not because they have pressed strongly for such recognition, but because so much of the population finds it more comfortable to declare itself awestruck than to acquire the knowledge that might dissipate the awe.

Like it or not, the responsibility for a remedy to this palpable and increasingly dangerous defect in the foundations of republican existence lies principally with the university. For all its failings, and despite its supposed economic decline relative to the other Western industrial powers, the United States remains unique in its ability to provide a large proportion of its young people with some kind of higher education. In the face of all just criticism of the race and gender biases of this society, the fact remains that a bright and energetic young person who is determined to get a college education can probably get one, no matter how "marginalized" or "disempowered" his or her social background. The campus culture not only shapes the maturation of most middle and upper class youth; it functions in a similar way for a large proportion of the "underclass"—at least those of its young people who manage to get through adolescence without major damage and with some hope left intact.

There is, however, an inevitable corollary to the praiseworthy demographics of American undergraduate education. The majority of students entering college are inadequately prepared for it across the board. In science and mathematics they are as a rule wretchedly prepared, so that undergraduate teaching responsibilities in the sciences involve a great deal of remedial work. Even worse, many students in lower-level science courses are not only ignorant of science but are ignorant as well of the fundamental frame of mind, the attitudes, the intellectual rhythms needed if one is to acquire useful knowledge. Thus, for better or worse, university scientists, as a body, have, in addition to their research goals and their duty to train new generations of scientific professionals, the responsibility of inculcating basic scientific literacy in an enormous number of students who are unprepared, recalcitrant, and skeptical about the whole business. Superb teaching may overcome these obstacles; but excellence of that kind, like excellence of any kind, is rare and appears only fitfully in the population of college teachers.

In the face thereof we now confront the emergence of a new body of criticism of science, one that holds, when we get down to cases, that Western science is in fundamental ways blind or blinkered, that it is corrupted by its subtle bigotry and by its servile accommodation of power, that it is the artifact of a worldview liable, any day now, to be overthrown. The critics, by their stance, their language, and the terms on which they engage the scientific views they criticize, offer another dispensation as well. They declare, in effect, that parallel, even superior ways of knowing—those of feminism or postmodernism or deep racial wisdom—are available for the evaluation of scientific questions, and that from the heights of such alternative epistemologies the newly enlightened can discern the fundamental errors and weaknesses of traditional science—*without* troubling to become well informed about the substance and the inner logic of the scientific enterprise. Thus, with the aid of an unrelenting moralism that cloaks itself in political and social virtue, a moralism seconded with reigning platitudes of activists and cowed, beleaguered college presidents alike, the critics enthrone a doctrine and a methodology for thinking about science that is at once scornful and ignorant.

We have been at pains to say why we think most of this is sheer puffery, and why the residue of genuine insights is negligibly small. Nevertheless, the odor of rectitude, as it is now defined in many areas of university life, can insulate a swarm of silly errors from criticism—indeed, it can silence the critics. Remarking on Robin Fox's critique of the extreme relativism prevalent among ethnographers and cultural anthropologists, Bernard Ortiz de Montellano has this to say:

> I think that Fox is understating the amount of political correctness in certain fields. For example, the American Anthropological Association's last meeting had multiculturalism as a theme and ran a number of inane PC sessions on the topic. However, it rejected a session I proposed on "Multiculturalism and Pseudoanthropology." This session included the dean of Olmec and Maya studies and the only African-American Meso-American archaeologist, critiquing Ivan Van Sertima's idea that Nubians/Egyptians were here at the time of the Olmecs, which is received wisdom among African American circles . . . The foremost authority on Nubia was going to talk on "The Afrocentric Misuse of Nubia"; . . . Eugenie Scott . . . was going to give a case study of how multiculturalism is being used in the Berkeley schools; and I was going to give my talk on melanin. This problem is not just in anthropology. I have found extreme reluctance and avoidance of the topic of the Portland Baseline Essay and Afrocentric pseudoscience generally at the AAAS Education directorate, . . . National Science Foundation Science Education, and even at the National Academy of Sciences Committee on National Science Standards.[11]

As we have noted, the criticism of science that we find so thin and unconvincing has been the making of a number of academic careers. To some degree at least, the attitudes it reflects and encourages have radiated into the general atmosphere of academic life. We encounter increasing numbers of students, graduate as well as undergraduate, whose primary contact with science has been through the work of feminist or cultural constructivist critics, and who are convinced, moreover, that they have imbibed doctrines that are wise (as well as stylish). Even more sadly, we have run into bright and ambitious black students who have become emotionally committed to some of the most risible Afrocentric myths—those of the black Egyptian superscientists, for instance, or of white, government scientists having created HIV secretly, in the laboratory, as a tool for genocide. We know of mathematics departments where the most straightforward pedagogic housekeeping task—that of giving placement exams to insure that students are assigned courses commensurate with their background and ability—is complicated by the insufferable intervention of ideologues, who insist that such tests are inherently "culturally biased" or "gender biased," an intervention whose probable consequence is to make life miserable for the poor undergraduates who are shoehorned, courtesy of their would-be benefactors, into courses they aren't ready to handle.

At the level of primary or secondary education, where even in more halcyon times science education was often a stepchild, matters stand even worse. The inanities of Afrocentric "science" now have free rein in a number of urban, predominantly black school districts. Of course, simple charity urges us to see this as a desperate, if horridly ill-considered, response to a desperate educational and social situation; but that situation is not to be ameliorated by the teaching of nonsense. As Ortiz de Montellano points out, however, many professionals in science education, who should, presumably, see the danger of this situation and take action to defuse it, sit on their hands and even devise relativistic excuses for letting it continue and worsen. It is never easy to estimate how much of this nonresponse is due to ignorance of what is really going on and how much to simple cowardice.

Even more startling, however, is the attitude of school authorities in some upscale, politically "progressive" districts. Eugenie Scott, executive director of the National Center for Science Education, reports (despite the apparent disapproval of the American Anthropological Association) that multicultural antiscientism fulminates in the progressive mecca of Berkeley, California.[12] According to Scott, some Berkeley textbook committees are now trying to bar history and social science books that assert (innocuously, one would think) that Native American populations arrived in this hemisphere from Asia toward the end of the last ice age. Native American myths, they point out, contain no such assertions; why, therefore, should the confabulations of scientists be privileged over the "narratives" that the indigenes tell about themselves? It is ironic that Scott, an anthropologist who has devoted her recent career to fighting the influence of presumably right-wing "scientific creationists," should now find herself trying to ward off similarly appalling nonsense that is backed by a large faction of what now passes for the left. In some ways, it is even more frightening nonsense. For, whereas the well-educated folk of the Berkeley community were eager to take up arms against the intrusion of "creation science" into the schools, and insisted that the facts of evolution be taught without apology, a certain hesitancy gripped many of them when scientific nonsense intruded under the banner of multicultural-ism, convoyed by the new relativism of the postmodernist thinkers.[13]

The Inanition of Public Discourse

If, as seems obvious, scientific and technical issues will become increasingly and urgently relevant to public policy in the decades ahead, how well will such matters be debated in this country? Obviously, we cannot hold high hopes. The historic record of American education in making the general public conversant with basic science has always been poor, except for a brief flurry of serious effort in the post-Sputnik era. Superstition, whether about astrology, ancient astronauts, or alien abductions, has always had easy and profitable going. Fringe medicine and outright quackery, long endemic in American culture, have taken on new and ominous vigor, thanks in part to the dizzily rising costs and increasing impersonality of ordinary health care. The contrast between the incomparable virtuosity of professional American science and the general, public disregard of scientific substance, whether from complacency or hostility, grows ever more pronounced. It is one of the great social paradoxes of history.

Those on the left of the political spectrum are concerned, and rightly so, about the abridgment of democratic procedure and debate inherent in a system that delegates all responsibility for important policy matters to a technocratic elite. These misgivings are manifestations of a significant dilemma. How are such decisions to be made in a manner that takes serious and accurate account of technical and scientific matters without abrogating popular rule or reducing it to a mere symbol? Dozens of pressing issues, from AIDS to alternative energy sources, are complicated by this question. How do we permit a wide public to have a serious voice in such deliberations without inviting in gullibility, ignorance, and mere faddishness—without inviting in the PR operators? The easy answer, of course, is to educate the great mass of citizens in such a way that thinking accurately about science is possible, if not quite second nature. The countervailing obstacle, however, is that widespread ignorance of science in and of itself prevents the development of an educational system that could dispel it.

It is clear that many of the left-wing thinkers whose ideas we find so unsatisfactory are, at bottom, obsessed by the same essential concern as the one we are now trying to address. How do we democratize scientific and technological decision making? How do we give the heretofore powerless some measure of control over the decisions, technological and otherwise, that so profoundly shape their lives? The unfortunate trajectory of academic radicalism has carried it to a position where this question is not so much answered as dispelled by a fog of philosophical conceits. Andrew Ross, Sandra Harding, Simon Schaffer, Stanley Aronowitz, Carolyn Merchant—even Jeremy Rifkin—are all, in their various ways, insisting that the mountain come to Mohammed. In this they are seconded by a squad of educational bureaucrats. Since science seems so difficult, so inaccessible and intimidating, when viewed on its own terms, the radical critics take the daring step of insisting that science can't be what it claims to be, no matter how well it backs up its claims with experiment and applied technology. Outwardly or covertly, they insist on supplanting standard science with other "ways of knowing" that, by their very nature, will be inclusive and welcoming. This is the true agendum of Harding's "strong objectivity," of Ross's insistence that New Agers and others on the far fringes of science will rally us against the omnivorous monster of technoculture.

The generosity of the democratic impulse when conjoined to this mode of thinking is instantly perverted to a kind of inverse intellectual snobbery, a form of coarse populism that is willing to exile the most stringent kinds of analytical thought and jettison the reliable devices of empiricism in the name of opening the doors of knowledge and driving the haughty priests of science from the temple. The theorizing done on behalf of this project is thus a species of incantation, a ritual rather than an argument. It does not conceive the need to examine science closely on the terms set by the logic of science because, in itself, that kind of examination would concede too much to the temperament and mind-set of the scientist. It would require precisely the kind of education, or self-education, that the critics, in the name of some kind of popular participation and empowerment, want desperately to prove superfluous.

A recent address by Carolyn Merchant gives voice to this sentiment in a fashion that reveals clearly the characteristic combination of earnest concern, wistfulness, and outright superciliousness that marks recent leftist criticism of science.[14] Drawing upon the misleading characterizations of "chaos theory" that have controlled public consciousness of this topic, she further distorts those second-hand characterizations in order to view it as a license to abandon the predictive strategies of science, to ignore "computers," in favor of a warm, cozy, inclusive discussion in which all voices—especially those of the formerly disempowered—will now have weight. Chaos theory, she reasons, tells us that the predictive claims of science have been ill founded; so why not abandon, or at least demote them, in the name of communitarian solidarity and radical egalitarianism? Of course, Merchant's daydream ignores the simple reality that, popular images to the contrary, chaos theory as an actual science does not *diminish* the claims of science to provide accurate predictions, but rather *enhances* and extends them substantially. One would have guessed that a Berkeley professor, motivated to take an interest in such matters and having had scientific training, could easily have discovered the facts of the case—they are certainly readily available. But this would not have sorted well with the egalitarian and ultimately pastoral vision Merchant wants to vindicate. She seems to prefer a science that is unsure and a little bit helpless—it would be far more amenable to her ideology.

Back-door utopianism, so characteristic of academic-leftist critics of science, is a sad and, ultimately, a woefully impatient business. Behind it stands a Romantic discontent that echoes perilously certain sentiments that were once recognized as *reactionary*. How much it will add, in the end, to the burden of outright superstition and ignorance that has always plagued the American democratic experiment is difficult to say. It is plain, however, that the underlying disaffection is hostile to enlightenment as such, and not just to the Enlightenment. What is chiefly discouraging about its new ascendancy in academic life is the evidence it provides of a tradition of egalitarianism falling under the sway of obscurantism and muddle. We do not need to convict the paladins of the postmodern left of any particular superstitious foolishness, in the ordinary sense, to notice that they have an appalling tendency to condone such foolishness with a relativist nod and a deconstructionist wink.

Above all, the net effect of all this is to debase still further the already corrupt coinage of public debate. The damage wrought by denatured language is all too apparent. Public health officials struggle to gain the credibility that should rightfully be theirs, and have to fight continually to be heard over a hubbub of voices stridently denouncing the arrogance of Western scientific and materialist paradigms, and offering to replace them with "alternative modes of healing" that promise to make us better faster and cheaper than stodgy old M.D.'s. At the root of it all is an ancient amalgam of quackery and self-delusion, but now, the fashionable shibboleths of postmodern academic discourse have become available to array the old foolishness in up-to-date scholarly language. Discussion of environmental questions is now, at least to some degree, hostage to a rhetorical style and a technique of public relations in which unrelenting ecobabble plays an increasingly peremptory role. The locutions "environment friendly" and "environmentally sane" cover a broad range of styles and practices that, as we have seen, can be neither of those. This, too, is a language fostered in large measure by the peculiar intellectual gamesmanship of the academic left, and it is as often as not employed for purely political purposes.

As we have seen, practical measures for making discussion of scientific issues effectively more democratic by what should be the straightforward process of extending scientific literacy are continually subverted by the intrusion of "identity politics" into the pedagogy of science. In the case of "Afrocentric" science education, the phenomenon is nothing less than garish, although it remains strangely immune to criticism. It is clear that black youngsters who aspire to scientific careers will be in deep trouble if their early education is dominated by the Afrocentrism espoused by Ivan Van Sertima, Hunter Havelin Adams, and their co-workers.

The feminist critique of science is subtler and, superficially, less provocative in style; but it may, ultimately, have even more widely exclusionary results. Young women—or men, for that matter—who try to embark on scientific vocations with the explicit aim of reconstituting science along the lines advocated by Harding, or Haraway, or Keller, or Longino, are on a course leading, we submit, to frustration and disillusion. We are not imagining such young people: we encounter them regularly in our classes!

Science does *not* work the way the critics say it works, and the program of reforms mooted by the critics will turn it into something other, and less than, science. Enthusiastic recruits to the cause of "feminist" science will have to face this contradiction sooner or later—most likely sooner. They may then come to take the view, shared by most women scientists of our acquaintance, that feminism, whatever its strengths as a moral stance and a social program, is not a methodology for doing science: it does *not* offer any privileged insights into scientific questions. That will be their victory. But if they attempt to hold fast to the most emphatic tenets of feminist dogma—for instance, Sandra Harding's assertion that "women" can't be "scientists" under the present order, because society constructs these as mutually exclusive categories, and therefore that scientific practice must be reconstituted along radical-feminist lines before women can participate[15]—they will quickly find themselves effectively excluded from serious scientific work.

On the other hand, many women with scientific talent may not get even that far. They will be discouraged from the outset by the litany of the most prominent critics to the effect that science, as it is practiced, is *innately* antagonistic to women, that it reflects and embodies a system of "patriarchal" domination and "violent" subordination of nature. Thus, to the degree that this sort of thing actually happens, feminism will find itself in the position of frustrating an original, legitimate, and honorable feminist goal: that of augmenting the proportion, as well as the absolute number, of women in science.

We must also note that the left itself—not only the peculiar ideological tribe we have dubbed the "academic left," but the far broader and deeper tradition of egalitarian social criticism that properly deserves such a designation—is, potentially, one of the ironic victims of the doctrinaire science-criticism that has emerged, just as it has long been the victim of the worst kinds of Marxist, Leninist, and Stalinist cant. It is quite legitimate, for instance, to assert that socialist views, as such, have a place in the important debates about environmental questions. Without here endorsing—or rejecting—a socialist point of view, we appreciate that it exists, that it is distinguishable from alternative political visions, and that it has a certain force. The underlying argument is that free-market capitalism, with its enthronement of profit and its tendency to insulate crucial economic decisions from democratic oversight, is, per se, an obstacle to changes we should make in our uses of technology if we are to develop sound environmental practices. In itself, this is a view that can be argued intelligently and that cannot simply be dismissed without specific, and historically informed, criticism. If, however, "eco-socialists" are forced to carry the ideological baggage of the academic left—the relativism of the social constructivists, the sophomoric skepticism of the postmodernists, the incipient Lysenkoism of feminist critics, the millenialism of the radical environmentalists, the racial chauvinism of the Afrocentrists—then they will, in effect, greatly accommodate their opponents, and facilitate the rapid dismissal of their own soundest points, since those will be embedded in a tissue of unscientific and antiscientific nonsense. Scientists, and the scientifically well informed, will simply not accept any form of "socialism" whose agenda include the subversion of legitimate science.

The Role of Scientists

If a jeremiad is to be more than a prolonged lament, it should, by custom, conclude with a call to arms, with a list of actions to be taken by people of goodwill. Thus are the diagnosed evils to be remedied and the disasters foreseen somehow avoided. Our list comes up, however, disconcertingly sparse; and the actions we can recommend are, on the whole, unheroic. We address ourselves chiefly to scientists, engineers, physicians, and other scientifically well-informed people who are members, by actual affiliation or by inclination, of the academic community. What ought they to do, as formal or informal educators, about the bizarre war against scientific thought and practice being waged by the various ideological strands of the academic left?

Obviously—at least we hope that by now it's obvious—we are not calling for a purge of the institutions of higher learning in this country. We don't advocate supplanting one regime of "political correctness" by another, even more odiously high-handed one. Having made that disclaimer, however, we can in good conscience urge that certain forms of vigilance are appropriate, troublesome as they may be to preoccupied teachers and scholars. First and most important is the necessity of seeing to it that whatever is labeled as "science education" in our colleges and universities *deserves* that designation. Science courses must teach science. It's as simple as that. They should have substantive scientific content, validated by perfectly well-known and legitimate modes of scientific inference. As educators, scarred in battle and wearing a few tarnished medals, we have experience of the attempts to label shaky theorizing and tendentious quibbling as "science" for the sake of introducing it into the curriculum.

"Creation science" is an example that most of us are familiar with, although institutions of higher education (outside of sectarian colleges of dubious legitimacy) were rarely the targets of that campaign. We have also the example of various New Age confections peddled at some community colleges and in extension programs, under the pretext that they represent "science" of some kind. But the influence of the academic left is of a different kind, since it is seconded by the support, often enthusiastic, of many established senior members of the academic community. To date, it has merely nibbled at the fringes of the "hard" sciences, although, as we have seen, such heretofore honorable fields as anthropology and psychology have been gravely contaminated. We cannot help feeling, however, that there will be many more calls for "feminist" courses in biology and "Afrocentric" courses in mathematics. How much force such campaigns will have is hard to predict. In any event scientists and science educators must, on their professional honor, be prepared to resist the insertion into the science curriculum of courses whose content is tailored to the demands of *any* ideological faction. It will be alleged, of course, that conventional science, like all knowledge, is inevitably "ideological," and that the proposed "reforms" are therefore just as legitimate and considerably more self-aware than the traditional kinds of scientific education. This contention is, however, simply *wrong*, as we hope we have at least suggested: in fact, we think it comes very near to being nonsense. Scientists ought to reject it out of hand. This is not really a matter that demands intellectual acuity—the theories of the left-academic critics are not, on the intellectual plane, particularly intimidating. Rather, it is a question of not letting one's social ideals, which may well find much to admire in feminism and in the quest for racial justice, overrule one's professional judgment and simple common sense. Unthinking sentimentality, be it remembered, is the great enemy of genuine compassion.

Beyond this, there is the matter of courses, seminars, symposia and the like, that claim to address scientific matters while falling outside the official boundaries of science departments. We urge scientific professionals to scrutinize these offerings, whether or not invited to do so, to participate in them if possible, and with appropriate skepticism. We urge them not to fear making judgments, not to hesitate, for the sake of someone else's imagined good social intentions, to make their misgivings public. The academic left, after all, is fiercely judgmental and highly vociferous, though all the while it is eager to denounce judgmental behavior in its opponents. Moreover, some of the instinctively deferential habits of academics will simply have to be put aside. One can't assume, in these matters, that possession of an advanced degree or a professorship equates to intellectual legitimacy. Most of this book has been devoted to a critique of work done by academics whose nominal credentials are quite impressive. That has not prevented them from propounding wrongheaded, even fatuous, theories about matters in which their knowledge ranges from shallow to nonexistent. This is a disconcerting fact of contemporary academic life. It should stimulate energetic action and argument, not resignation or quietism. It should be as strong a goad to attending faculty meetings, and paying attention to curricular proposals, as the threat of state-mandated program cuts or of new parking regulations.

In our experience, scientists who try to engage the radical critics of science in direct debate are in for a frustrating time of it. This is not because their foes are expert rhetoricians. In fact the critics of science are decidedly reluctant to contend with actual scientists in the flesh. They prefer a hermetic, self-referential atmosphere for the promulgation of their ideas; often they appear to regard the presence of well-informed scientists, unless invited for specific "technical" contributions, as intrusive. This is a characteristic of the sectarian mentality: it usually implies a skittishness on the part of the critics, one that may derive from inward knowledge that their theories rest as much on wishful thinking as on learning. In any case scientists oughtn't to be reluctant to stick their two cents in. They should insist—always within the bounds of courtesy, of course—on being included in debates and presentations that center on science and the relations between science and culture. If they are nevertheless excluded, as will sometimes happen, they should be prepared to make a bit of noise about it. They will be on very firm ground in today's academy, where few proceedings are allowed to proceed without proper representation of the affected class, and where noise is no longer taken, as it used to be, as evidence of an unfortunate class origin.

We realize, of course, that the greatest disincentive to participation in such controversies is the time and effort it takes, costs that will add greatly to the burden of sustaining a serious research program and meeting one's instructional and professional responsibilities. Intellectually, these quarrels tend to be tiresome. Nature is the scientist's worthy adversary (we use the figure in defiance of the fact that science critics will sniff it out as evidence that we are slaves to the Western-patriarchal paradigm of dominance and control). Academic leftists, on the other hand, tend to be unfocused bores, and a certain deliberate, cheerful simple-mindedness is needed to hear them out sufficiently to catch the drift of the arguments and to formulate an apposite response. It is an unlikable chore, but one that a good many of us ought to be doing, out of loyalty to our own disciplines and to—forgive the pretentiousness of the word—civilization.

Finally, there is the unpleasant, but serious, question of careerism, which, conjoined to ideological enthusiasm, has been responsible for much of the misbegotten scholarship we have considered. It has been traditional, in the academy, for the humanists to let the scientists do their own hiring, firing, and promotion, and for the scientists to reciprocate. If representatives from one camp are obliged by local custom to rule on the other's actions in such matters (as in the case of tenure committees), they usually assent pro forma, without making any inquiries beyond the superficial ones, for example, to ascertain that the letters of recommendation and so forth are adequately fulsome or lack, at least, the hint of bones rattling in closets. The proliferation of would-be science critics and epistemologists among left-wing humanists raises serious questions about this cozy arrangement. If an aspiring scholar is to be judged on work affecting to make deep pronouncements on questions of science, scientific methodology, history of science, or the very legitimacy of science, it strikes us that scientists should have some say in evaluating it. This holds even if the candidate resides academically in the English department or the art department or the sociology department. It will be objected that scientists, as a hermetic, self-protective guild, ought not to sit in judgment of those who are studying *them*. But academic leftists, postmodernists, deconstructionists, and the like have *their* self-protective guilds, and experience shows that they are not at all reluctant to rally round their own. Elementary fairness requires that a broader spectrum of opinion should be brought into the process. If an assistant professor of English is to stake his bid for tenure on work that, for example, purports to analyze quantum mechanics as an ideological construct, then he has no right to complain if a professor of physics is brought into the evaluation to say whether he evidences any real understanding of quantum mechanics. More broadly, if scientists perceive that a spate of nonsense concerning science has been coming out of the mouths of their young humanist colleagues, then they have the right to raise questions about the mechanisms that give a fair wind to such shaky scholarship.

It will be argued immediately that this is an asymmetric, and therefore inequitable, proposition. If physicists are to judge scholars of English, why shouldn't English professors judge physicists? The fallacy here is that the asymmetry originates from the pretensions, legitimate or otherwise, of members of the English (or sociology or cultural studies or women's studies or African-American studies) department to qualification on scientific questions. If, say, a member of the mathematics department were to engage in the (most unlikely) scholarly project of analyzing the rhetorical and stylistic elements of certain mathematics papers, it would be entirely legitimate for literary scholars to pronounce judgment on the work, and for the promotion process to take that judgment fully into account. To put it bluntly, it is humanists of the academic left who have transgressed the boundaries—as they are eager in most circumstances to proclaim. That's their privilege; but they are not (or should not be) exempt from customs duties!

Finally, there is the question of reconquering lost territory for the scientific approach. As we have noted, some fields, long recognized as scientific in principle, have fallen victim to antiscientific relativism. Anthropology is one example; other, partial examples could be found within psychology and sociology. Plainly, there is no direct way to enjoin our counterparts in these fields to abandon the pleasures of subjective narrativity for the fuddy-duddy rigors of empirical and statistical research. Still, "hard" scientists should find some way of supporting those of their colleagues in these areas who are willing to honor the principle that the right to make knowledge claims, in a university, has to be earned by the methodologically sound sweat of one's brow. It's fine to argue about competing methodologies; it is not fine to congratulate oneself on having abandoned method.

We must conclude on a note of melancholy. At this point in history, for anyone who has read it honestly, the status of science as a reliable, profound, and productive source of knowledge ought to be beyond serious question. That vague but grandiloquent challenges nevertheless recur incessantly remains, after all our attempts to understand, a source of sad perplexity. That many of these challenges now issue from a community that consists, regardless of ideology, of people who have presumably enjoyed a first-class education and who have, all their adult lives, played a central role in the larger intellectual world deepens our misgivings. We would have been much happier if this book had been unnecessary. We may be misguided; we may have made mistakes; our erudition may be more deficient than we already know: but we are not dishonest. Honestly, then, we care deeply about our students, and honestly we treasure that collegial life of the mind which no external insult—not of age nor loss nor straitened circumstances—has in our time been able to diminish. For us to believe that a book of this kind is needed means at very least that, in making our inquiries and absorbing a large and distressing literature, we have had to abandon the complacent feeling that the republic of intellectual inquiry is secure from internal decay.

Finally, and with an ironic nod to Andrew Ross, we would like to acknowledge with deepest gratitude the gifts of all the science teachers we have had throughout our lives. That includes colleagues, senior and junior, and, most important, some of our students. This book could not have been conceived, let alone written, without them.

Notes

1. The Academic Left and Science

1. This is not to suggest for a moment that our subjects would spurn the opportunity to manipulate mass consciousness, if they were to get it. All too many of the claims of the academic left flow from the wish to dominate public discussion.

2. One fine distinction we do *not* want to make is that between *postmodernism* and *poststructuralism*. The former is the more inclusive term, implying as it does a range of stylistic attitudes and judgments that reject "modernism" (including much of the Enlightenment), as well as the specifically philosophical positions associated with "structuralism" and its presumably triumphant successor, "poststructuralism." No writer we identify here as a postmodernist is done an injustice thereby. We were tempted, but only tempted, to rely upon Patrick Scrivenor's definitions, which begin: "POSTMODERN: *adjective*. So new it's out of date already" (in *Egg on Your Interface: A Dictionary of Modern Nonsense*).

3. This is not to suggest that other sorts of institutions are more lovable. Indeed, if institutions were to be ranked on a scale of delinquency, the most at the top, universities would fall far below governments and multinational corporations.

4. It would be a mistake to believe that conservative thinkers are no longer generating nonsensical opinions about science. A recent Op-Ed piece in the *New York Times* by historian John Lukacs ("Atom Smasher is Super Nonsense," June 16, 1993) is as ignorant and presumptuous as any of the writings by nominal leftists that we examine in this book.

5. There are honorable thinkers on the left—and elsewhere—who advocate strongly a multicultural approach (e.g., to education) that looks toward the development of fruitful interchange between the various ethnic and historical strands woven into our culture. (See for instance, Robert Hughes, *Culture of Complaint*.) They are as appalled as are we by the chauvinism and blindness to fact of the Afrocentric movement. In the academic trenches, however, the call for multiculturalism is largely answered by a proliferation of increasingly intransigent and insular Afrocentric curricula.

2. Some History and Politics

1. See, e.g., Peter Gay, *The Party of Humanity*.

2. For an incisive essay on Maistre and his foreshadowing of modern thought, see Isaiah Berlin, *The Crooked Timber of Humanity*.

3. See Leo Loubere, *Utopian Socialism*.

4. Timothy Ferris, "The Case against Science," 17. (The title of Ferris's piece refers not to its own content, but to that of a book under review, Brian Appleyard's frankly reactionary *Understanding the Present*). See also Ian Hacking, "His Father the Engineer," for an equally skeptical view of the same work. Appleyard at least has the virtue of wearing his nostalgic social and religious views on his sleeve. The curious thing about so much of the "left-wing" critique of science is its unconscious and unwitting echo of this kind of right-wing grousing.

5. See Philip S. Foner, *History of the Labor Movement in the United States*, vols. 3 and 4.

6. Robert K. Murray, *Red Scare*.

7. For an exhaustive history of the American Communist party during the Depression, see Harvey Klehr, *The Heyday of American Communism*. For a more personal view of the lives and motivations of American Communists, see Vivian Gornick, *The Romance of American Communism*.

8. For an account of the Communist party and its intellectual critics, from the Hitler-Stalin pact to the 1950s, see William L. O'Neill, *A Better World*.

9. Victor S. Navasky, *Naming Names*.

10. Taylor Branch, *Parting the Waters*.

11. See Harold Fromm, "Scholarship, Politics, and the MLA," for a shrewd assessment of the current state of affairs.

12. "Traditionalists," of course, see the spectrum of opinion offered by the media through a different lens. From their perspective, the right end of it has been eliminated. The bias they see is opposite to the one we describe. For example, in an angry letter to the *New Republic*, in which an unnecessarily snide account has been given of a recent cultural conference organized by Patrick Buchanan, John Zmirak writes: "Technology is on our side, shattering the hegemony of the Three Networks, the One Newspaper and the public schools. All three institutions are hopelessly discredited outside of a few square miles in Washington, New York and Los Angeles" (pp. 4–5).

13. A woman scientist friend, who is also a feminist interested in our argument, took umbrage at "academic left," describing herself as one of those, yet sharing our negative view of the science critiques we examine here. She undertook to help us out of the terminological trap; after much thought, she proposed the signifier "New Rage Academics." We demurred.

14. Vincent P. Pecora, "What Was Deconstruction?"

15. *Lingua Franca*, September–October 1992, 21.

3. The Cultural Construction of Cultural Constructivism

1. Stephen J. Gould, "The Confusion over Evolution."

2. We have already mentioned Bryan Appleyard's *Understanding the Present* (note 4 for Chapter 2). Ian Hacking's review, "His Father, the Engineer," observes that "traditionalist" views like this flourish in the same soil as postmodern or cultural constructivist arguments. Both genres address a constituency poorly educated in science and, for that reason, already inclined to view it with hostility.

3. Harvie Ferguson, *The Science of Pleasure*, 238.

4. Or, to put it otherwise and as is demonstrated by example in Chapter 5, where we take up feminist anti-science: constructivist critical thought has neither identified major flaws in existing, specific science nor provided better alternatives to existing scientific interpretations.

5. See Alan Chalmers's excellent book *Science and Its Fabrication* for an extensive rejoinder to the constructivist viewpoint, temperate in tone and sympathetic to the political motives of constructivist theorists but adamant in insisting on the shallow and unconvincing nature of cultural constructivism in general and in its most vaunted examples.

6. Paul Feyerabend, "Atoms and Consciousness."

7. For the record, one of us is also a member of that organization.

8. Here are some names that go *unmentioned* in this "history": Poincaré, Planck (?!), Weyl, Schrödinger (??!!), deBroglie, von Neumann, Pauli. Dirac's name flashes past briefly in another chapter (p. 119), but the context for it is a paraphrase of the views of a Marxist philosopher (who is himself paraphrasing). This is not the kind of history of quantum mechanics where the intriguing role of noncommuting self-adjoint operators on complex Hilbert space gets much of an airing.

9. The assertion that deterministic causality is still viable within the phenomenal world of quantum mechanics may come as a surprise to people far better informed than Aronowitz, but recent work in the mathematical foundations of the subject seems to support it strongly. A beautiful result of Dürr, Goldstein, and Zanghi shows that a large part of classical quantum theory emerges naturally from a dynamical model which is deterministic through and through, and in which the only "hidden" variables are thoroughly classical variables like position and velocity. See their "Quantum Equilibrium and the Origin of Absolute Uncertainty" and "Quantum Mechanics, Randomness, and Deterministic Reality" (a short exposition of these results without mathematical details). The model is so simple that we cannot resist the temptation to summarize it for readers with a little knowledge of mathematics and physics:

Consider a universe consisting of N particles the set of whose possible respective coordinates form a configuration space (modeled on 3N-dimensional Euclidean space). Let q_k denote the position of the kth particle (as a triple of local coordinates) and m_k its mass. Thus $\mathbf{q} = (q_1, \ldots, q_N)$ is the configuration vector. We assume as well a complex-valued wave function, Ψ, defined on the configuration space. The dynamics are then given by the familiar Schrödinger equation

$$i\hbar \partial\Psi/\partial t = -(\hbar^2/2)\Delta\Psi + V\Psi$$

(where V denotes a potential energy function) together with the ordinary differential equation

$$dq/dt = \hbar Im(grad\Psi/\Psi)$$

(Here the Laplacian Δ and the gradient are given in terms of a Riemannian metric scaled by the masses of the particles.) The mathematically literate reader will readily see that, given an initial Ψ_0 at one particular time, Ψ evolves purely deterministically (as usual), and thus, with initial values for q as well, the whole system evolves deterministically. As it turns out, however, with modest and unproblematic assumptions on the initial values of Ψ_0 and $q(o)$, "small" subsystems consisting of "small" numbers of particles will behave so that, statistically speaking, the standard quantum mechanical formalism applies.

This work amplifies and makes rigorous rather old ideas of the philosopher/physicist David Bohm, vindicating him and bluntly contradicting the "Copenhagen interpretation" of Bohr and Heisenberg. Aronowitz mentions Bohm briefly and skeptically (p. 17) but seems unaware that Bohm's program is not just philosophical but involves a specific strategy, now seen to be quite fruitful, for doing mathematical physics.

We emphasize, of course, that this particular work only rederives the classical (i.e., nonrelativistic) quantum mechanics of a collection of point-masses in a potential field. The philosophical point is strongly clear, however, since the standard debates use this case as a touchstone. The moral, if one must be whimsical, is that Occam's Razor may now cut through the leash that heretofore bound us to Schrödinger's Cat, that half-and-half beast weirder by far than Centaur, Sphinx, and Hippogriff. In other words: (a) Einstein was right: God does not play dice with the cosmos. (b) On the other hand, since individually and collectively we are not God, nor Laplace's Demon, nor any other demiurge of comparable intellectual power, we must inevitably regard the universe as something of a crap game. As far as (b) is concerned, note well that even if the universe were purely Newtonian in the best eighteenth-century tradition, and the perplexities of quantum mechanics entirely avoided, the crap game would still be inevitable. This follows from the work of Poincaré on classical mechanics and from that of his latter-day disciples, which goes under the fashionable name "chaos theory."

10. Stanley Aronowitz, *Science as Power*, 346.

11. Aronowitz, "Science under Capitalism."

12. Foremost in Einstein's mind when he was developing the ideas that matured as the theory of special relativity were certain anomalies in the mathematics of the Maxwell equations for electromagnetic radiation. The Maxwell equations, unlike Newton's principles of mechanics, are not invariant under "Newtonian" change of coordinates, but rather under what is called Lorentz transformation, a situation that bewildered late-nineteenth-century physicists. If the fundamental correctness of the Newtonian picture was assumed, then the very experimental success of Maxwell's equations could be accounted for only under the assumption that they were invariant under Newtonian transformation! Einstein's staggeringly brave and brilliant solution to the mystery was to insist on recasting mechanics so that its laws become Lorentz invariant. The entirely counterintuitive result is that quantities like length, mass, and duration lose their absolute character. See Ronald Clark, *Einstein: The Life and Times*, and Gerald Holton, *The Thematic Origins of Scientific Thought*, for complete discus-

sions of Einstein's relation to the Michaelson-Morley experiment.

13. Michael Sprinker, "The Royal Road," 138.

14. See Hans Eysenck, *Decline and Fall of the Freudian Empire*, for an accurate and unrelenting critique of the claims of Freudianism and its variants to scientific status as either theory or clinical practice.

15. Among the Gould collections that exhibit this analytic sanity through most of their length we recommend: *The Mismeasure of Man*, *The Panda's Thumb*, and *Bully for Brontosaurus*.

16. Thomas Kuhn, *The Structure of Scientific Revolutions*.

17. For further examples of such distortion, see Paul R. Gross, "On the Gendering of Science."

18. Bruno Latour, *Science in Action*, 99 and 258.

19. Ibid., 186ff.

20. We assert—not that there is any great intellectual accomplishment involved—that our argument was invented independently, since we devised it before becoming acquainted with Latour's views on the matter. We are puzzled, however, about why Latour thinks it necessary or instructive to resort to flow-chart-like diagrams (ibid., 193–94) in making such a simple point. May we suggest that this is an example of showmanship, rather than of profundity?

21. Ibid., 209, 259.

22. Latour, *Science in Action*, 237ff.

23. Ibid., 90. The reference here is to the mathematician Georg Cantor's discovery that the (infinite) cardinality of a one-dimensional line segment is equal to that of a two-dimensional square. Latour cites Cantor's comment—"I see it but I don't believe it!"—on the note about this theorem that he sent to Dedekind, the mathematician who initiated the rigorous study of the real continuum. This remark has to be understood as we understand the same exclamation coming from a sportscaster; but in this case, it is an instance of sly self-praise. Cantor is celebrating his own cleverness in coming to such a "counterintuitive" conclusion. Latour regards this result as "scarcely conceivable." On the contrary, it is, once one has caught the spirit of the game, a pretty easy result to grasp, and may easily be taught to undergraduates—or even bright high school students—just beginning the study of rigorous mathematics. One of us, in fact, has just done that very thing. See Karel Hrbacek and Thomas Jech, *Introduction to Set Theory*, 121 (Theorem 2.3).

A more deeply counterintuitive result was produced a few years afterward by Peano, who showed that there is a *continuous* path that passes through every point of the square at least once—the so-called Peano Space-Filling Curve. This is a somewhat more tricky example and requires a bit more background to understand.

24. Ibid., 85.

25. Ibid., 237–39. "The Reynolds number . . . works only as long as there are hundreds of hydraulic engineers working on turbulences." Latour seems to be saying here that the abstract and general formulation of fluid dynamics in which Reynolds numbers appear, and in which their general applicability is made logically clear, is just so much window dressing that disguises a widespread social convention and the politicking that leads to its acceptance. On the other hand—pinpoint clarity is not

one of Latour's virtues—he may be saying that in a society that paid little attention to fluid mechanics, Reynolds numbers would not be considered important. The latter does not seem to be a particularly choice insight. Could he *really* mean that if all hydraulic engineers went on strike, baseballs, boats, and bullets would start to behave differently?

26. Ibid., 242. The reader of *Science in Action* might find it instructive to measure Latour's snide aphorisms against some of the better-known examples of prediction arising from intensive engagement with deep abstract models and their logical consequences. Halley's prediction of the reappearance of the now-eponymous comet is perhaps universally familiar; but we also have, for example, the staggeringly counterintuitive predictions of special and general relativity, the unbelievable precision of the quantum electrodynamics predictions that arose from the work of Feynman and Schwinger, the uncanny correctness of the "correlation at a distance" prediction that followed from John Bell's analysis of the so-called Einstein-Podolsky-Rosen paradox. Effrontery is far too mild a word for Latour's wisecracks on this point.

27. Steven Shapin and Simon Schaffer, *Leviathan and the Air Pump*, 283.

28. Ibid., 301.

29. Ibid., 344.

30. Ibid., 139.

31. Shapin and Schaffer identify Spinoza along with Hobbes as an opponent of Boyle's particular views of the nature of air, as well as a builder of a priori rationalistic systems, and thus an opponent of "experimental" philosophy. Spinoza clearly was an a priorist—and much better than Hobbes as a pure dialectician in this vein. On the other hand, as a leading expert on the science of optics—and a professional lens-grinder—Spinoza must also count as an "experimentalist." Still, given Spinoza's "outcast" status, as well as his subversive views on religion, he should have been cast further into the outer darkness even than poor old Hobbes! Yet he was closely associated with Henry Oldenburg, the Royal Society's inaugural secretary.

32. See Andrew Hodges, *Alan Turing*.

33. These problems—constructing a square equal in area to a given circle or a cube of volume twice that of a given cube using only the ideal compass and straightedge of synthetic Euclidean geometry—go back to classical antiquity. By means of deep algebraic arguments, they were shown, in the nineteenth century, to be insoluble.

34. Shapin and Schaffer, *Leviathan and the Air Pump*, 135. Also cited (100) is Hobbes's question, "For who is so stupid, as both to mistake in geometry, and also to persist in it, when another detects his error to him?" but not the ironic answer, "Thomas Hobbes!"

35. In taking this position, we contradict not only Shapin and Schaffer, but some later work on the same subject from a similar social-determinist perspective. "Why was Hobbes excluded from membership in the Royal Society?" asks James R. Jacob in "The Political Economy of Science in Seventeenth-Century England," 532n. "Quite simply," he answers, "we can now see that the social and political views of leaders of the Society like Boyle and Wilkins diverged sharply from those of Hobbes in certain fundamental respects." Again, out of eagerness to derive everything from "politics," a scholar has overlooked the contrast between Hobbes's high opinion of his own mathe-

matical abilities and his manifest mathematical incompetence. Hobbes's contemporaries, however, were not deceived on this point.

36. See, for example, Marc Bloch's justly celebrated essay "The Advent and Triumph of the Water Mill," reprinted in *Land and Work in Medieval Europe*.

37. Gar Alperovitz, *Atomic Diplomacy*.

4. The Realm of Idle Phrases

1. John Patrick Diggins, *The Rise and Fall of the American Left*, 356–57.

2. Ibid., 363.

3. Heather MacDonald, "The Ascendancy of Theor-ese," 360.

4. Diggins, *The Rise and Fall of the American Left*, 356.

5. See David Lehman's fine book, *Signs of the Times*, for a devastating account of de Man's character.

6. See Thomas Sheehan, "A Normal Nazi," 30–35, in the *New York Review of Books*, for a discussion of Heidegger's zeal as a Nazi, including a brief but telling discussion of Derrida's machinations in his attempt to occlude the obvious. See also the subsequent exchange in the same journal (Feb. 11, 1993, and Mar. 25, 1993), of which Derrida's own letters form the unintentionally comic centerpiece. A continuation of the exchange (Apr. 22, 1993) by a posse of Derrida's American camp-followers, including such stalwarts of the academic left as Judith Butler, Jonathan Culler, Stanley Fish, Gerald Graff, Fredric Jameson, J. Hillis Miller, Joan Scott, and Hayden White, adds immeasurably, if inadvertently, to the high drollery.

7. See Lehman, *Signs of the Times*, for an account of the attempts—remarkable for their moral and intellectual opacity—of Derrida and deconstructionist faithful to defend de Man. See also Heather MacDonald, "The Holocaust as Text," for a merciless but well-earned scourging of de Man's student, Shoshanna Felman, for her bizarre attempt to get her mentor off the hook.

8. Foucault's most influential work is probably to be found in his historical studies: *Madness and Civilization, The Birth of the Clinic, Discipline and Punish*, and *The History of Sexuality*. It goes without saying that despite Foucault's celebrity, many professional historians are highly critical of his selective and impressionistic methodology. Many of his admirers now concede that his histories are probably best regarded as "novels" of a singular type, works which, for better or worse, say more about their author than about the truths of history.

9. George Steiner, *In Bluebeard's Castle*, 131.

10. Jacques Derrida, "Structure, Sign, and Play in the Discourse of the Human Sciences," 267. (This citation appears in Ernest Gallo's short article, "Nature Faking in the Humanities," to which we are much indebted.) Note that Derrida follows this up with a peroration on algebra that is as opaque as it is silly.

11. Jacques Derrida, *Acts of Literature*, 208. We cannot resist the impulse to point out that in Derrida's usage the word *topology* seems to be virtually synonymous with *topography*—at least the index regards them as identical. This recollects an experience of one of us (N.L.) at the age of eighteen. When being interviewed by an insurance executive for a summer actuarial job he was asked: "What kind of mathematics are you

interested in?" "Topology," he replied. "Well, we don't have too much interest in topography," said the insurance man. Obviously a deconstructionist *avant la lettre*.

Defenders of deconstruction and other poststructuralist critical modalities will no doubt wish to point out that *topos* (pl.: *topoi*) is a recognized term within literary theory for a rhetorical or narrative theme, figure, gesture, or archetype, and that therefore it is permissible, without asking leave of the mathematical community, to deploy *topology* to designate the analysis of textual *topoi*. One's suspicions are reignited, however, when the term *differential topology* suddenly appears. (In mathematics, *differential topology* is used to denote the study of the topological aspects of objects called "differential (or smooth) manifolds," which are, roughly speaking, higher-dimensional analogues of surfaces in three-dimensional space.)

When we first encountered this usage in literary theory, we guessed that sooner or later some ambitious but naive young scholar would conflate the mathematical sense of the term *topology* with its meaning within postmodern theory. Second thoughts, however, inclined us to believe that no literary scholar, no matter how impressionable or callow, could fall into such error. We were right the first time. See note 12.

12. Tim Dean, "The Psychoanalysis of AIDS," 107–86. Dean is a graduate student in English. His guru is the psychoanalyst Jacques Lacan. Lacan is cited as declaring that topology is a conception "without which all the phenomena produced in our domain would be indistinguishable and meaningless." Since the citation is from a piece of Lacan's called "Desire and the Interpretation of Desire in *Hamlet*," it is pretty clear that someone has been getting literary *topoi* mixed up with the subject matter of mathematical topology. Whether the confusion is Lacan's or Dean's, we can't say. If Dean's, it is probably free of arrant fakery; but on the other hand, it then crosses the line from mere confusion into the realm of actual stupidity.

Dean's propensity for inadvertently comical confusion of categories is underlined by the associated footnote, which asks: "What are we to make of the fact that the development of topological science is historically coincident with the emergence both of 'the homosexual' as a discrete ontological identity and of psychoanalysis?" (It is also, we note, coincident with the era in which people played football without a helmet. Would it be too cruel of us to wonder whether Dean is overly fond of some similar activity?) For the record, non-Euclidean geometry involves changing what are usually called the *metric* properties of space. The topological properties, and thus topological mappings, remain unaffected. Also for the record, Dean's piece appears in *October*, which is published at the Massachusetts Institute of Technology. We doubt that it is vetted by the mathematicians of that magnificent institution.

13. Jean Baudrillard, *Le xerox et l'infiniti*.

14. J.-F. Lyotard, *The Postmodern Condition*, 45.

15. Alexander J. Argyros, *A Blessed Rage for Order*, 234. Argyros himself is concerned with the philosophical implications of contemporary mathematics and science, in relation to the postmodern ideological positions promulgated by Derrida and others. He is critical of many of these positions, although he finds some useful. He is also slightly guilty of bluffing his way through mathematical points. In his discussion of Lyotard, for example, he states: "Lyotard has correctly diagnosed the failing prestige of

linear, or continuously differentiable, functions." Of course, a mathematical function
—linear, continuously differentiable, or otherwise—is not the kind of entity that
carries prestige, although a mathematician may be. In any event, the mathematics of
continuously differentiable functions, along with those mathematicians who work on
it, is in no particular danger of suffering a decline in prestige. More important, Argyros
seems to think that continuously differentiable functions are, in general, linear,
which is grossly untrue.

What is really involved in this remark is Argyros's naive acceptance of Lyotard's
naive enthusiasm for what is called "applied catastrophe theory." This is a point of
view, advocated by such mathematicians as René Thom and Christopher Zeeman, for
making mathematical models of phenomena in physics, chemistry, biology, and even
economics, which exhibit sharp "jumps" or discontinuities. Although it is based on
beautiful mathematics concerning the topology of function spaces (spaces of continuously differentiable functions, as it happens), applied catastrophe theory has not, over
the years, provided empirical scientists much help in the way of insight or technique.
There have been stringent criticisms of the approach from other mathematicians.
(See R. S. Zahler and H. J. Sussmann, "Applied Catastrophe Theory.")

16. J. Crary and S. Kwinter, ZONE 6: Incorporations.

17. Here is a brief bill of particulars concerning the mathematical sins of the cited
authors of pieces in ZONE 6: Incorporations:

Deleuze ("Mediators," 283): Gives an utterly incoherent, essentially meaningless
account of the notion of "Riemannian manifold" (Riemannian "space" in the text) in
an attempt to make it relevant to film criticism.

Simondon ("The Genesis of the Individual," 301ff): Tries to use the notion of
"metastable" point of equilibrium without much in the way of an honest definition,
the point being to find some philosophical resolution to the problem of "wave-
particle" duality in quantum mechanics. This reads like so much metaphysical hand-
waving and doesn't make any particular sense in terms of mathematics or physics.

Eisenman ("Unfolding Events," 425): Gives an utterly incoherent account of the
"seven catastrophe theorem" of René Thom. Presumably, the idea is to exploit rhetor-
ically the fact that "unfolding" is a technical term in the mathematics of "catastrophe
theory."

Stone ("Virtual Systems," 614): Misrepresents both the chronology and the content of the Heisenberg uncertainty principle in order to create a strained metaphor.

Turner ("Biology and Beauty," 412): Gives a misleading paraphrase of the Gödel
incompleteness theorem (a common sin of amateur logicians—and cultural theorists).

De Landa ("Non-Organic Life," 137): Correctly points out that a minimum of a
potential function is a dynamical "attractor" (of the simplest sort) but then (p. 163,
note 27) claims, essentially, that such a situation does not represent an attractor. In
the latter discussion, there is some muddle about "linear" versus "nonlinear" mathematics. Nevertheless, de Landa's paper, which is essentially a piece of scientific
journalism pointing out the insights into empirical science provided or promised by
recent developments in mathematics, is pretty clear and straightforward, by no means

a typical spate of postmodern hyperbole. For a nonexpert, de Landa has done a good and honest job, although one might wish for a more careful delineation of how much of this is really speculative.

18. Robin Fox, "Anthropology and the 'Teddy Bear' Picnic," 51–52.

19. Ibid., 53.

20. Ibid., 49.

21. Alan Ryan, "Princeton Diary." See also Ryan's illuminating, judicious, and fair-minded article "Foucault's Life and Hard Times."

22. Frank Lentricchia, "Reading Foucault—II," 57.

23. Ibid., 51.

24. Bogdan Denitch, address to Socialist Scholars Conference.

25. Argyros, A Blessed Rage for Order, 9.

26. Elizabeth Wilson, "The Postmodern Chameleon," 187.

27. Kate Soper, "Postmodernism, Subjectivity, and the Question of Value," 128.

28. Kate Ellis, "Stories without Endings."

29. Ibid., 47.

30. Vincent Pecora, "What Was Deconstruction?" 65.

31. Stanislav Andreski, Social Sciences as Sorcery, 129–30.

32. An example of the hostility of some social scientists to the authority of the "exact" sciences is to be found in the quixotic embrace of the notorious crank Immanuel Velikovsky by a few notable social theorists. Velikovsky (Worlds in Collision) held that various mythical catastrophes, e.g., the Noachian flood, really represented folk memories of a time when the planet Venus was born from the gas-giant Jupiter and initially careened around the solar system before settling into its present (near-circular!) orbit. His defenders among the sociologists (see De Grazia, 1978) were motivated by a desire to reveal "hard" scientists as closed-minded and intolerant, and to display a case where historical and ethnographic studies ostensibly revealed an astronomical truth denied to mere astronomers. This, we submit, was an important harbinger of postmodern relativism and antiscientism.

33. See J. W. Grove, "The Intellectual Revolt against Science," for an interesting account of this phenomenon.

34. See, for example, the right-wing libertarian journal Critical Review, vol. 5, no. 2, 1992, a special issue devoted to postmodernism.

35. See the profile of the "cultural studies" phenomenon at the NILS, and of Andrew Ross in particular, in Anne Matthews, "Deciphering Victorian Underwear."

36. Andrew Ross, "New Age Technicultures," 535.

37. Andrew Ross, Strange Weather, 29.

38. Ibid., 60.

39. David Ruelle, Chance and Chaos. See also Ivar Ekland, Mathematics and the Unexpected, for an equally intelligible treatment (by a first-rate mathematician) of much the same material. On the whole, Ruelle is probably preferable to Ekland because of its tone (Ruelle is urbane where Ekland is somewhat histrionic in a way we judge to be slightly misleading), completeness (Ekland scarcely mentions quantum mechanics, which is quite important in any historical and philosophical discussion of the physical significance of chaos theory; Ruelle treats the matter thoroughly), and

because of Ruelle's record of having been "present at the creation" of chaos theory. This last gives added weight to Ruelle's sensible advice to keep matters in proportion when tempted to proclaim the chaos revolution. Ruelle's brief description of the typical mathematics seminar is also a gem of sociological observation. On the other hand, Ekland concludes his book, oddly enough, with a beautiful chapter of literary criticism—on Homer, to be precise—which, in the context, is neither inappropriate nor presumptuous.

40. Harmke Kamminga, "What Is This Thing Called Chaos?"

41. For a typical, relatively innocuous example of this sort of thing see the concluding pages of Alan Beyerchen's article "What We Now Know about Nazism and Science." Other examples abound.

42. Steven Best, "Chaos and Entropy," 188.

43. Ibid., 204.

44. See Stephen J. Gould, "Integrity and Mr. Rifkin." This is a scathing analysis of Rifkin's pretentiousness and incompetence, by a distinguished scientist and educator whose pro-environmental and politically left-wing views are well known. We shall have more to say on Rifkin in another chapter.

45. Curiously enough, something similar happens in a recent paper of the distinguished economic historian and theorist of Third World development Immanuel Wallerstein ("The TimeSpace of World-Systems Analysis," section 4, pp. 17–20). Wallerstein is under the impression that somehow or other, quantum mechanics and nonlinear topological dynamics have overthrown the claims of the physical sciences to yield reliable knowledge about the world, and that physicists are frantically throwing paradigms overboard. (Apparently he hasn't heard the contrasting rumors, far more plausible although still worthy of skepticism, that physics is closing in on a "theory of everything.") In short, most of the distortions of Best's paper are reproduced, although without the political rodomontade, in Wallerstein.

The latter's motives are clear. He is eager to believe that the social sciences are now at the top of the knowledge hierarchy: "But all of a sudden, the physical scientists seem to be looking towards the historical social sciences for models" (Wallerstein, 20). We, on the other hand, have noticed no such thing; but it's understandable that a social scientist would wish it to be so. As gauge of Wallerstein's actual knowledge of mathematical physics, we offer the following: "Einstein spent his life searching for the unified field theory, the single equation that would encompass all reality. He only achieved $E = mc^2$, which explained a large part of, but not all of, the physical world" (Wallerstein, 19). If ignorance is an excuse for bizarre condescension, Wallerstein is excused.

46. Best, "Chaos and Entropy," 203.

47. We might put things in a kinder light by taking Best's assertion as a refutation of a favorite "Creationist" argument. The latter asserts that since life—of ever-increasing complexity—represents increasing "orderliness," its existence violates the "Law of Entropy," and hence divine intervention must account for it. This is fallacious reasoning, of course; and Best's point may be taken as a sketchy indication of the error. Since, however, serious scientists have long been aware of the fallacy, this can hardly count as a "postmodern" development.

Also: we don't want to leave the impression that formation of snowflakes and similar phenomena are *perfectly* understood. Nor do we for a moment deny that there are exciting developments in the statistical mechanics of self-organization in nonequilibrium systems. (See, for an enthusiastic account, Gregoire Nicolas, "Physics of Far-from-Equilibrium Systems.") As with flowers that bloom in the spring, however, "dialectics" and the postmodern zeitgeist have nothing to do with the case.

48. Best, "Chaos and Entropy," 225, note 5.

49. "Thus one can readily compare Heisenberg's theory of uncertainty and chaos theory with Derrida's concept of undecidability, or Bohr's emphasis on the discontinuous movement of subatomic particles and Foucault's emphasis on the discontinuous breaks from one *episteme* to another." Ibid., 212.

50. N. Katherine Hayles, *Chaos Bound*, 181.

51. The Feigenbaum number, approximately 4.66920, is a universal constant (the term *universal* is off-putting to some postmodern enthusiasts of chaos theory, including Hayles) that arises in connection with any parameterized dynamical system where "chaos" arises via a cascade of "period doublings" as the parameter is increased. See Ruelle, *Chance and Chaos*, 67–70, or Ekeland, *Mathematics and the Unexpected*, 137–38, for details.

52. Argyros, *Blessed Rage for Order*, 238. Argyros's own comparisons between mathematics and literary philosophy, while still stretching analogies to what we consider a questionable degree, are more cautious and tentative than those of Hayles. His mathematical exposition is also far more systematic and coherent, although it is far from flawless.

53. N. Katherine Hayles, "Gender Encoding in Fluid Mechanics." The standards Hayles sets in *Chaos Bound*—for erroneous accounts of physics, hallucinatory history of mathematics, and substitution of arbitrary analogy for logic—are fully honored here. In short, don't expect to learn much about Reynolds numbers or divergence-free vector fields.

54. Both these mainstays of postmodern skepticism toward science seem to have lost heart for this particular battle. Rorty is now, it seems, chiefly concerned with old-fashioned politics and in this regard advocates a moderate, pragmatic, and unideological brand of social democracy. (See Rorty, "For a More Banal Politics.") A telling and convincing quote from Rorty, "Love and Money":

> All the talk in the world about the need to abandon "technological rationality" and to stop "commodifying," about the need for "new values" or for "non-Western ways of thinking," is not going to bring more money to the Indian villages. . . . All the love in the world, all the attempts to abandon "Eurocentrism" or "liberal individualism," all the "politics of diversity," all the talk about cuddling up to the natural environment, will not help. Maybe technology and centralized planning will not work. But they are all we have got. We should not try to pull the blanket over our heads by saying that technology was a big mistake, and that planning, top-down initiatives, and "Western ways of thinking" must be abandoned.

In the very same number we find the remark of Paul Feyerabend already cited (Chapter 3) whose essential thrust is equally cautionary: "Movements that view quantum mechanics as a turning-point in thought—and that include fly-by-night mystics, prophets of a New Age, and relativists of all sorts—get aroused by the cultural component and forget predictions and technology."

55. From the article on Rudolf Carnap in the *Encyclopedia Britannica* (1965 edition): "Carnap, Rudolf (1891–), one of the chief representatives of logical positivism . . . From 1926 to 1931 he taught philosophy at the University of Vienna where he belonged to the famous Vienna circle which first proclaimed the logical positivist doctrine."

56. Bertrand Russell, *Autobiography*, 341.

57. Hayles, *Chaos Bound*, 267.

58. The degree to which this kind of association has penetrated the academy, affecting undergraduates as well as scholars, is illustrated in an op-ed piece by a student, Jeffrey Robinson, that appeared in the *Rutgers Targum* ("Lessons in Holistic Learning," Apr. 19, 1993). The piece calls for an undergraduate curriculum that puts a double emphasis on "technological" education and on "multicultural" and "holistic" learning. Says Robinson:

> Our undergraduate education is failing us by not providing holistic learning. From kindergarten through the years of undergraduate education we are taught with a monochronic process. Western society has been taught to think in a *linear, one-dimensional, somewhat mechanistic* fashion which focuses on one sub-problem and solves it before it moves on to the next. Eastern societies tend to process polychronically, reviewing all of the sub-problems as a whole and solving with a process different from our own.
>
> The advantage of such a polychronic process lies in the way in which the solution is found. Multiple problems are worked on simultaneously and ideally with multiple approaches. This results in solutions that incorporate many systems of thought . . . Diversity of thoughts is essential to 21st Century problem solving. This holistic approach many times [*sic*] is cross disciplinary and cross cultural. (Emphasis added.)

It would be intriguing to learn the source of Mr. Robinson's ideas, which draw upon slogans in support of multiculturalism, New Age catchphrases, and the shibboleths of postmodern scholars. The disparaging reference to "linearity," in particular, echoes a widespread attitude, one of whose consequences is the enthusiastic, if mathematically illiterate, fascination with "nonlinearity" that abounds in postmodernist theory. Here is a further example from a militant feminist tract by Patti Lather, (*Getting Smart*, 104–5): "These [scientific] practices were rooted in a binary logic of hermetic subjects and objects and a linear, teleological rationality . . . Linearity and teleology are being supplanted by chaos models of nonlinearity and an emphasis on historical contingency." Lather's authority for this assertion is the unlikely duo of James Gleick and Michel Foucault. As it happens, a similar confusion shows up in the paper by Wallerstein, cited in note 45 above (Wallerstein, "The TimeSpace of World-Systems Analysis," 18).

59. See Thomas and Finney, *Calculus and Analytic Geometry*, 165, for a charming picture of this fractal phenomenon. Of course, calculus books (e.g., earlier editions of Thomas and Finney) have ignored this sort of thing until now. The fad for matters chaotic has induced publishers of elementary texts to decorate their offerings with such doodles. Similarly, the new version of William E. Boyce and Richard C. Di Prima (*Elementary Differential Equations*, 1992), a cut-and-dried traditional text for science and engineering majors, now contains a short discussion (section 2.12) of the evolution of the attractors of the iterations of the logistic function $f(x) = 4kx(1 - x)$, as k increases from 0 to 1. Argyros, *Blessed Rage for Order*, also contains a picture of this phenomenon with a misleading caption but a reasonably adequate explanation in the text. Strange doings for philosophers of literature!

60. Michael Rosenthal, in "What Was Postmodernism," suggests, with more relief than regret, that the high-water mark of postmodernism has passed and that it is likely soon to be regarded as a mere relic of the discontents of the eighties. It is ironic that this appears in a journal recently enthusiastic for the dogmas of the postmodern left. Rosenthal acknowledges that both the stance and the terminology of postmodernism, as well as the term itself, will persist in the books and articles of radical academics for years to come. In this insight, he is supported by the fact that a considerable number of books and journals advertised in this issue of *Socialist Review* prominently announce themselves with the term *postmodern*. Then, too, one of us (NL) recently dropped by the main scholarly book outlet near an Ivy League university to find that, of the featured "new releases" in all humanities fields, 20 percent contained the term *postmodern* in the title or subtitle.

5. Auspicating Gender

1. See Alison Galloway et al., exchange of letters in *Science*.
2. Sue V. Rosser, *Teaching Science and Health from a Feminist Perspective*.
3. Marianne Hirsch and Evelyn Fox Keller, eds., *Conflicts in Feminism*.
4. Ann Garry and Marilyn Pearsall, eds. *Women, Knowledge, and Reality*.
5. Sandra Harding, *The Science Question in Feminism*, 25. To these adjectives we might have added several more, e.g., white, male, European. The fashion for grouping them as mutually reinforcing pejoratives seems not yet to have peaked.
6. This is not the place to enlarge upon the meaning of the assertion that science "works," and that in so doing it is epistemologically unique. That is, obviously, the central issue for texts in the philosophy of science; and it is of course obvious, as well, to many nonphilosophers who are able to think honestly about the question. We note, however, that a particularly eloquent, as well as brief, statement on the matter is to be found in Jonathan Rauch's "Kindly Inquisitors" (pp. 29–30), otherwise a journalistic, but philosophically serious, treatment of the contemporary attack on free thought.
7. Mary Anne Campbell and Randall K. Campbell-Wright, preprint of the article. This paper was read originally at a meeting of the Mathematical Association of America, a professional association concerned primarily with mathematics education at the post-secondary level. (The other main organizations of professional mathemati-

cians are the American Mathematical Society, which is chiefly concerned with current research in all mathematical fields, and the Society for Industrial and Applied Mathematics, whose name is self-explanatory.) It has been also submitted as a contribution to a proposed volume on how to teach science so as to attract women into scientific careers.

8. Ibid., 6.

9. Here is a sample problem from Carroll's *Symbolic Logic*:

> Derive logical conclusions from the following premises:
> (1) Puppies, that will not lie still, are always grateful for the loan of a skipping-rope.
> (2) A lame puppy would not say "thank you" if you offered to lend it a skipping-rope.
> (3) None but lame puppies ever care to do worsted work.

Another example, more redolent, perhaps, of C. L. Dodgson's class and time:

> Derive logical conclusions from:
> (1) No Gentiles have hooked noses.
> (2) A man who is a good hand at a bargain always makes money.
> (3) No Jew is ever a bad hand at a bargain.

Still, we can read Carroll with pleasure and ignore the stereotyping foolishness of such problems as the second one.

10. This is from a problem in Stanley Grossman, *College Algebra* cited by Campbell and Campbell-Wright, "Toward a Feminist Algebra."

11. One of us (NL) will cheerfully read any proposed solution to this problem that our readers might come up with.

12. Campbell and Campbell-Wright, "Toward a Feminist Algebra," 11.

13. Ibid., 12.

14. Ibid., 10.

15. Mathematical insight can generate whimsical and instructive metaphors, sometimes quite elaborate. See, for instance, the graphic work of the artist/mathematician Anatolii T. Fomenko, (*Mathematical Impressions*, 1990). Fomenko's elaborate visual fancies are inspired by such things as "the action of the fundamental group on the higher homotopy groups" (32) and "proper Morse functions on 3-dimensional manifolds" (42). However, this is a case of the mathematics generating the "visual metaphor," not of a "social metaphor" generating mathematics.

16. Athena Beldecos, Sarah Bailey, Scott Gilbert, Karen Hicks, Lori Kenschaft, Nancy Niemczyk, Rebecca Rosenberg, Stephanie Schaertel, and Andrew Wedel.

17. The words *privileged* and *oppressor*, appearing in the same sentence and within the same line of type, should be added to a list of useful clinical tests.

18. Scott Gilbert et al., "The Importance of Feminist Critique," 173.

19. For evidence, see the review (of a new book on the Human Genome Project): Sydney Brenner, "That Lonesome Grail." This piece exemplifies, in addition, the approval and sympathy with which many distinguished scientists (among whom Bren-

ner is one) tend to respond to "social criticism," even when it bears upon science, in contrast to the tough-minded, analytical treatment they accord arguments of scientific substance.

20. Although eventually, when the full story of the functions of signal, binding, and receptor proteins that play a role in fertilization is known, McClung's metaphor might turn out—ironically and by accident—to have been right.

21. The "theory," became "fact" very quickly, of course. It was critical for Morgan's great work (for which he received a Nobel Prize) that founded modern genetics. It has been triumphantly confirmed countless times, most recently by identification of the location on the "Y" chromosome of male mammals at which the testis-determining factor—the molecule that initiates the chain of developmental switches that produce the male—is encoded.

22. See Harding, The Science Question, 113.

23. Nor did, presumably, his two distinguished predecessors in this sort of experimentation—A. D. Mead and T. H. Morgan. See Frank R. Lillie, The Woods Hole Marine Biological Laboratory. None of the papers of these workers betray any obviously greater admiration for the sperm than for the egg.

24. Scott F. Gilbert, Developmental Biology, 490ff. The key experiments, reported by P. R. Gross and G. H. Consineau, established the existence and function of maternal messenger RNA. The data are represented on p. 493 by a graph taken from the 1964 paper.

25. Eric H. Davidson, Gene Activity in Early Development.

26. We can't resist naming here this ancient and dangerous rhetorical device, in which natural objects are portrayed (inappropriately) as having human feelings.

27. It is worth pointing out that to "control for bias" is an ancient house rule of empirical science, and a direct inheritance from the Enlightenment. It is one of the hallmarks of the "good science" that the postmodernist critics of science disparage.

28. Discover, June 1992. This issue was deemed a finalist for the National Magazine Award (in the category of single-topic issues) by the American Society of Magazine Editors.

29. Ibid., Meredith F. Small, "What's Love Got to Do with It?"

30. Anne Fausto-Sterling, "Is Nature Red in Tooth and Claw?"

31. Anne Fausto-Sterling, Myths of Gender.

32. Sandra Harding, "Feminist Justificatory Strategies."

33. The reader who doubts this should have a look at what physicists say in books, rather than in space-limited research articles. Among dozens of recent possibilities, here are four well-known ones: Freeman Dyson, Infinite in All Directions, 1988; Stephen W. Hawking, A Brief History of Time, 1988; Roger Penrose, The Emperor's New Mind, 1989; and David Layzer, Cosmogenesis, 1990.

34. Philip Lenard was a Nobel Prize–winning German experimental physicist of emphatic anti-Semitic views, notorious for his hatred of Einstein and relativity theory. The cited quote, from his Deutsche Physik, is reproduced in Clark, Einstein, 525–26.

35. Harding, The Science Question in Feminism, 43.

36. Ibid., 47.

37. Observe, please, Harding's magically malleable logic: Assertion: Physics is tainted by sexism. Assertion: Physics is the (illegitimate?) model for all science. Ergo: All science is sexist. *But*: if we search the actual content of physics, we have trouble finding the sexism. *Not to worry!* Since physics *shouldn't* be paradigmatic, it *isn't.* The rest of science can therefore and nonetheless be declared sexist. This, we suppose, is what is meant among the avant-garde by "nonlinear" thinking.

38. Margarita Levin, "Caring New World," 106.

39. Marcy Darnovsky, "Overhauling the Meaning Machines," 66. A similarly extravagant encomium is also to be found in Robert M. Young, "Science, Ideology and Donna Haraway."

40. Darnovsky, "Overhauling the Meaning Machine," 82.

41. Harding, "Feminist Justificatory Strategies," 193.

42. A similar ambiguity afflicts the chapter "The Women's Movement Benefits Science" in Harding's most recent book (*Whose Science? Whose Knowledge?*). The shadowy "benefits" don't seem to have much to do with particular science that anyone actually works at.

43. Harding, "Feminist Justificatory Strategies," 197.

44. Ever since the publication of Nien Cheng's moving book on the Cultural Revolution (*Life and Death in Shanghai*, 1988), texts sprinkled with the word *struggle* smell of the auto-da-fé. (The public tribunals at which innocent people were denounced were known as "struggles.")

45. Garry and Pearsall, *Women, Knowledge, and Reality*, 198.

46. Kenneth Minogue, "The Goddess That Failed."

47. Sir Charles Sherrington, *Goethe on Nature and on Science.*

48. Quoted ibid., 38. This relatively early fragment appears in the complete (thirty-volume) edition of Goethe by Cotta (Stuttgart, 1858).

49. Evelyn Fox Keller, "The Gender/Science System."

50. Evelyn Fox Keller, broadcast interview by Bill Moyers, 1991.

51. Evelyn Fox Keller, "Long Live the Difference between Men and Women Scientists."

52. Thomas Kuhn, *The Structure of Scientific Revolutions.*

53. Ibid., 206.

54. See Larry Laudan, *Science and Relativism.*

55. Evelyn Fox Keller, "Feminism and Science."

56. This, although she surely knows that the word *dogma* was used in this connection tongue-in-cheek, an inside joke. Even if not, however, she *must* know, as a scientist herself, that no scientist would describe seriously his or her own idea as a *dogma.* Dogma, as every freshman taking biology or chemistry or physics for the first time is told, is exactly what science is *not* supposed to be.

57. Evelyn Fox Keller, *A Feeling for the Organism.*

58. Garry and Pearsall, *Women, Knowledge, and Reality*, 186.

59. Gould has within the hearing of one of us (NL), and before an audience quite sympathetic to the radical social critique of science, unambiguously repeated the assertion that there is "no such thing as 'feminist' science." (Gould, "Capitalism and the Environment.")

60. Helen E. Longino, "Can There Be a Feminist Science?"

61. Ibid., 206.

62. That is, there is a great deal of important work ahead for feminist criticism of science.

63. Longino, "Can There Be a Feminist Science?" 208.

64. Helen E. Longino, *Science as Social Knowledge*, 133–61.

65. Gerald Edelman and Vernon Mountcastle, *The Mindful Brain*. See also Edelman, *Bright Air, Brilliant Fire*.

66. Edelman's theory of neuronal group selection ("TNGS"), elaborated in a remarkable series of scientific books and papers, is heavy going even for working scientists, if they are not simultaneously expert—as he is—in molecular and developmental biology, neurobiology, and cognitive science. He has reduced the difficulties, however, even for lay readers, in a summary volume: "Bright Air, Brilliant Fire." Here he deals again, explicitly, with the problem of consciousness. For our present purposes it is important to say that in this book as elsewhere Edelman insists that the theory is—as many of its predecessors and competitors are not—rigorously *biological*. It is standard science, to the highest possible standard. To be sure, it can be dubbed an "interactionist" description of the brain's—and the mind's—ontogeny and phylogeny; but the component interactions are among genes, molecules, cells, and *time*—evolutionary time and its embedded history. To recruit this body of fact and hypothesis as justification for an ideology comes close to trivializing it.

67. Longino, "Can There Be a Feminist Science?" 211.

68. Ibid., 214.

6. The Gates of Eden

1. David Day, *The Eco Wars* (foreword).

2. Tom Athanasiou, "Greenhouse Blues," 107.

3. George Steiner, *In Bluebeard's Castle*, 62.

4. We hope that these concoctions are tested on *something*. It would do the environmentally sensitive no good at all to shampoo with what turned out, upon its first tests in the marketplace, to be an herbally fragranced depilatory.

5. Dave Foreman, *Confessions of an Eco-Warrior*.

6. Morris Berman, *The Reenchantment of Nature*.

7. See also Morris Berman, *Coming to Our Senses*. The publisher, Bantam Books, quite properly labels this one of its "New Age" series, but curiously includes among its companion volumes books by such emphatically *non*-New Age types as Richard Feynman, Heinz Pagels, Lewis Thomas, Michio Kaku, and Douglas Hofstadter. This is grossly unfair to both camps!

8. Berman, *The Reenchantment of Nature*, 31.

9. Carolyn Merchant, *The Death of Nature*.

10. Ibid., 295

11. Jeremy Rifkin, "Beyond Beef," 185.

12. We are aware that *hysterical* is a politically incorrect word; but it is hard to find a substitute for it in this case.

13. Anna Bramwell, *Ecology in the Twentieth Century*.

14. Another, typical, example of the urge to lay the sins of a supposed patriarchal technocracy at the feet of Francis Bacon can be found in Susanne Scholz's, "The Mirror and the Womb," a critique of Bacon's presumed misogyny: "From its very beginnings, the aim of scientific knowledge in the western world has been the domination of nature. This is as obvious in the thought of Francis Bacon as in the philosophies he set out to criticize."

15. Edward Goldsmith, "Evolution, Neo-Darwinism, and the Paradigm of Science," 73.

16. Michael Fumento, *Science under Siege*, 272.

17. The difference between "*the* greenhouse effect," which is a constant property of Earth's atmosphere, and a possible excessive *enhancement* of that effect is not in the least trivial. As regards policy, as well as physical chemistry, it is absolutely fundamental. Commentators who have not quite mastered the science and, unfortunately, a few who have, give the impression that the greenhouse is something new, something imposed by human activity on a previously invariant atmosphere. Such statements are false and inherently alarmist. They imply, moreover, a much greater certainty about future surface temperatures on this planet than is justified by available knowledge. A good example of this careless usage is to be found in (then Senator) Al Gore's well-meaning and fervent book (*Earth in the Balance*, 5). Despite its merits, this work manages to create false impressions, e.g., that an overwhelming majority of scientists believe that we are now passing or have already passed the point of no return toward a *catastrophic* global warming due to a greenhouse effect (37–39).

18. A grumble, widespread among atmospheric physicists whose predictions of global warming have so far failed, is that their expected climatic shift has been prevented—or frustrated—by recent volcanic eruptions (e.g., that of Mt. Pinatubo in the Philippines). Vulcanism puts sulfate aerosols and CO_2 into the atmosphere in huge quantities. The aerosols, added to the gases, nucleate cloud formation; and the more cloud cover, the more incoming solar radiation is reflected back into space. In the end there is less energy available to be transformed, via the greenhouse effect, into surface heat. It is perhaps not necessary to remind readers that we have no control over volcanoes.

19. One, again, among hundreds of possible examples, is the recent "News and Views" essay of Mark Chandler (of the NASA Goddard Institute for Space Studies) in *Nature*, on current understanding of the relationships between atmospheric CO_2 and oceanic heat transport in the regulation of planetary surface air temperatures (Mark Chandler, "Not Just a Lot of Hot Air"). Chandler's closing sentence is worth quoting in full: "Global circulation models, working from the fundamental physical equations, may one day do the job, but not until atmospheric water and ocean features are simulated more realistically. *Recognizing that the devil is in the details would ensure that the road to hell is not just paved with good conventions*" (emphasis added).

20. See J. W. Waters et al., "Stratospheric ClO and Ozone from the Microwave Limb Sounder on the Upper Atmosphere Research Satellite."

21. Observe, for example, the recent, much-advertised conversion of fast-food packaging from plastic to paper. The outcomes will include a sharp acceleration of

tree-cutting for paper, an increase of energy utilization (in pulping the logs and making the paper), and a *decrease* in biodegradability of the discard (because the "paper" used is actually a composite, lined with a plastic heat-retainer). In short, this "remedy" is with high probability a net *loser*, from any seriously "environmentally conscious" point of view, as, for example, McDonald's management is the first to admit in the right company. An interesting explanation for the failure of specific interventions to influence the behavior of very complex systems is given by the physicist F. David Peat, in an effort to come to grips with the startling implications of Bell's theorem (which demonstrates nonlocality in the quantum universe). Applying the insight macroscopically, while it encourages him in a certain amount of rhapsodizing on nonlinearity, nevertheless suggests why hasty interventions so often cause trouble:

> Such systems have a wide variety of behaviors that range from stability and rigidity to change to sudden bifurcation points and extreme sensitivity. The very complexity of these systems may make them impossible to model in any satisfactory way. In addition, any human intervention often leads to unpredictable results . . . Such "solutions" can have unexpected effects and . . . when it comes to highly complex and interrelated systems, the cure may even be worse than the problem itself! (F. David Peat, *Einstein's Moon*, 163.)

22. See, e.g., Gary Taubes, "The Ozone Backlash."

23. E. O. Wilson, *The Diversity of Life*. We applaud the arguments of this distinguished and humane scientist, and await with curiosity the response to them of the academic left, some of whose members have accused him repeatedly of sexism and racism, among other, lesser crimes, for his role as a founder of sociobiology.

24. Taubes, "The Ozone Backlash," 1581.

25. Although to do so will place us irretrievably—in the minds of some of our colleagues—among the Reaganites and dinosaurs, we recommend on this point chapter 10 of *Trashing the Planet*, by the straight-shooting Dixy Lee Ray and Lou Guzzo (pp. 123ff).

26. See Gregg Easterbrook, "Green Cassandras," 23, reporting on the Rio de Janeiro Earth Summit: "There, discussion of palpable threats to nature mixed in equal proportion with improbable claims of instant doom. Environmentalists, who would seem to have an interest in separating the types of alarms, instead encouraged the confusion on doctrinal grounds, namely that environmental news should be negative."

27. Merchant, *Radical Ecology*, 236. Merchant derives her "postmodern" view of science, i.e., that "mechanistic" and "atomistic" science is being supplanted by "holistic" science, in part from opinions of physicist David Bohm, in his last years a philosopher and something of a mystic (pp. 93–94). Bohm's meditations on quantum mechanics and its implications (see, for example, his nontechnical essay "Postmodern Science and a Postmodern World), as Merchant understands them, lead to this position. It is thus an irony that a rigorous working-out of Bohm's ideas about physics (see chapter 3, note 9 herein) lead to a strong reaffirmation of the "mechanistic" and "atomistic" picture of reality. Merchant is also eager, of course, to claim that chaos

theory provides metaphysical leverage for her notions of the superiority of "organistic" science over "mechanistic" (96–97). Therein she echoes Steven Best, giving as little evidence as does he of significant understanding of the theories she cites. Mercifully, she is less prolix.

28. Editorial by G. B. Gori.

29. Philip Abelson, "Testing for Carcinogens with Rodents," and the same editor's more recent "Toxic Terror; Phantom Risks."

30. Lewis, *Green Delusions*, 81.

31. Merchant, *The Death of Nature*, 236.

32. See, for an example of the *scientific* arguments against general circulation models (GCMs) that predict disastrous global warming during the next century, and of arguments for a *protective* role of rising CO_2 in the current atmospheric greenhouse, the views of the climatologist Patrick J. Michaels, brought together in *Sound and Fury*.

33. Stephen Schneider, statement in *Discover* magazine, October 1989, quoted in Michaels, *Sound and Fury*, 161.

34. For instance, as insatiable supplicants for increased federal funding, and as infighters, with one another, over a share of the pie—or pork.

35. As has been argued, not unconvincingly, by many serious scientists in the business—see Michaels, *Sound and Fury*.

36. J. Imbrie et al., "Milankovitch Theory Viewed from Devil's Hole."

37. See "Environmental Politics Is Making the Kitchen Hotter," a story by Marian Burros on Jeremy Rifkin's alliance with the celebrity-chef Wolfgang Puck. We hope that the reader understands that except for seafood, wild game, and a few genuinely wild berries and greens, *all* the food we eat is the product of extensive genetic modification; and that this has been true almost from the beginning of agriculture and animal husbandry. There is no reason to believe that genetic modification by recombinant DNA techniques is any different in effect or more dangerous than the ancient practices of selective breeding for favorable varieties. See also John Seabrook, "Tremors in the Hothouse," for an account of this quarrel and Rifkin's role in it as it centers on tomatoes modified by recombinant DNA techniques.

38. Steven Best, "Chaos and Entropy," 204–5.

39. Stanley Aronowitz, "Science under Capitalism" lecture.

40. Steven J. Gould, "Integrity and Mr. Rifkin," 236.

41. Jeremy Rifkin, *Beyond Beef*, 158.

42. Jonathan H. Adler, in the *American Spectator*, April 1993; Rifkin's quotation is from Lester R. Brown et al., in the Worldwatch Institute's *State of the World 1990*.

43. Bill Freedman, *Environmental Ecology*. This, of course, does not address the problem of species loss from the deforestation of the tropical rainforest.

44. Ibid., 104. See also Michaels, *Sound and Fury*, 144.

45. Ibid., 120ff.

46. Ibid., 165.

47. For one, and only one, of hundreds of possible recent examples of such research, see H. Wayne Polley, et al., "Increase in C_3 Plant Water Use Efficiency and Biomass over Glacial to Present CO_2 Concentrations."

48. Michaels, *Sound and Fury*.

49. Val Plumwood, "Nature, Self, and Gender," 18.
50. Janet Biehl, *Rethinking Ecofeminist Politics*, 6.
51. Berman, *The Reenchantment of Nature*, 278.
52. From Blake's personal papers, quoted by Charles Rosen, "The Mad Poets," 35.
53. Martin Lewis, *Green Delusions*, 247.

7. The Schools of Indictment

1. Herman Melville, *Moby-Dick*, chap. 35, "The Quarter-Deck."
2. Thomas Kuhn, *The Structure of Scientific Revolutions*.
3. Here is but one example: the pathway of development of those cells that function as the body's guardians against non-self—including pathogenic invaders—could not be worked out until the identity and life history of such cells were known. Such a life history cannot be obtained merely from *looking* at the cells: they all look alike. A means of identifying the stages of development—these cells are lymphocytes of thymic origin (hence "T-cells")—had to await the solution of a separate, older, technical problem. In due course it came down to the question of how to produce an antiserum, every antibody molecule of which would be the same as every other—a "monoclonal" antibody. Without the aid of such antibodies, which can function as markers for (otherwise invisible) surface molecules (antigens) of cells, there would be no way of knowing exactly what is happening within the system of cellular immunocompetence as it varies normally or in disease. The technical problem of preparing monoclonal antibodies was solved, eventually, by an ingenious and elegant application of then-recent knowledge about the behavior and genetics of cell mixtures in culture. Only after that development, and others that evolved from an increasingly sophisticated cell biology, was it possible to make a start on the molecular-level analysis of T-cell development and maturation. Only the results of such an analysis could provide insight into a catastrophic breakdown of the body's defenses against invaders and of its tumor-surveillance systems. Progress was very rapid after 1975; but even so, it took five or six years before a comprehensive picture of the molecular-level events of T-cell-mediated immunity began to emerge. Such a picture is, of course, absolutely essential for any understanding of a disease that wipes out immunity, since disease happens in the first instance, here as elsewhere, at the cellular and molecular levels.
4. Anthony Fauci et al., "Acquired Immunodeficiency Syndrome," 92.
5. As is well known, there was considerable bad blood between these two eminent researchers. It is not for us to judge their conflicting priority claims. There are questions of history and ethics here; but no doubt at all about the validity of the science that emerged.
6. To date, this category of AIDS sufferers, expanded by the number of babies born to infected mothers, accounts for about 10 percent of the total. Readers with some life science background but unfamiliar with modern immunology will find an excellent source in Golub and Green, "Immunology: A Synthesis." The concluding chapters on immunodeficiency disease are highly accessible.
7. In the rest of the world, matters are grimmer. According to Daniel Tarantola and Jonathan Mann, "Coming to Terms with the AIDS Pandemic," the spread of the

disease is explosive now in Southeast Asia, where, by the turn of the century, more than 40 percent of cases will be found.

8. Everyone keeps score, everyone for his own purposes; and the published scores are sometimes suspect, as we demonstrate here; but it seems clear that while the rate of new HIV infections may have begun to decline nationally, it continues to rise among *young* homosexual men, i.e., between the ages of thirteen and twenty-nine. See, for an example of the sermonizing this elicits, an editorial in the *Richmond Times-Dispatch* for March 8, 1993.

9. Recently, the decade-old puzzle of the long "latency" of AIDS, the remarkably long interval between infection and the appearance of symptoms, has yielded to research. It is now clear that the virus is alive and well, but hiding, for a long period after initial infection, in the lymph nodes and in other lymphoid tissues. See Giuseppe Pantaleo et al. and Janet Embretson et al., back-to-back articles in *Nature* on the progressive activity of HIV in lymphoid tissue during the clinically latent stage of the disease. One more little—but deadly serious—"puzzle" of nature solved.

10. *Science* magazine for May 28, 1993, has for its title "AIDS: The Unanswered Questions." This was as fair and authoritative a summary of the issues as was available at the time of writing this book. The reader interested in a well-documented summary of the pathogenesis of AIDS, as understood in the summer of 1993, is referred to the paper by Robin A. Weiss ("How Does HIV Cause AIDS?") in that issue.

11. The wildfire spread of the disease in parts of Africa, by what is almost certainly heterosexual contact, has yet to be fully explained, but it may be a property of a variant HIV. It becomes increasingly likely, however, that there are other factors, such as endemic diseases, venereal and otherwise, that produce genital irritation, scars, and lesions, in whose presence transmission of the virus is greatly facilitated, whereas in their absence transmission via heterosexual, vaginal intercourse is highly unlikely.

12. Oddly enough, even the most abusive oppositionist arguments favor *research*; what is taken to be at stake here, inconsistently, is not the power and effectiveness of scientific research, but who is to be in charge of it.

13. As an example—and a notorious one at that—we have the militant Chicago politician Steve Cokely, who insists that Jewish doctors are deliberately infecting black babies with the AIDS virus, under the pretext of vaccinating them against disease! This charge is sickening, but by no means unusual.

14. We might call it the school-condom school, except that to do so would encourage unnecessarily the extreme right, in whose members the distribution of condoms in the public schools triggers fits of rage. Such distribution will protect few heterosexual males for whom sexual intercourse is the only risk factor. It might be of some value to sexually active young women among whose partners may be bisexual or IV drug users. It will protect no one from the consequences of needle sharing. On the other hand, as a contraceptive for teenagers who are—and who will continue, come what may, to be—sexually active and sexually promiscuous, it is possibly better than nothing. Conservative rage is against the implication that sex sans sacrament is okay; the radical delusion is that "education," in the form of slogans and free penile sheaths, is even a partial solution to the problem.

15. "Notebook," *New Republic*, August 17 & 24, 1992, 8.

16. See, typically, Peter H. Duesberg, "HIV Is Not the Cause of AIDS," *Science* 241, July 1988, 514–16, and for a rebuttal, W. Blattner, R. C. Gallo, and H. M. Temin, "HIV Causes AIDS," *Science* 241, July 1988, 515–17.

17. Robert Root-Bernstein, "Re-thinking AIDS," *Wall Street Journal*, editorial, March 17, 1993. Root-Bernstein's full argument is in his book *Rethinking AIDS: The Tragic Cost of Premature Consensus.*

18. See, for example, Tom Bethell, "AIDS Reporters Snooze," in an issue of *Heterodoxy* that gives sympathetic coverage to Duesberg and his followers.

19. The argument also has an undeniable appeal to a small faction of gay militants, to whom it holds out the hope that, despite the AIDS epidemic, it might be possible to return to the omnivorous sexual habits that pervaded the most "liberated" gay communities in the seventies. The argument is that if HIV is not, per se, causative of AIDS, then the danger of infection can be ignored, provided other damaging behaviors and bad health habits are avoided.

20. Let us take, for the single example to which we can devote space, the particular set of claims about drug abuse as fatally predisposing or even as the main cause of AIDS. This one has been a favorite, oddly, on the political left as well as on the right; and it is still taken very seriously by some scientists. M. S. Ascher and colleagues (see M. S. Ascher, H. W. Sheppard, W. Winkelstein, and E. Vittinghoff, "Does Drug Use Cause AIDS?") have investigated the question by means of a direct epidemiological test. They analyzed the large data source of a population-based cohort, the San Francisco Men's Health Study. The cohort is 1,034 single men aged twenty-five to fifty-four at the time of recruitment, from a neighborhood where the AIDS epidemic had flourished in its early years, but otherwise *without* regard to sexual preference, lifestyle, or, at the time of recruitment, to serology (i.e., presence or absence of HIV antibodies in the blood). In this statistically rigorous investigation, the variables included sexual preference, drug use, and serostatus: the endpoints were AIDS cases and deaths. An added, unique feature of the work, however, was the availability, during eight years of the study, of an additional endpoint: $CD4^+$ T-lymphocyte counts for its participants. Thus the study's variables could be related, not only to appearance of the syndrome known as AIDS, but specifically to an objective criterion: the disease's principal (and unquestionably causative) pathologic feature: depletion of $CD4^+$ cells.

The results of this study are as clear as epidemiology ever gets to be. Environmental or behavioral factors proved, in this group, to be statistically insignificant, while seropositivity—the presence of antibody to the putative AIDS virus—and its progression over time were highly significant. The $CD4^+$ data are impressive. People free of the virus had normal cell counts throughout the interval, or even slightly elevated counts if they were heavy drug users (a matter of separate, potential interest); among those infected with HIV and remaining alive (a good many of them died of AIDS, of course), there was a steady and cataclysmic decline in cell number over the entire period.

21. I.e., vaginal intercourse.

22. This is the outcome of the "Concorde Trial" in Europe, reported in the April 3, 1993, issue of *The Lancet*.

23. By association of the most irrational sort, lesbian women are also tagged, in some quarters, as AIDS bearers. Obviously, committed lesbians are at less risk for AIDS than other categories of sexually active persons.

24. E.g., middle-class college students.

25. Larry Kramer, *Reports from the Holocaust*, 178.

26. Ibid., 103.

27. Kay Diaz, "Are Gay Men Born That Way?" 43.

28. Daniel Harris, "AIDS and Theory," 16.

29. Ibid., 19.

30. Cindy Patton, "Inventing AIDS," 71.

31. Jacques Derrida, "Rhétorique de la drogue." Quoted in Alexander G. Düttmann, "What Will Have Been Said about AIDS," 102.

32. John O'Neill, "Horror Autotoxicus," 265–66. Note that *autoimmunity* does not mean what the author seems to think it means. In ordinary clinical use, this word refers to a disease state in which the immune system fails to recognize self, and therefore attacks it. Autoimmunity is never "enjoyed."

33. Tim Dean, "The Psychoanalysis of AIDS," 84ff. Dean's verbal mush, appearing as it does in the wake of Harris's debunking *Lingua Franca* article, is self-consciously defensive. Dean accuses Harris of "a thoroughly reactionary, anti-intellectual stance." Dean should be reminded that nothing is more anti-intellectual than shooting one's mouth off when one is overwhelmingly ignorant (see note 12, chapter 4).

34. Harris, "AIDS and Theory," 19.

35. Steven Epstein, "Democratic Science? AIDS Activism and the Contested Construction of Knowledge."

36. Ibid., 60.

37. David L. Kirp, "R_x Populi."

38. Benjamin Wittes and Janet Wittes, "Group Therapy." The regulation in question is Section 131 of the National Institutes of Health Revitalization Amendments (pending). While it is true that Section 131 is concerned principally with the inclusion of women and minorities in clinical trials, it represents a political impetus of precisely the kind Epstein and others are talking about, i.e., the "democratization" of clinical research.

39. They calculate the practical consequences of such a requirement in a hypothetical test of a stroke-reducing treatment for hypertension patients. Those would include increasing the sample size from, say five thousand persons, followed for five years, to forty thousand or more. The cost implications are staggering: Wittes and Wittes report that a recent hypertension trial involving five thousand subjects cost $50 million. Costs for such activities as clinical trials rise at least in proportion to sample size: in fact they usually rise faster, since there are emergent complications. Would the additional information emerging be useful? Possibly; but not significantly more so than that emerging from a standard test on a designed and limited cohort (i.e., limited to the primarily affected group).

40. See Dennis L. Breo, "At Large," 19.

41. Peter Singer, *Animal Rights*.

42. See Tom Regan, *All That Dwell Therein*.

43. At Rutgers, for instance, there is an Animal Rights Legal Center affiliated with the Rutgers (Newark) Law School. At the University of Virginia, on the day of this writing, the concrete walks have chalked all over them, in foot-high letters, such maxims as "FUR KILLS."

44. Morris Berman, *Coming to Our Senses*, 82.

45. *Hypatia—A Journal of Feminist Philosophy* 6, no. 1 (Spring 1991). Special issue on ecological feminism.

46. Carol J. Adams, "Ecofeminism and the Eating of Animals," 140.

47. Deborah Slicer, "Your Daughter or Your Dog?" 108–24. Note "rethinking the status of fur": *pour encourager les autres*—those potential recruits or contributors who have not yet determined upon a defensible means of disposing of the mink coat. To *sell* it, of course, even if at a loss, would be shaming.

48. This does not include Peter Singer himself, whose utilitarian propensity for moral bookkeeping intrudes and complicates the question.

49. An article in the left-wing journal Z (Joan Dunayer, "Censorship: Faculty Who Oppose Vivesection") gives some evidence of the affinity of the contemporary academic left for the animal rights cause. For the most part, this is a standard anti-vivisectionist screed that asserts categorically the uselessness of animal research for drug development, creation of new treatment modalities, or devising diagnostic technology. (The article purports to raise certain questions about the academic freedom of anti-vivisectionists. There may be a point here but the evidence presented is too skimpy and selective to allow any judgment.) In fairness to Z, we note that the magazine published subsequently a devastating rejoinder to Dunayer's piece, in the form of a lengthy letter to the editor from Matt Spitzer (Z, June 1993, 3).

50. The following are a very few samples of well-documented articles on the de-predations of animal rights activists and on the nature and effects of their arguments: Herbert Pardes et al., "Physicians and the Animal-Rights Movement"; Rudy M. Baum, "Biomedical Researchers Work to Counter Animal Rights Agenda"; Jerod M. Loeb et al., "Human vs. Animal Rights." An editorial account of a "Consensus Conference" of the National Institutes of Health, devoted to exploration of non-animal models in the study of diabetes, cardiovascular disease, and pulmonary dis-eases, is in Paul R. Gross, "Animals Still Crucial to Research."

The clearest and most fundamental statements of the opposing point of view that are not *simply* activist propaganda tracts are the popular philosophical books of Singer (*Animal Rights*) and Regan (*All That Dwell Therein* and *The Case for Animal Rights*). It must be noted, however, that distinguished moral philosophers can be deeply con-cerned about animal rights and yet reject decisively the simple—and shaky—utilitarianism of such arguments. See, for example, Cora Diamond's often quoted "Eating Meat and Eating People."

51. See Todd Gitlin, "The Rise of 'Identity Politics.'"

52. Charles S. Finch, "The African Background of Medical Science," 151ff. By way of comparison to Europe, see Renate Blumenfeld-Kosinski, *Not of Woman Born*, which demonstrates that in medieval and Renaissance Europe, successful cesarean sections were extremely rare.

53. See Bernard Ortiz de Montellano, "Melanin, Afrocentricity, and Pseudo-

science," which briefly lays out the evidence that the ancient Egyptians, like the modern residents of that country, were of mixed ancestry, with a small contribution from sub-Saharan Africa.

54. Beatrice Lumpkin, "Africa in the Mainstream of Mathematics."

55. Actually, 256/81.

56. These are published by the Multnomah School District of the Portland, Oregon, public schools (1990).

57. Hunter Havelin Adams, *Baseline Essay in Science*. (Outline, page 7 of the general introduction.)

58. Ibid., 41.

59. See Bernard Ortiz de Montellano, "Multicultural Pseudoscience."

60. "The Dogon, with no apparent instrument at their disposal, appear to have known these amazing facts [about the Sirius system] for at least 500 years!" (Adams, *Baseline Essay in Science*, S-60.)

61. See Bernard Ortiz de Montellano, "Magic Melanin" and "Melanin, Afrocentricity, and Pseudoscience," for an account of the bizarre doings of the Melanin scholars.

62. The latter (Kenneth R. Manning, *Black Apollo of Science*), however, albeit very well written and handsomely published, offers a rather superficial view of Just's research accomplishments, theoretical positions, and eventual politics.

63. We note with great satisfaction that an *art critic* has devoted substantial space to a properly devastating critique of Van Sertima, H. H. Adams, and their cronies (Robert Hughes, *Culture of Complaint*). As scientists, we owe him enormous thanks; and we wonder why the scientists left the job to someone else.

64. Adams, *Baseline Essay in Science*, vi. De Broglie is quoted as follows: "Many scientists of the present day, victims of an ingenuous realism, almost without perceiving it, have adopted a certain metaphysics of a (sexist), materialistic, and mechanistic character and have regarded it as the very expression of scientific truth." De Broglie was, of course, addressing the implications of quantum mechanics, which had rendered problematical the "billiard-ball" determinism of classical physics. Even if we allow that the use of parentheses, rather than editorial square brackets, is a simple misprint, Adams's imputing concern about "sexism" to de Broglie, who died before the term was coined, is a solecism. That simply was not de Broglie's concern in the cited passage. It is possible to infer that Adams is lured into this childish error by his reading of Ruth Bleier, *Feminist Approaches to Science*.

65. Bell Hooks, "Columbus," 25.

66. Bernard Ortiz de Montellano, "Chariots of the [Black] Gods?" (preliminary version).

67. Sandra Harding, *Whose Science? Whose Knowledge?* 223–27. We note that Harding's approving reference to the work of Van Sertima et al. comes in for similar praise, in its turn. The favorable review of *Whose Science? Whose Knowledge?* by Sue V. Rosser ("The Gender Equation") singles out Harding's citation of African scientific achievement as particularly admirable. The appearance of Rosser's featherweight piece in *The Sciences*—a normally serious journal of the New York Academy of Sciences—says quite a bit about the sympathy that feminism enjoys in the scientific

community—and about the moral leverage that feminist scholars are now able to apply in many areas.

68. A recent, well-received book reflecting Montellano's hard-science background is *Aztec Medicine, Nutrition, and Health.*

69. In addition to the works already cited, Ortiz de Montellano has pursued these questions in "Afrocentric Creationism" and "A Critique of the Portland Schools Baseline Essay."

8. Why Do the People Imagine a Vain Thing?

1. See, for example, Jeri Laber, "Bosnia."
2. George Steiner, *In Bluebeard's Castle,* 65.
3. Ibid., 65.
4. See, for example, Montaigne's essay "On Cannibals."
5. J. Hillis Miller. Quoted in David Lehman, *Signs of the Times,* 40–41.
6. The story of the Reagan family's involvement with astrology is universally known. By way of symmetry, here is Camille Paglia, the celebrated literary critic, in the canceled preface—now reborn in "Sex, Art, and American Culture," 107—to her *Sexual Personae:* "I am partial to cyclic theories of reality, as in Hinduism or astrology and the ever-turning Wheel of Fortune." On the other hand, her exquisitely formulated view of the postmodern pantheon—Derrida, Foucault, Lacan—redeems her in full: "French rot! Gibberish." Readers who have not yet encountered Paglia's comments on this subject can meet her at her best in the long essay "Junk Bonds and Corporate Raiders," reprinted in the same volume (170–248).
7. See, for example, P. Gerdes, *Marx Demystifies Calculus,* which reprints, with enthusiastic commentary, much of Marx's writing on mathematics. The book is apparently intended as some kind of manual for students who have been bewildered by the attempts of mathematics professors to teach them the bourgeois foundations of calculus. No description can do justice to its absurdity—it must be sampled directly. On the other hand, there seems to be an embryonic movement to put this stuff into the universities as part of "multicultural" reforms. We trust this won't get too far, although stranger (but not much) things have happened.
8. See, for example, V. I. Lenin, *Materialism and Empiro-Criticism.*
9. For a devastating critique of "matriarchalist" feminism and the superstitious and merely wishful strain in ecofeminism generally, see Janet Biehl, *Rethinking Ecofeminism.* As to what there is to criticize, consider the following: "Rationalism is the key to the linked oppressions of women and nature in the West" (from Val Plumwood, "Nature, Self, and Gender").

9. Does It Matter?

1. As one among many possible examples, see Anne Fausto-Sterling, "Is Nature Really Red in Tooth and Claw?" This piece is earnestly devoted to knocking down a straw man: an established biology the author characterizes as fixated on *competition,* while it suppresses the facts and significance of *cooperation* (i.e., mutualism and

symbiosis) in nature. Fausto-Sterling's political purpose of this is clear: "A strong case can be made for viewing nature as a socialist cooperative . . . only a change in social ambience will permit these already existing ideas to be incorporated into the mainstream of biological thought." Anybody who thinks of life in a complex ecosystem in a major landscape, say, the continental shelf, a coral reef, or a tropical rainforest, *either* as a system of competing capitalist enterprises *or* as a socialist cooperative has been staying up late with the wrong literature. We know of no serious contemporary natural history, ecology, or evolutionary biology so afflicted in its structure.

2. Arthur Potynen, "Oedipus Wrecks."

3. Alan Chalmers, *Science and Its Fabrication*, 125.

4. Indeed, such responses as are here printed, to already-famous attacks on "patriarchal" science, have been criticized when made *viva voce* as tasteless and unfair "because women's studies is a struggling young discipline, still finding its feet in a hostile academic world."

5. For a clear-eyed analysis of this phenomenon at its most intense, see Martin Jay, "The Academic Woman as Performance Artist."

6. Lewis S. Feuer, "From Pluralism to Multiculturalism."

7. The media, including such trade organs as the *Chronicle of Higher Education*, seem to have settled upon the description of this body (the N.A.S.) as "traditionalist." While that certainly fits many of the members, it clearly annoys others, who have some claim to being thinkers quite as advanced and "progressive" as their equivalents in the MLA.

8. Jacob Weisberg, "The NAS: 'Who Are These Guys, Anyway'?", 34–39.

9. These exact words were used by an energetic young postmodernist, in the hearing of one of us, to impugn the warning given by a distinguished cardiologist to a colleague who was being pressured to seek help for a serious heart condition from the local holistic healing emporium.

10. The full stories of several such media distortions, compounded in equal parts of avarice (for sensation and its consequent ratings), politics, and scientific ignorance, is beginning to be told. Michael Fumento's new (1993) book, *Science under Siege*, is full of them, meticulously documented. Any reader who still doubts that the huge environmental public-interest groups are as interested in public relations, and as casual about the scientific truth, as the most profit-driven multinational corporation should study Fumento's account of the disgraceful Alar scare, to which the main contributors were the EPA, CBS's *60 Minutes*, the Nader organizations, the National Resources Defense Council, and a public relations firm hired by the NRDC.

11. Bernard Ortiz de Montellano, personal communication.

12. Eugenie Scott, "Multiculturalism—The Good, the Bad, and the Ugly" (lecture).

13. Stanley Aronowitz, having deconstructed the scientific enterprise in *Science as Power*, has gone on to co-author a tract on how to reform education in accordance with postmodern philosophy and the academic left's canon of political virtue (Aronowitz and Giroux, *Postmodern Education*). For a self-appointed expert on science, Aronowitz has wondrous little to say about the reform of *science* education. The book avoids this question, except to convey a generalized distrust of scientists and to suggest

that ecological questions be introduced into science courses. We should be thankful for this reticence.

14. Carolyn Merchant, address, University of Virginia, 1993.

15. Sandra Harding, seminar talk at Center for the Critical Analysis of Contemporary Culture, 1992.

Supplementary Note

The literature with which we are concerned is large and scattered. New items, and older ones of which we have become aware too late for mention in the text, continue to demand our attention. We feel obliged to take note of a few samples here. In the following paragraphs, numbers in square brackets refer to relevant pages of this book. Sources are identified in the supplementary list at the end of the references.

Andrew Ross [89–92] (now at New York University) has added to his bibliography on science with "The Chicago Gangster Theory of Life" (1993), a tantrum in the form of an essay that denounces science in general and genetics in particular. There is the predictable scourging of E. O. Wilson's sociobiological thought as some kind of capitalist-patriarchalist conspiracy. We note, with some puzzlement, however, that Ross, through ignorance or unkindness, fails to credit Wilson's well-known and effective efforts on behalf of biodiversity, a cause that seems close to Ross's heart.

Meredith F. Small has just published *Female Choices*, a celebration of the bonobo that enlarges to book length her claim that this primate can and should serve our species as a nonsexist role model [125]. A brief, caustic review, "Oh, Those Bonobos!" by the biologist and feminist Helena Cronin, has appeared in the *New York Times*.

Freeman Dyson's recent article, "Science in Trouble," includes acute remarks on the obstacles put in the way of a safe and beneficial biotechnology by Jeremy Rifkin's endless stream of environmentalist lawsuits [170].

The postmodern sage Avital Ronell has joined the list of thinkers who view AIDS as a product of the corrupt metaphysics of Western post-Enlightenment discourse with "A Note on the Failure of Man's Custodianship." Sample sentences from this effusion: "The co-factors that have produced the destruction of internal self-defence capabilities still need to be studied in a mood of Nietzschean defiance toward the metaphysico-scientific establishment. For surely AIDS is in concert with the homologous aggression that is widely carried out against the weak within the ensemble of political, cultural, and medical procedures. It is not far-fetched to observe that these procedures take comparable measures to destroy any living, menacing reactivity, and thus have to be considered precisely in terms of the disconcerting reciprocity of their ensemble" [60]. Ronell easily outshines, in sheer loopiness, the examples [191–93] to which our readers have already been introduced.

As to "medical procedures," we are saddened to discover that one source of instruction for undergraduates in the history and practice of obstetrics is Alexandra Dundas Todd's *Intimate Adversaries*, the burden of which is that male physicians have turned the natural, healthy process of childbirth into a disease and that their destructive (to women) contribution, rooted in (patriarchal) science, should never have replaced midwifery. There is, of course, no serious discussion in this airy volume of such matters as puerperal fever, infection, eclampsia, and the like, nor of the role of twentieth-century male physicians in the advocacy of "natural" childbirth. This book is a prototype of what happens when social-scientific disciplines (here sociology and anthropology) are recruited wholly to the service of an uncompromising sexual politics.

Among a number of comprehensive refutations of the strong form of cultural contructivism (as regards natural science) are several impressive recent books. Steven Weinberg's new *Dreams of a Final Theory* contains devastating remarks on the cultural constructivist theory of scientific knowledge [42–70] (Weinberg, 184–90). We note with amusement that he, too, found the post–World War II cargo cults pertinent to his analysis [40–41]. Sociologist Stephen Cole provides a detailed refutation in *Making Science*; and *Scrutinizing Science*, a recently reprinted compendium of papers edited by Arthur Donovan, Larry Laudan, and Rachel Laudan on the philosophy and history of science, is a rich resource for those interested in empirical efforts to *test* the major post-positivist claims on guiding principles ("paradigms") and theory choice.

Supplementary Notes to the 1998 Edition

p. 9 The notion that right-leaning intellectuals, academic and otherwise, keep their distance from Creationism turns out to have been premature. Within the last few years, and with the support of such academics as Phillip Johnson (University of California at Berkeley), Michael Behe (Lehigh University), and Dean Kenyon (San Francisco State University), religiously linked anti-evolutionism has become a fixture of the campus scene. A pseudo-scholarly journal, *Origins and Design*, has been established to promote this point of view. Conservative journalist Tom Bethell (see also p. 282, n. 18) is a supporter. (See *National Center for Science Education Reports*, Winter, 1996.)

p. 20 Laplace's famous quip, "I did not find the hypothesis necessary," seems, according to recent scholarship, to be apocryphal. Laplace might have said it (but didn't); somebody did say it (but not Laplace).

p. 34 To the panoply of communications instruments by which the academic left propagates its ideas, we may now add World Wide Web sites. Of course, it's the rare upwardly mobile thinker, these days, who doesn't have his own homepage, so the left has no monopoly.

p. 39 Perusal of recent numbers of *Lingua Franca* reveals that the tone and scope of such announcements hasn't changed much in four years.

p. 54 Recent Newton scholarship shows that Newton's jealous editing of the historical record was even more extensive than has been commonly thought. While his claim to have invented calculus during the late 1660s seems to hold up, he apparently backdated much of his work on celestial mechanics in order to suppress the role of Robert Hooke. Hooke has priority for much of what we now think of as Newton's Laws of Motion, and Newton focused clearly on the relation between an inverse-square, central force formulation of gravitation, and Keplerian orbital ellipses at the urging (sometimes as tart criticism) of Hooke in the late 1670s. Newton nevertheless deserves full credit for the key mathematical insights necessary to make this formulation rigorous. See A. Rupert Hall, *Isaac Newton: Adventurer in Thought* (Cambridge: Cambridge University Press, 1992), and Michael Nauenberg, "Hooke, Orbital Motion, and Newton's *Principia*," *American Journal of Physics* 62, no. 4 (April 1994): 333–50, and John Aubrey's "brief life" of Hooke.

p. 55 In the early 1700s, Newton did address himself to the urgent and practical navigational problem of determining longitude; his scheme involved lunar observations interpreted through the astronomical theory of the *Principia* and was never of practical use. (Dava Sobel has written a short but eloquent account of the great longitude problem and the achievement of the clockmaker, John Harrison: *Longitude* [New York: Penguin Books, 1995].) However, since all this occurred decades after Newton's foundational work in mechanics, it can in no sense have motivated that work.

p. 57 Bruno Latour is on record as denying that he is a leftist. He also denies that he is a constructivist, or indeed an *anything* but the inventor of his own system (dare we call it actor/network theory?) of verbally pyrotechnical science-studies and one of the "Darwins of science" (see "Who Speaks for Science?" in *Sciences*, March/April 1995, 7). But his fame, at least in the United States, is due to his having been taken up by the academic left and the cultural constructivists. Perhaps *they* have really misunderstood him as, presumably, his French compatriots have not (since he claims to have been attacked there by the left). Misunderstanding him would not be difficult. The internecine wars of science studies (for examples of which, see Andrew Pickering, ed., *Science as Practice and Culture* [Chicago: University of Chicago Press, 1992], especially Part 2) suggest that practitioners of this subdiscipline disagree, and misunderstand one another, about everything *except* that science is a social construct, decisively shaped by social forces and "interests," rather than by nature or reality, and that it needs to be taken down a peg or two. The misfired attempt to secure Latour's appointment at the Institute for Advanced Study (which preceded the writing of this book, but was unknown to us at the time) was led by two prominent members of that tribe, anthropologist Clifford Geertz and historian Joan Wallach Scott. See David Berreby, "That Damned Elusive Bruno Latour," *Lingua Franca*, October 1994, 22–32 and 78.

p. 58 The hostility toward Latour among philosophers of science and others concerned with scientific practice (including some sociologists) has not lessened over the years. See Gross, Levitt, and Lewis, eds., *The Flight from Science and Reason* (Baltimore: Johns Hopkins University Press, 1997); N. Koertge, ed., *A House Built on Sand* (forthcoming from the Oxford University Press); and a volume forthcoming from A. Sokal and J. Bricmont.

p. 61 Some criticism has emerged of our reference to the Traveling Salesman Problem—the problem of finding an optimal route through a network—as anachronistic. The confusion arises because these days, many people in theoretical computer science think of the "Traveling Salesman Problem" as a synonym for the so-called P = NP Problem (with whose technical significance we will not burden the reader). However, at an earlier period, the TSP had a more practical connotation, that is, of actually finding efficient routes through networks. The scheduling problem for the Aramis scheme is obviously a generalization of *that* kind of problem.

p. 62 Latour's difficulties with mathematics are made hilariously clear in his now-notorious paper, "A Relativistic Account of Relativity," *Social Studies of Science* 18 (1988): 3–44. See John Huth, "Latour's Relativity," in *A House Built on Sand*.

p. 63 Pentimento: to call the Hobbes-Boyle dispute "resounding" is perhaps to concede too much to the Shapin-Schaffer thesis. The dispute was loud and vituperative, but essentially a historical sideshow.

p. 66 The distinguished historian of science Margaret C. Jacob advises us that she considers Halley's reputation as an atheist to have been a canard. We think the question is still open; at any rate, Halley's views fell far short, in point of piety, of Newton's enthusiasms.

pp. 67–68 Our refusal to accept the Shapin-Schaffer thesis on the socially determined nature of the vacuum dispute of 1661, and on its significance for enthroning the "experimental life," has provoked considerable agitation among those of the science studies community who accord classic status to *Leviathan and the Air Pump*. But few of them have addressed the rhetorical sleight-of-hand through which Shapin and Schaffer avoid the question of Hobbes's eccentric—indeed, mad—pretensions to mathematical genius, and their bearing on his extensive quarrels. A number of additional points can be made to illustrate just how far astray this omission has led gullible readers of their book. (For what follows, Alexander Byrd, "Squaring the Circle: Hobbes on Philosophy and Geometry," *Journal of the History of Ideas* 57, no. 2 [1996]: 217–32, is a good general reference; S. Probst, "Infinity and Creation: The Origin of the Controversy between Thomas Hobbes and the Savilian Professors Seth Ward and John Wallis," *British Journal of the History of Science* 26 [1993]: 271–79, is also valuable.)

1. The invective between Wallis and Hobbes over mathematical questions began as far back as the early 1650s. In particular, Hobbes's claim to have "squared the circle" was widely circulated and widely scorned during this decade, as was his supposed "duplication of the cube." With respect to these claims, Wallis's colleague Seth Ward had this irony to offer in 1654: "Geometry hath now so much place in the Universities, that when Mr. *Hobbs* shall have published his Philosophicall and Geometrical Pieces [i.e., his circle-squaring and cube-duplication efforts], I assure my selfe, I am able to find a great number in the University, who will understand as much or more of them then he desires they should, indeed too much to keep up in them that Admiration of him which only will content him." Hobbes's repeated, but obtuse, attempts to answer his critics merely deepened the scorn, which is typified by Huyghens's remarks that, as mathematician, Hobbes was "ridiculous" and "childish." Equally damning was Hobbes's reluctance to recognize significant results obtained by others, particularly Wallis. Hobbes uncomprehendingly rejected Wallis's characterization of the conic sections of classical geometry as second-degree plane curves. This, as even a mathematical tyro knows, is an achievement of surpassing elegance and beauty, and Hobbes's refusal to come to terms with it can only be called stupid.

2. Shapin and Schaffer took note of Hobbes's much-quoted remark "For who is so stupid, as both to mistake in geometry, and also to persist in it, when another detects his error to him?" to emphasize his commitment to the deductive method. We noted (264 n. 34) the irony of this remark: Hobbes himself was exactly that stupid, on many

occasions. We were not the first to take note—Wallis, with similar ironic intent, cited the very same words on the title page of his *Due Correction for Mr. Hobbes* (1656). Another remark of Wallis, anent Hobbes: "How little he understands the mathematics from which he takes his courage." All of this bolsters our chief point: The reputation of Hobbes *as a mathematical scientist* was, by 1661, so deservedly low among important thinkers that his not having been taken seriously as a participant in scientific debate hardly needs explaining. What's puzzling is Wallis's eagerness to continue the endless exchange, even though his friends warned him that it was futile to try to wring concessions from a crank. This, however, was a matter of egotism and irascibility, not a case of political and social forces shaping scientific dispute.

3. We have been attacked (see R. Hart, "The Flight from Reason" in *Science Wars*, ed. A. Ross [Durham: Duke University Press, 1996], 259–92) for putting forth our observations on Hobbes's mathematical folly as new scholarly discoveries on our part. This is absurd. The point, which we should have thought obvious, is that these very well known tales were not only swept under the rug by Shapin and Schaffer but also ignored by hordes of supposed scholars eager to heap praise on their work. (Hart's critique on this point is bizarre, to say the least, in other respects as well; he characterizes the proposition that Hobbes was wrong in these mathematical disputes, and Wallis correct, as mere "conventional wisdom." Would that conventional wisdom were always so wise!) Whether this obvious criticism was neglected out of mere ignorance or, equally likely, out of a desire to laud the superiority of a certain brand of social analysis of science, inconvenient facts notwithstanding, the science studies community failed in an elementary duty to scrutinize sweeping claims. It should have been warned by a peculiar elision in the text of *Leviathan and the Air Pump*. On the very last page of the book, which is the conclusion of the Appendix presenting Simon Schaffer's translation of Hobbes's *Dialogus Physicus*, these words appear in brackets: "Here follows a proof, which we omit, of the duplication of the cube." In other words, in the very middle of the slanging match over the air pump, Hobbes reverted once again to his absurd mathematical pretensions, to the great amusement, no doubt, of his more-than-knowledgeable contemporaries. Can there be any question at all why Hobbes's version of deductive methodology based (so he said) on Euclid's geometry, was dismissed? As early as 1652, Seth Ward had pointed out, with respect to Hobbes's attempt to claim mathematical validity for his physical theories, "that he [i.e., Ward, writing in the third person] is sure he [i.e., Hobbes] hath much injured the Mathematicks, and the very name of Demonstration, by bestowing it upon some of his discourses, which are exceedingly short of that evidence and truth which is required to make a discourse able to bear that reputation." Although the Appendix to *Leviathan and the Air Pump* is chock-full of footnotes on all sorts of matters pertaining to *Dialogus Physicus*, there is not one word about this strange omission, or about the significance of what is left out.

4. We note that the work of Shapin and Schaffer—individual and joint—has recently come under telling attack by a number of scholars, of whom we mention Mordechai Feingold, "When Facts Matter" (*Isis* 87, no. 1 [1995]: 31–39; Cassandra Pinnick, "What's Wrong with the Strong Programme's Case Study of the 'Hobbes-Boyle' Dispute," in *A House Built on Sand*; and Alan E. Shapiro, "The Gradual Recep-

tion of Newton's Theory of Light and Color, 1672–1727," *Perspectives on Science* 4, no.1 [1996]: 59–140.) We also find it remarkable that Shapin was able to write a well-publicized book, *The Scientific Revolution* (Chicago: University of Chicago Press, 1996), which has not one reference to the invention of the calculus and its significance. All these points raise questions as to why *Leviathan and the Air Pump* and the constructivist philosophy that it incarnates were elevated to the exemplary academic status they have enjoyed.

   p. 69 In his essay "The Pioneer Defended," *New York Review of Books*, December 21, 1995, 54–58, on G. Geison's *The Private Science of Louis Pasteur* (Princeton: Princeton University Press, 1995), M. F. Perutz has called attention to the eagerness of social scientists and historians with sketchy scientific backgrounds to sit in judgment of the ethics of scientists and (more to the point) on the science itself, upon the authority of constructivism and other relativist notions. Says Perutz, "The entire approach emphasizing 'relative' truth seems to me a piece of humbug masquerading as an academic discipline; it pretends that its practitioners can set themselves up as judges over scientists whose science they fail to understand." The conclusion of this fascinating argument, illustrating the tortuous justifications offered for the kind of derogatory science-history to which Perutz was responding, and the bluntness of his response, can be found in an exchange of letters in the *New York Review of Books*, February 6, 1997, 41–42.

   p. 79 A. Plotnitsky, a deconstructive literary theorist with some mathematics and physics background, has attempted to show that the bizarre "Einsteinian constant" comment of Jacques Derrida, which has by now attracted much scornful attention from physicists, actually makes sense in the context of relativity theory ("'But It Is above All Not True': Derrida, Relativity, and the 'Science Wars,'" *Postmodern Culture*, published electronically). His effort is admirable for its ingenuity and even more so for its presumptuousness. Its accuracy is another matter, and its honesty is decidedly another matter.

   It has also been pointed out, quite correctly, that the term *differential topology* does not appear in Derrida's *Acts of Literature*; what shows up is *differantial topology*, as a translation of "topique *differantiale*." Let us assume, then, that Derrida did not authorize the translator's locution (but remember, the Wolin affair demonstrated that he can be very finicky about translation of his writing). The honor of the coinage then goes to translator Derek Attridge, a well-known American deconstructionist. We leave it to the judgment of the reader to assess the probability that the near-identity of Attridge's phrase with the mathematical one is merely adventitious.

   In any case, there has been an enormous amount of comment from our critics on this one specific matter, which, remember, was a casual illustration of an incidental point: literary theorists sometimes use jargon borrowed from scientific terminology to create the impression of rigor and congruity with "cutting edge" science. We think the point still stands, even if Attridge, rather than Derrida, is responsible for the example. Plenty of other instances can be found. For example this, from postmodern social theorist John Law, an ally of Bruno Latour, in an attempt to characterize "actor network theory":

Another . . . way of tackling the issue is to think topologically. Topology concerns itself with spatiality, and in particular with the attributes of the spatial
which secure continuity for objects as they are displaced through a space. The
important point here is that spatiality is not given. It is not fixed, a part of the
order of things. Instead, it comes in various forms. We are most familiar with
Euclideanism [sic]. Objects with three dimensions are imagined to exist precisely
within a conformable three dimensional space without violence so long as they
don't seek to occupy the same position as some other object. And so long as
their coordinates are sustained, they also retain their spatial integrity. . . .
    All of this is intuitively obvious. . . .
    . . . But studies of exotic societies suggest that there are other spatial possibili-
ties—and so does actor network theory.

(John Law, "Topology and the Naming of Complexity" [draft], published by the Cen-
tre for Social Theory and Technology, Keele University, at <http://www.keele.ac/depts/
stt/staff/jl/pubs-JL3.htm>). If the success of this analogy depends upon Prof. Law's
grasp of topology, the prospects are not bright.

    In any case, we take comfort in the fact that our critics have such difficulty finding
substantive grounds upon which to impeach our arguments that they are forced to
hammer away at this trifle.

    p. 80 Fairness obliges us to note that the well-known cultural critic Frederick Turner
is an honest admirer of natural science (e.g., Frederic Turner, *Natural Classicism*
[Charlottesville: University Press of Virginia, 1992]), an emphatic antagonist of
postmodernism, and, ultimately, an advocate of traditional aesthetic and cultural standards.

    p. 81 For a fascinating account of the purposeful and systematic rejection of sci-
ence by large parts of anthropology, and a thoughtful analysis of what will be needed
to reverse the process, see Lawrence A. Kuznar, *Reclaiming a Scientific Anthropology*
(London: AltaMira Press, 1997).

    p. 83 Since our book was written, Duke University literary critic Frank Lentricchia
has stunned the community of literary critics by an eloquent public recantation, in
which the fatuities of postmodernism and "theory" are scornfully denounced. (See F.
Lentricchia, "Last Will and Testament of an Ex-Literary Critic," *Lingua Franca*, Sep-
tember/October 1996, 59–67.)

    pp. 89–90 Prof. Andrew Ross was understandably discomfited by the analysis in
this volume. His response was to organize and edit a special number of the cultural
studies journal *Social Text* (no. 46/47 [1996]) under the rubric "Science Wars," whose
contributors included a number of people criticized in this book, along with their
sympathizers. Their intention, in large measure, was to reply to our critique. That
intention misfired to a degree when Ross and his coeditors fell into the trap set by our
friend, Alan Sokal, accepting his hilarious hoax article "Transgressing the Boundaries: The Transformative Hermeneutics of Quantum Gravity" (*Social Text*, 215–52)
at face value. This volume was subsequently republished in book form as *Science Wars*,
but *without* the Sokal piece. It sports, however, a few new essays. See the subsequent
Social Text 49 (1997) for further commentary.

Roger Hart's *Science Wars* piece, "The Flight from Reason" (see note to pages 67–68 above) alleges that in quoting the long passage (from "New Age Techniculture" and *Strange Weather*) likening the science-versus-pseudoscience distinction to the high-brow-versus-lowbrow cultural face-off, we misrepresented the views of Andrew Ross (R. Hart, in *Science Wars*, 290 n. 37). Hart points out that the paragraph following (unquoted by us) begins with Ross conceding, "I do not want to insist on a literal interpretation of this analogy" (*Strange Weather*, 26); this, claims Hart, is exculpatory. Hart's claim is either silly or sophistical. The paragraph we quoted got Ross into hot water with many scientists and philosophers, not because it pushed the analogy between "science" and "taste cultures" too far, but because it clearly insisted that the distinction between science and pseudoscience is a boundary-policing power-play, by which establishment scientists maintain their social and political hegemony. The next paragraph, although it hedges slightly on the "taste culture" analogy, does *nothing* to qualify, modify, or soften the assertion that the science-pseudoscience distinction is essentially dirty work perpetrated by scientists anxious about their status. To quote further from that paragraph:

> A more exhaustive treatment would take account of the local, qualifying differences between the realm of cultural taste and that of science, but it would run up, finally, against the stand-off between the empiricists's claim that non-context-dependent beliefs exist and that they can be true, and the culturalist's claim that beliefs are only socially accepted as true. Ultimately, the power of science rests upon making and maintaining that distinction, and we ought to recognize that science's anxiety about authenticating its belief in truths is, in the truly [sic] Foucauldian sense, a question of power. Consequently, it is not such a great leap from seeing that categories of taste are also categories of power employed to exclude the unwanted to seeing that the power of scientific ideology rests upon its unwillingness to question the role of the powerful institutions of sponsors whose interests are not only heavily mortgaged in the demarcation debate [between science and pseudoscience], but who are also well served by the hireling scientists who referee it.

Should the reader guess that this kind of immaterial quibble might be entirely characteristic of Hart's critical equipment, we would not spurn the presumption.

p. 96   The recent claims, both mathematical and philosophical, of Ilya Prigogine have been subjected to withering scrutiny by the physicist and specialist in statistical mechanics Jean Bricmont. (See his essay "Science of Chaos or Chaos in Science?" in *The Flight from Science and Reason*, 131–75.)

p. 97   We should make clear that when we speak of the process of snowflake formation, we mean to include the entire system: warm water-vapor, cold air, etc. It is the entropy of the whole system that increases, of course.

p. 98   N. Katherine Hayles responds to this critique with a rejoinder ("Consolidating the Canon," in *Science Wars* [book version], 226–37). We recommend this to our readers as a specimen of Hayles's logical rigor and scrupulous handling of evidence. Says Hayles, "One of the grotesque exaggerations in which Levitt and Gross indulge is the fantasy that the cultural and social studies of science are responsible for cuts in funding for basic scientific research" (*Science Wars*, 234). Readers of the present text are invited to scrutinize it, line by line or even word by word, for *any* support of Hayles's statement.

p. 100 For an extensive and devastating analysis of Hayles's "Gender Encoding in Fluid Mechanics," see P. A. Sullivan, "An Engineer Dissects Two Case Studies: Hayles on Fluid Mechanics, and MacKenzie on Statistics," in *A House Built on Sand*.

p. 101 The lingering doubts about the correctness of general relativity on the cosmic scale still linger. Recently, for instance, there has been renewed interest in the cosmological constant and similar modifications.

p. 110 In stating that "the only widespread, *obvious* discrimination today is against white males," we may have created, inadvertently, the impression that we think there is widespread, obvious discrimination against white males in science. For that confusion, we apologize. Such is not the case, by and large. Science departments have, *on the whole*, avoided the kind of race-and-sex-conscious hiring and retention policies that have been commonplace in other areas of academic life, just as they have been more successful in avoiding exclusionary practices of the traditional kind. (We cannot, and do not, deny that there have been exceptions of both kinds.) We note also, however, that there are precious few science departments that have not been told by a dean or provost, in the decade now ending, "We don't have a faculty line for you, but, if you find a qualified woman or minority candidate . . ." What effect this has had in these days of chronic understaffing, we do not pretend to know. Whether this practice constitutes discrimination against white males is a question that we leave to the judgment of our readers.

pp. 113 and 272–73, n. 7 "Toward a Feminist Algebra" has now appeared as an essay in S. V. Rosser, ed., *Teaching the Majority* (New York: Teachers College Press, Columbia University, 1995), 127–44. We hope, somewhat forlornly, that its influence will be limited.

pp. 116–17, 205 and 253 A current experiment in teaching mathematics "multiculturally," as well as with a feminist slant, is worth noting. It is a course, offered at the State University of New York at Plattsburgh, designed to fulfill the mathematics requirement for non-science majors. Here is a list of its "objectives" as given in the syllabus:

After taking this course the student will be able to:

1. Describe the political nature of mathematics and mathematics education.
2. Describe gender and race differences in mathematics and their sociological consequences.
3. Examine the factors influencing gender and race differences in mathematics.
4. Critically evaluate Eurocentrism and androcentrism in mathematics.
5. Describe the role the culture plays in the development and learning of mathematics.
6. Give examples of the historical role of women and people of color in mathematics.

7. Critically evaluate research on the relationship of gender and culture to mathematics and mathematics education.

In the context of this course, the descriptions asked for are already inflexibly determined; they are built into the reading list and lectures. The instructor's clear task is to make sure the answers come out in a way that will satisfy the right-thinking on all points. Just in case someone might be tempted to slip into unorthodoxy, the course requires that a journal be kept so that the instructor can keep weekly track of the progress of the student's opinions. Arts and crafts are not neglected; the student is to make an Incan quipu, a quilt square (presumably illustrating nonmale mathematics), or an African board game. A typical lecture assigns readings on "fighting Eurocentrism," the role of mathematics in "building a democratic and just society," becoming "critically numerate," and "justice and equity, and mathematics instruction."

pp. 117–22 The argument that there is bias in developmental biology *for* sperm and *against* eggs continues to be made, publicly and with strong emphasis of its exemplary character, despite its perfect emptiness. The usual system of circular quotation of a politically favorable story that happens not to be true (but what matter?) is in full career. E. Fox Keller, H. Longino, N. Tuana, E. Martin, B. Spanier, and others cite one another and, of course, the Biology and Gender Study Group, as authoritative sources on the matter. So far as we can see, the only source in the recent scientific literature is the one cited by the BGSG (Schatten and Schatten), and that is a gloss on earlier research of Schatten and Mazia, the conclusions of which give no support to the statement that an aggressive egg "grabs" the spermatozoon, with the latter contributing nothing to the interaction. In fact the Schatten-Mazia conclusions are to the opposite effect: the morphological change signaling effective interaction starts with the acrosome reaction of the sperm head. Martin refers to certain claims that the force exerted by the spermatozoon's flagellum is "weak," and others in this game report with glee and astonishment that many spermatozoa are malformed, or swim in circles and don't seem to know where they are going. These "observations" are textbook material in embryology and in the clinical literature of fertility and sterility. They have nothing to do with the spermatozoon's "incompetence" or the egg's "aggressiveness" in fertilization. An effort to compare this popular chapter in the book of feminist science-critique with the facts, biological and historical, has been made by one of us in "Bashful Eggs, Macho Sperm, and Tonypandy," in *A House Built on Sand*; but the effort may well be in vain. You can't keep a good story down.

p. 141 For further analysis of Keller's misrepresentation of McClintock's work and scientific style, see our essay in the AAUP house-organ, *Academe* (N. Levitt and P. R. Gross, "Academic Anti-Science," *Academe*, November–December, 1996, 38–42) We might strengthen an observation somewhat: There are no "developments" in the history and philosophy of science that prove, or even plausibly suggest, a social construction of the final product of empirical science. Finally, we note that the recent cloning of Dolly has been the signal for a remarkable outpouring of commentary from self-appointed as well as professional ethicists. Leaving the latter aside for now, it is clear from the asseverations of the former that most of them have no idea of the technique of nuclear transplantation, which dates back to the late 1950s and early 1960s, or of what the Edinburgh group did to modify it so that it works in mammals (rather than in amphibians, in which Briggs and King, but especially John Gurdon, using donor nuclei from differentiated cells, showed thirty years ago that it works pretty well). The socio-political bombast has included quite a lot of commentary on DNA that displays no understanding—still—of the roles of genes, chromosomes, nuclei, and the egg cytoplasm in the reproduction and development of multicellular animals (like us).

pp. 146 and 230 An interesting new twist in the history of feminist dogmatism concerning the "social construction" of gender roles has just emerged. One of the most celebrated cases supposedly proving that sexual identity is plastic at birth and that environmental cues push it in either a male or female direction irrespective of physiology was that of an unfortunate male infant whose penis was severed in a surgical accident. The decision of physicians and psychologists at the time (some thirty-five years ago), acting under a version of this theory already firmly entrenched, was to castrate the child, surgically create the external female genitalia, and raise the child as a girl, with hormone treatments administered chronically to foster the development of female secondary sex characteristics. The supposedly successful adjustment to this gender reassignment has long been celebrated by feminist theorists as confirming the "constructed" nature of masculinity and femininity.

It has recently been revealed, however, that, unfortunately for the theory, the child never accepted a "female" identity, engaging repeatedly and defiantly in "male" behavior (e.g., trying to urinate in a standing position). Finally, the unhappy parents confessed the truth, and the child swiftly adopted an emphatically male identity—with the aid of further reconstructive surgery. He is now married. We wonder what Professor Longino makes of this particular case.

pp. 153, 178 Irony suffuses the role assigned to Francis Bacon by recent feminist theorists. Supposedly, it was Bacon who bequeathed modern science a covert set of values encoding the male (scientific) desire for conquest, domination, and penetration of the female (nature). But see Aubrey's "brief life" of Bacon:

> He was a παιδεραστής. His Ganameds and Favourites tooke Bribes; but his Lordship always gave Judgement *secundum æquum et bonum.* His decrees in Chancery stand firme, i.e. there are fewer of his Decrees reverst then of any other Chancellor.
>
> His Dowager married her Gentleman-usher Sir Thomas (I thinke) Underhill, whom she made deafe and blinde with too much of Venus.

Does this mean that contemporary science ought to be regarded as a branch of "queer studies"?

p. 157 The difficulty of estimating the Earth's carrying capacity for human population, to which we refer, gets a certain amount of scoffing from the Chicken Littles of the left and the Micawbers on the right. The former insist that we already know the carrying capacity and have exceeded it; the latter that the whole question is a preoccupation of eggheads and purveyors of class envy. For a scientifically sound assessment of the question, establishing both its difficulty and its gravity, see the recent volume by Joel E. Cohen, head of the Laboratory of Populations at Rockefeller University, *How Many People Can the Earth Support?* (New York: W. W. Norton, 1995). Most recently, demographers have concluded that the world's population will *not* double in the twenty-first century; but that it *will* grow from the current 5.8 billion to about 10 billion in 2050. Population growth is slowing down in the West and in Eastern Europe, but it is very rapid in the Middle East, sub-Saharan Africa, and North Africa, where a tripling of population is expected by 2050. Worldwide, there will be a major demographic shift toward middle- and old age (Reuters, London, June 19, 1997).

The idea that all our neighbors in the solar system are sterile has lately come under challenge with the discovery of what might be interpreted as traces of microbial life in meteorites of Martian origin, and with the confirmation that the Jovian satellite Europa is covered by a huge ocean of liquid water under a crust of ice, which suggests an interior source of heat.

pp. 158–74 Recently, the scientific consensus has moved in the direction of greater certainty of a significant global warming due to anthropogenic emission of greenhouse gases. See S. Schneider, *Laboratory Earth: The Planetary Gamble We Can't Afford to Lose* (New York: Basic Books, 1997); T. R. Karl, N. Nicholls, and J. Gregory, "The Coming Climate," *Scientific American*, May 1997, 78–83; and John Houghton, *Global Warming: The Complete Briefing* (Oxford: Lion Books, 1994). The fact remains, however, that we have as yet no unequivocal signal, that is, a measurement or a series of climate events that all or most atmospheric scientists agree *is* the signal, of global warming of anything like the magnitude predicted by the available climate models. The atmosphere has in fact cooled, according to the most reliable measurements (by satellites), about 0.1°C since 1979. It is almost certain that anthropogenic greenhouse gases added to the atmosphere at the present rate will eventually warm the planet; but at what rate, by how much, with what sort of geographic variation, and with what specific climatic consequences remain questions in need of *much* better answers than we have. Were not politics involved (which would of course be impossible), there would be much more research and much less posturing. Because it was a project supported by the Competitive Enterprise Institute, a collection of essays entitled *The True State of the Planet* (ed. Ronald Bailey [New York: Free Press, 1995]) is dismissed out of hand by ecoradicals and ecopoliticians; but its authors include distinguished scientists and economists and it is worth reading on environmental issues like this.

p. 161 The persistence of the habit, among some environmental activists, of excluding hydroelectric power from the "solar" category was recently illustrated by the comments of Ralph Nader during a broadcast (PBS *Frontline*, April 22, 1997) on the ongoing debate about nuclear power. Nader mentioned windmills and photovoltaic sources as solar alternatives to nuclear, but not hydroelectric—except to the extent that he called for exploitation of tidal power as well, identifying this one type of hydroelectric power as "solar" when, in fact, the ultimate source of energy for tidal power is not the flux of solar energy, but rather the kinetic energy of the earth's rotation.

p. 163 See M. Fumento, *Science under Siege* (New York: Quill, William Morrow, 1993) for a well-documented account of how much and to whose benefit the dioxin danger has been overplayed.

pp. 165 and 247 Recently, the well-confirmed idea that Native Americans are descendents of immigrants from Asia has encountered increasingly vituperative rejection from Indian-rights activists. Most prominent among them has been Prof. Vine Deloria, whose book *Red Earth, White Lies* (New York: Simon and Schuster, 1996) explicitly rejects Western science and its conclusions about the peopling of the Western Hemisphere. Deloria's writing shows substantial influence of precisely those postmodernist and constructivist critiques of science that this book scrutinizes. See also M. Gladwell's curious account (*New Yorker*, November 11, 1996, 36–38) of a speech by Alan Sokal and the audience reaction thereto.

p. 174 Of late, there is some empirical evidence associating the ozone hole with temporary decreases in the populations of plankton and some fish species in the Antarctic. It is not yet clear that this represents a serious threat to the biota. Empirical evidence is of course the *sine qua non*; but it must always be evaluated. Commonly lacking in the public discussion of environmental worries, themselves reasonable, is reasonable evaluation. The *Boston Globe* carried, for example, a full lead story on a putative epidemic of malformed frogs (Scott Allen, July 28, 1997, B1–B3). Among the proposed culprits are pesticides and excess ultraviolet radiation due to the ozone hole. The article indicates obliquely that such "explanations" are contradicted by available data: the malformations are seen where pesticides are not and that there is no evidence of increased UV radiation at those loci. But only the final paragraph of this alarming story addresses the primary "evidence":

> But researchers' first concern is simply determining the extent of the abnormalities. The data still consists [sic] mainly of anecdotal reports in most states. Until last week, the main information on deformities in Vermont had come from a one-day outing last October in which state officials found that 13.1 percent of the frogs they collected had abnormalities.

pp. 179–82, 190 The recent development of a whole class of drugs called protease inhibitors, in addition to AZT, has led to many cases in which the progress of AIDS seems to have been arrested. (For example, Magic Johnson, who is under treatment with such drugs, has announced that all direct signs of the virus are absent from his system.) Whether this treatment will live up to its promise and be developed to the point where it is more widely available and applicable than now, or whether, on the contrary, it will ultimately be defeated by the mutability of HIV genes, remains to be seen. In any case, it has produced remarkable results in the clinic and by the hard evidence of viral load reduction. It is the first medical intervention that has allowed AIDS investigators and clinicians to speak of "management" of the disease, if not of a "cure."

pp. 182–87 Issues involving the origin of the AIDS epidemic and the existence of "high risk" populations retain their power to stir up strong feelings. The role of the unfettered sexual behavior of members of the gay community in the late '70s and early '80s in spreading AIDS is still a painful topic to many gay activists. Gabriel Rotello's new book, *Sexual Ecology: AIDS and the Destiny of Gay Men* (New York: Dutton, 1997), emphasizes the link, and pleads for a new and more restrained sexual ethic among gay men to prevent another flare-up of the AIDS epidemic. In reviewing Rotello's book for the *Nation* ("Epidemic Arguments," May 5, 1997, 27–28), Martin Duberman, an activist as well as a distinguished social historian, accepts this judgment on the origins of the epidemic, but clearly feels uncomfortable having to do so. On the other hand, the same issue of the *Nation* carries a review ("Reality Bites," 31–33) of Katie Roiphe's new *Last Night in Paradise* (New York: Little, Brown, 1997) by Emily Gordon, who is clearly incensed at Roiphe's (correct) assumption that AIDS is not a major threat to the heterosexual, non-drug-using, "general community" in this country. Roiphe's heresy, which is to say, Michael Fumento's, is not to be allowed a disinterested examination, even in 1997, by those committed to gay identity politics and the like. On the other hand, see Chandler Kerr, "The AIDS Exception: Privacy vs. Public Health," (*Atlantic Monthly*, June 1997, 57–67), for a penetrating analysis that confronts these pieties. It's a safe guess that four or five years ago Kerr's piece could not have been published in any such mainstream forum.

pp. 184–86 The Duesberg arguments still manage to command inordinate attention, even though they are rejected, almost unanimously, by serious AIDS investigators. Extensive debates on the topic have appeared in *Science* and the *New York Review of Books*. (See, for instance, a series of pieces on Duesberg by Jonathan Cohen, "The Duesberg Phenomenon," *Science*, December 9, 1994, 1642–49, and the correspondence between Duesberg and his critics in several succeeding issues. For its part, *The New York Review of Books* published Richard Horton's "Truth and Heresy about AIDS" [May 23, 1996, 14–20], followed by an exchange between Duesberg and Horton [August 8, 1996, 51–52].)

pp. 186, 189–95 Steven Epstein has expanded his analysis of the relation between AIDS scientists and the gay community into a book, *Impure Science: AIDS, Activism, and the Politics of Knowledge* (Berkeley: University of California Press, 1996), in which he claims that the active role of the gay community in monitoring and redesigning clinical AIDS research amounted, in some sense, to reinventing science. Although some of Epstein's claims are advanced with Foucauldian and postmodernist flourishes, there is some truth to them. Large numbers of activists representing a cohesive, well-educated gay community did, in fact, a considerable amount of homework to learn some of the science and methodology involved, and did put pressure on the research community and the regulatory agencies. Their efforts altered various traditional protocols so that largely untried drugs could by obtained by HIV-infected persons and so that drug trials could be done without placebo methodology. Sociologically, this is an important phenomenon, but it is *not* true that it effected revision of the canons of scientific validity. We might summarize this complicated ethical and methodological situation as follows:

1. Numbers of HIV-infected individuals did get access to speculative treatments that more rigid observance of standard cautions and research designs would have denied them. If nothing else, this provided the psychological comfort of hope in the face of what is generally considered a hopeless disease.

2. Clinical researchers learned a great deal about the diplomacy, tact, and flexibility needed to deal with an aroused and militant community in order to secure its cooperation in testing therapy. Scientists reconciled themselves to the necessity of settling for less efficient experimental designs in order to secure the cooperation of subjects who would have rebelled against a more stringent system.

3. Compromising the canons of methodology did, in all probability, retard the acquisition of reliable knowledge about AIDS and modes of treatment, although how severe this delay will prove to have been is of course unknown.

4. The emphasis on treatments suitable for an urban, relatively affluent, knowledgeable community probably diverted some work that might have been more germane to the situations in Africa and Asia, where a vastly larger population is infected by the virus or stands at risk. (It is noteworthy that Epstein pays no attention to the AIDS situation outside North America.)

5. The origins and development of the protease inhibitor therapeutic strategy owes little to "science" created by gay activists and almost everything to relentless traditional research in molecular biology and virology laboratories and in the theory of infectious disease.

An interesting commentary on Epstein's book can be found in S. J. Heginbotham's review, "The Power of HIV-Positive Thinking," *Sciences*, May/June 1997, 38–42.

p. 198 The influence of animal rights activist groups, at least in the area of medical research, seems to be on the wane. Part of the reason is the adamant opposition of AIDS victims, their families, and articulate AIDS activists, to any interference with medical research in this area. On a more frivolous note, fur coats and the like have enjoyed a significant revival in popularity, while the antifur movement seems to have lost steam and is no longer quite so fashionable as it once was. However, not all signs are good. Recently, the young winner of a high school science competition was denied his prize for a time on the grounds that his research had been cruel to fruit flies! In Massachusetts, a veterinarian (Dr. Richard Rodger) was attacked on a Cape Cod golf course by a charging Canada goose. The good doctor defended himself by swiping at the goose with his golf club, injuring her. He is now in pretrial hearings, charged with cruelty to animals. Judge and District Attorney have recommended that he plead guilty, go on probation, and pay $3,500 in restitution to the Humane Society. At the time of writing, the accused has rejected such plea agreements in favor of a simple "not guilty" (*Cape Cod Times*, June 19, 1997, 1).

If that needs a topper, consider the efforts of the animal-rights monitors on the set of the popular film *Men in Black*; their task was to see to it that none of the many cockroach "extras" used in the film were killed, harmed, or inconvenienced. Of course, once filming ended for the day, exterminators arrived to see to it that no escaped bit-players made themselves permanently at home (David Shenk, "Star Treatment for *Men in Black*'s Six-legged Extras," *New Yorker*, July 21, 1997, 25–26). On a less whimsical note, consider the continuing indulgence of the putatively serious press towards animal rights rhetoric; for example, Joy Williams, "The Inhumanity of the Animal People," *Harper's Magazine*, August 1997, 60–67. (The Williams piece was only ambiguously and partially countered in the same issue of *Harper's* by Wedeline L. Wagner, "They Shoot Monkeys, Don't They?" 27–30.) And, a few months ago, the once austerely scientific *Scientific American* ran an extensive debate on the ethics and efficacy of research on animal subjects. (See *Scientific American*, February 1997, in particular Neal D. Barnhard and Stephen R. Kaufman, "Animal Research Is Wasteful and Misleading," 80–82, and Jack H. Botting and Adrian R. Morrison, "Why Animal Research is Vital to Medicine," 83–85.) Needless, perhaps, to say, no new evidence for the "wastefulness" of research on animals, or of its having "misled" medical science, has been forthcoming since more than a decade ago, when one of us (P. R. G.) organized the report of a consensus conference of the National Institutes of Health, at which all the animal rights groups had their passionate say.

p. 210 One of us (N. L.) has had an opportunity to give a guest lecture on science and culture to a class supposedly studying the cultural strands that make up American society. A suggestion that the tales told by Ivan Van Sertima in his various books should not be taken too seriously brought forth loud cries of indignation from the numerous black students in the class. It was a depressing experience.

p. 223 The recent success of the film Apollo 13 and the surging popularity of the comic strip Dilbert offer the hope that admiration for techno-nerds is not entirely dead in the hearts of their fellow Americans.

p. 246 The attempt by the National Academy of Sciences National Research Council to set up nationwide standards for science education was heavily influenced initially by social-constructivist "philosophers" of science appointed to the panel assigned to formulate them. Fortunately, the objections of alarmed scientists were eventually respected and the standards rewritten to eliminate these eccentricities.

Similarly, the Smithsonian Institution's costly exhibit Science in American Life was initially created under the influence of a number of constructivist historians and sociologists of science whom the curators of the institution, responsive to academic fashion, procured for the task. The protests of the American Chemical Society, which provided the funding, were ineffectual until the exhibit opened. The indignation of a few members and organizations of the scientific community (especially the American Physical Society) eventually brought about some changes. But Paul Forman, a Smithsonian historian of science and knight of postmodern science studies, exults in the insignificance of those changes. See Paul Forman, "Assailing the Seasons," Science, May 2, 1997, 750–52.

Both these incidents are recounted and analyzed in G. Holton, "Science Education and the Sense of Self," in The Flight from Science and Reason, 551–60.

p. 251 In their book Professing Feminism: Cautionary Tales from the Strange World of Women's Studies (New York: Basic Books, 1994), N. Koertge and D. Patai speculate that large doses of feminist theory, especially as regards the nature of science, demoralize and discourage young women. Their thesis is borne out by the indignant rejection of such "feminism" by mathematician Mary Beth Ruskai, "Are 'Feminist Perspectives' in Mathematics and Science Feminist?" in The Flight from Science and Reason, 437–42. Ruskai regards such "perspectives" as condescending at best and discerns a specifically anti-feminist bias in them.

p. 252 The suggestion, which we regarded as quite modest, that appropriate scientists should be among those consulted in judging, for promotion and tenure purposes, work of humanists or social scientists that purports to deliver judgments on the content or methods of science has provoked more howls of protest than anything we have said. Langdon Winner, a member of the science studies program at Rensselaer Polytechnic Institute, has told the readers of his review of this book ("Sheriffs of Scientific Correctness," Technology Review, February 1995, 74) that we propose that all faculty with left-wing views should be excluded from university life. Similar charges were made by Basil O'Neill in the (London) *Times Higher Education Supplement* ("Here Be Dragons," July 1, 1994, 23) and by Berkeley philosopher Elisabeth Lloyd ("Science and Anti-Science: Objectivity and Its Real Enemies," in L. H. Nelson and J. Nelson, eds., *Feminism, Science, and Philosophy of Science* [Dordrecht: Kluwer, 1996]). We need not refute this; we trust our text already does so, quite explicitly. It is interesting to note that Winner's review for *Technology Review* was solicited and published only after another (largely favorable) review, commissioned and paid for, was received and then spiked. We regard this incident, along with some similar ones, as the highest tribute to our powers of persuasion. As regards the remarkably lengthy indictment by E. A. Lloyd, see P. R. Gross, "Evidence-Free Forensics," in *A House Built on Sand.*

p. 266, n. 12 E. Roudinesco's newly translated biography *Jacques Lacan* (New York: Columbia University Press, 1997) makes it clear that the late psychoanalyst had an obsessive interest in topology—surfaces, knot theory, and the like—and tried to incorporate it into the foundations of his theoretical work. Alas, his mathematical talent seems to have been minimal. His efforts vitiated, rather than enhanced, the rigor of his theories. They can only be called addlepated.

References

Books

Adams, Hunter Havelin III. *African-American Baseline Essays—Science Baseline Essay: African and African-American Contributions to Science and Technology.* Portland: Multnomah School District 1J, Portland Public Schools, 1990.

Alexander, Judd H. *In Defense of Garbage.* Westport, Conn.: Praeger, 1993.

Alperovitz, Gar. *Atomic Diplomacy: Hiroshima and Potsdam.* New York: Simon and Schuster, 1965.

Andreski, Stanislav. *Social Sciences as Sorcery.* New York: St. Martin's Press, 1972.

Appleyard, Bryan. *Understanding the Present: Science and the Soul of Modern Man.* London: Picador, 1992.

Argyros, Alexander J. *A Blessed Rage for Order.* Ann Arbor: University of Michigan Press, 1992.

Aronowitz, Stanley. *Science as Power: Discourse and Ideology in Modern Society.* Minneapolis: University of Minnesota Press, 1988.

Aronowitz, Stanley, and Henry Giroux. *Postmodern Education: Politics, Culture, and Social Criticism.* Minneapolis: University of Minnesota Press, 1989.

Ayer, A. J. *Language, Truth, and Logic,* 2d ed. New York: Dover Publications, 1946.

Baudrillard, Jean. *Le xerox et l'infiniti.* Paris, 1987.

Berlin, Isaiah. *The Crooked Timber of Humanity.* New York: Alfred A. Knopf, 1991.

Berman, Morris. 1989. *Coming to Our Senses: Body and Spirit in the Hidden History of the West.* New York: Bantam Books, 1989.

———. *The Reenchantment of Nature.* Ithaca: Cornell University Press, 1981.

Best, Steven, and David Kellner. *Postmodern Theory: Critical Interrogations.* New York: Macmillan/Guilford, 1990.

Biehl, Janet. *Rethinking Ecofeminist Politics.* Boston: South End Press, 1991.

Bleier, Ruth, ed. *Feminist Approaches to Science.* New York: Pergamon, 1986.

Blumenfeld-Kosinski, Renate. *Not of Woman Born: Representations of Caesarian*

Birth in Medieval and Renaissance Culture. Ithaca: Cornell University Press, 1990.

Boyce, William E. and Richard C. Di Prima. *Elementary Differential Equations,* 5th edition. New York: John Wiley and Sons, 1992.

Bramwell, Anna. *Ecology in the Twentieth Century: A History.* New Haven: Yale University Press, 1989.

Branch, Taylor. *Parting the Waters: America in the King Years, 1954–68.* New York: Simon and Schuster, 1988.

Chalmers, Alan. *Science and Its Fabrication.* Minneapolis: University of Minnesota Press, 1991.

Cheng, Nien. *Life and Death in Shanghai.* New York: Grove Press, 1988.

Clark, Ronald. *Einstein: The Life and Times.* New York: World Publishing, 1971.

Crary, Jonathan, and Sandford Kwinter, eds. *ZONE 6—Incorporations.* Cambridge: MIT Press, Zone Books, 1992.

Davidson, Eric H. *Gene Activity in Early Development.* New York: Academic Press, 1986.

Day, David. *The Eco Wars.* Toronto: Key Porter Books, 1989.

De Broglie, Louis. *Physics and Microphysics.* Translated by Martin Davidson. New York: Grosset and Dunlap, 1955.

DeGrazia, Alfred, et al. *The Velikovsky Affair: Scientism versus Science.* London: Sphere, 1978. [Reprint from *American Behavioral Scientist,* September, 1963]

Denitch, Bogdan. *After the Flood: World Politics and Democracy in the Wake of Communism.* Hanover: University Press of New England, Wesleyan University Press, 1992.

Derrida, Jacques. *Acts of Literature.* Translated by Derrick Attridge. New York: Routledge, 1992.

Diggins, John Patrick. *The Rise and Fall of the American Left.* New York: W. W. Norton, 1992.

Dyson, Freeman. *Infinite in All Directions.* New York: Harper and Row, 1988.

Edelman, Gerald M. *Bright Air, Brilliant Fire: On the Matter of Mind.* New York: Basic Books, 1992.

———. *The Remembered Present: A Biological Theory of Consciousness.* New York: Basic Books, 1989.

Edelman, Gerald, and Vernon Mountcastle. *The Mindful Brain.* Cambridge: MIT Press, 1978.

Ekland, Ivar. *Mathematics and the Unexpected.* Chicago: University of Chicago Press, 1989.

Eysenck, Hans. *Decline and Fall of the Freudian Empire.* New York: Viking Press, 1985.

Fausto-Sterling, Anne. *Myths of Gender: Biological Theories about Women and Men.* New York: Basic Books, 1985.

Ferguson, Harvie. *The Science of Pleasure.* London and New York: Routledge, 1990.

Fomenko, Anatolii T. *Mathematical Impressions*. Providence: American Mathematical Society, 1990.

Foner, Philip S. *History of the Labor Movement in the United States*. Vols. 3 and 4. New York: International Publishers, 1973, 1974.

Foreman, Dave. *Confessions of an Eco-Warrior*. New York: Harmony Books, 1992.

Foucault, Michel. *The Birth of the Clinic: An Archaeology of Medical Perception*. Translated by A. M. Sheridan Smith. New York: Random House, 1973.

———. *Discipline and Punish: The Birth of the Prison*. Translated by Alan Sheridan. New York: Random House, 1977.

———. *The History of Sexuality, Volume I: An Introduction*. Translated by Robert Hurley. New York: Random House, 1978.

———. *Madness and Civilization*. Translated by Richard Howard. New York: Random House, 1973.

Freedman, Bill. *Environmental Ecology*. New York: Academic Press, 1989.

Fumento, Michael. *The Myth of Heterosexual AIDS*. New York: Basic Books, 1990.

———. *Science under Siege: Balancing Technology and the Environment*. New York: William Morrow, 1993.

Garry, Ann, and Marilyn Pearsall. eds. *Women, Knowledge, and Reality: Explorations in Feminist Philosophy*. Boston: Unwin Hyman, 1989.

Gay, Peter. *The Party of Humanity*. New York: W. W. Norton, 1971.

Gerdes, Paulus. *Studies in Marxism*. Vol. 16. *Marx Demystifies Calculus*. Minneapolis: MEP Publications, 1984.

Gilbert, Scott F. *Developmental Biology*, 3d ed. Sunderland Mass.: Sinauer Associates, 1991.

Gleick, James. *Chaos: Making a New Science*. New York: Viking Press, 1987.

Golub, Edward S., and Douglas R. Green. *Immunology—A Synthesis*, 2d ed. Sunderland Mass.: Sinauer Associates, 1991.

Gore, Albert, Jr. *Earth in the Balance*. Boston: Houghton Mifflin, 1992.

Gornick, Vivian. *The Romance of American Communism*. New York: Basic Books, 1974.

Gould, Stephen J. *Bully for Brontosaurus*. New York: W. W. Norton, 1991.

———. *The Mismeasure of Man*. New York: W. W. Norton, 1981.

———. *The Panda's Thumb*. New York: W. W. Norton, 1981.

———. *An Urchin in the Storm*. New York: W. W. Norton, 1987.

Grossberg, L., C. Nelson, and Paula Treichler, eds. *Cultural Studies*. New York: Routledge, 1992.

Harding, Sandra. *The Science Question in Feminism*. Ithaca: Cornell University Press, 1986.

———. *Whose Science? Whose Knowledge? Thinking from Women's Lives*. Ithaca: Cornell University Press, 1991.

Hawking, Steven W. *A Brief History of Time*. New York: Bantam Books, 1988.

Hayles, N. Katherine. *Chaos Bound: Orderly Disorder in Contemporary Literature and Science.* Ithaca: Cornell University Press, 1990.

Hirsh, Marianne, and Evelyn Fox Keller, eds. *Conflicts in Feminism.* New York: Routledge, 1990.

Hodges, Andrew. *Alan Turing: The Enigma.* New York: Simon and Schuster, 1983.

Holton, Gerald. *The Thematic Origins of Scientific Thought: Kepler to Einstein.* Cambridge: Harvard University Press, 1988.

Hrbacek, Karel, and Thomas Jech. *Introduction to Set Theory*, 2d ed. New York: Marcel Dekker, 1984.

Hughes, Robert. *Culture of Complaint: The Fraying of America.* New York: Oxford University Press, 1993.

Keller, Evelyn Fox. *A Feeling for the Organism: The Life and Work of Barbara McClintock.* New York: W. H. Freeman, 1983.

Kimball, Roger. *Tenured Radicals: How Politics Has Corrupted Higher Education.* New York: Harper and Row, 1990.

Klehr, Harvery. *The Heyday of American Communism.* New York: Basic Books, 1974.

Kramer, Larry. *Reports from the Holocaust: The Making of an AIDS Activist.* New York: St. Martin's Press, 1989.

Kuhn, Thomas. *The Structure of Scientific Revolutions*, 2d ed. Chicago: University of Chicago Press, 1970.

Lather, Patti. *Getting Smart: Feminist Research and Pedagogy with/in the Postmodern.* New York: Routledge, 1991.

Latour, Bruno. *Science in Action: How to Follow Scientists and Engineers through Society.* Cambridge: Harvard University Press, 1985.

Laudan, Larry. *Science and Relativism.* Chicago: University of Chicago Press, 1990.

Layzer, David. *Cosmogenesis: The Growth of Order in the Universe.* New York: Oxford University Press, 1990.

Le Guin, Ursula K. *Always Coming Home.* New York: Harper and Row, 1985.

Lehman, David. *Signs of the Times: Deconstruction and the Fall of Paul de Man.* New York: Poseidon Press, 1991.

Lenin, V. I. *Materialism and Empirio-Criticism.* Moscow: Progress Publishers, 1970.

Lewis, Martin. *Green Delusions: An Environmentalist Critique of Radical Environmentalism.* Durham: Duke University Press, 1992.

Lillie, Frank R. *The Woods Hole Marine Biological Laboratory.* Chicago: University of Chicago Press, 1941.

Longino, Helen. *Science as Social Knowledge.* Princeton: Princeton University Press, 1990.

Loubere, Leo. *Utopian Socialism: Its History since 1800.* Cambridge, Mass.: Schenkman, 1974.

Lyotard, J. F. *The Postmodern Condition: A Report on Knowledge*. Translated by Geoff Bennington and Brian Massunn. Minneapolis: University of Minnesota Press, 1984.

Manning, Kenneth R. *Black Apollo of Science: The Life of Ernest Everett Just*. New York: Oxford University Press, 1983.

Merchant, Carolyn. *The Death of Nature: Women, Ecology, and the Scientific Revolution*. San Francisco: Harper and Row, 1980.

——. *Radical Ecology: The Search for a Livable World*. New York: Routledge, 1992.

Michaels, Patrick. *Sound and Fury: The Science and Politics of Global Warming*. Washington, D.C.: Cato Institute, 1992.

Moore, Deborah Lela. *The African Roots of Mathematics*. Detroit: Professional Publishing, 1992.

Murray, Robert K. *Red Scare: A Study in National Hysteria*. New York: McGraw-Hill, 1974.

Navasky, Victor S. *Naming Names*. New York: Viking Press, 1980.

O'Neill, William L. *A Better World*. New York: Simon and Schuster, 1982.

Ortiz de Montellano, Bernard. *Aztec Medicine, Nutrition, and Health*. New Brunswick: Rutgers University Press, 1992.

Paglia, Camille. *Sex, Art, and American Culture*. New York: Vintage Books, 1992.

Patton, Cindy. *Inventing AIDS*. New York: Routledge, 1990.

Peat, F. David. *Einstein's Moon*. Chicago: Contemporary Books, 1990.

Penrose, Roger. *The Emperor's New Mind*. New York: Oxford University Press, 1989.

Rauch, Jonathan. *Kindly Inquisitors: The New Attacks on Free Thought*. Chicago: University of Chicago Press, 1993.

Ray, Dixy Lee (with Lou Guzzo). *Trashing the Planet*. Washington, D.C.: Regnery Gateway, 1990.

Regan, Tom. *All That Dwell Therein: Essays on Animal Rights and Environmental Ethics*. Berkeley: University of California Press, 1982.

——. *The Case for Animal Rights*. Berkeley: University of California Press, 1983.

Rifkin, Jeremy. *Algeny*. New York: Viking Press, 1983.

——. *Beyond Beef: The Rise and Fall of the Cattle Culture*. New York: Dutton, 1992.

——. *Entropy: Into the Greenhouse World*. New York: Bantam Books, 1989.

Ross, Andrew. *Strange Weather: Culture, Science and Technology in the Age of Limits*. London: Verso, 1991.

Rosser, Sue V. *Teaching Science and Health from a Feminist Perspective*. New York: Pergamon Press, 1986.

Root-Bernstein, Robert S. *Rethinking AIDS: The Tragic Cost of Premature Consensus*. New York: Free Press, Macmillan, 1993.

Ruelle, David. *Chance and Chaos*. Princeton: Princeton University Press, 1992.

Russell, Bertrand. *The Autobiography of Bertrand Russell.* Boston: Atlantic Monthly Press, 1968.

Scrivenor, Patrick. *Egg on Your Interface: A Dictionary of Modern Nonsense.* Tolworth, Surrey: Buchan and Enright, 1989.

Shapin, Steven, and Simon Schaffer. *Leviathan and the Air Pump.* Princeton: Princeton University Press, 1985.

Sherrington, Sir Charles. *Goethe on Nature and on Science,* 2d ed. Cambridge: Cambridge University Press, 1949.

Singer, Peter R. *Animal Rights: A New Ethic for Our Treatment of Animals.* New York: Avon Books, 1975.

Snow, C. P. *The Two Cultures and the Scientific Revolution.* New York: Cambridge University Press, 1962.

Steiner, George. *In Bluebeard's Castle.* New Haven: Yale University Press, 1971.

Thomas, George B., and Ross L. Finney. *Calculus.* New York: Addison-Wesley, 1992.

Tuana, Nancy, ed. *Feminism and Science.* Bloomington: Indiana University Press, 1989.

Van Sertima, Ivan, ed. *Blacks in Science, Ancient and Modern.* New Brunswick, N.J.: Transaction Books, 1983.

————. *They Came before Columbus: The African Presence in Ancient America.* New York: Random House, 1976.

Velikovsky, Immanuel. *Worlds in Collision.*

Wilson, E. O. *The Diversity of Life.* Cambridge: Harvard University Press, Belknap Press, 1992.

Chapters

Adams, Hunter Havelin III. "African Observers of the Universe: The Sirius Question." In Van Sertima, ed., *Blacks in Science.*

————. "New Light on the Dogon and Sirius." in Van Sertima, ed., *Blacks in Science.*

Bloch, Marc. "The Advent and Triumph of the Water Mill." In *Land and Work in Medieval Europe,* translated by J. E. Anderson. New York: Harper and Row, 1972.

Bohm, David. "Postmodern Science in a Postmodern World." In Charles Jencks, ed., *The Postmodern Reader.* London: Academy Editions, 1992.

Clarke, John Henrik. "Lewis Latimer—Bringer of the Light." In Van Sertima, ed., *Blacks in Science.*

Delanda, Michael. "Non-organic Life." In Crary and Kwinter, eds., *ZONE 6—Incorporations.*

Deleuze, Gilles. "Mediators." In Crary and Kwinter, eds., *ZONE 6—Incorporations.*

Derrida, Jacques. "Structure, Sign, and Play in the Human Sciences." In R.

Macksey and E. Donato, eds., *The Structuralist Controversy*. Baltimore: Johns Hopkins University Press, 1970.

Düttmann, Alexander G. "What Will Have Been Said about AIDS?" In Christine Davis, ed., *Public-Sacred Technologies*. Toronto: Public Access, 1993.

Eisenman, Peter. "Unfolding Events." In Crary and Kwinter, eds., *ZONE 6— Incorporations*.

Finch, Charles S. "The African Background of Medical Science." In Ivan Van Sertima, ed., *Blacks in Science*.

Gilbert, Scott, et al. "The Importance of Feminist Critique for Contemporary Cell Biology." In Nancy Tuana, ed., *Feminism and Science*. Bloomington: Indiana University Press, 1989. (Reprinted from *Hypatia* 3, no. 1 [Spring 1988].)

Harding, Sandra. "Feminist Justificatory Strategies." In Garry and Pearsall, eds., *Women, Knowledge, and Reality*, 1989.

Keller, Evelyn Fox. "Feminism and Science." In Garry and Pearsall, eds., *Women, Knowledge, and Reality*, 1989.

———. "The Gender/Science System: or, Is Sex to Gender as Nature Is to Science?" In Tuana, eds., *Feminism and Science*.

Longino, Helen E. "Can There Be a Feminist Science?" In Garry and Pearsall, eds., *Women, Knowledge, and Reality*.

Lumpkin, Beatrice. "Africa in the Mainstream of Mathematics." In Van Sertima, ed., *Blacks in Science*.

Messiha, Khalil, et al. "African Experimental Aeronautics: A 2000-Year-Old Glider." In Van Sertima, ed., *Blacks in Science*.

Nicolas, Gregoire. "Physics of Far-from-equilibrium Systems." In Paul Davies, ed., *The New Physics*. Cambridge: Cambridge University Press, 1989.

O'Neill, John. "Horror Autotoxicus: Critical Moments in the Modernist Prosthetic." In Crary and Kwinter, eds., *ZONE 6—Incorporations*.

Shore, Debra. "Steel Making in Ancient Africa." In Van Sertima, ed., *Blacks in Science*.

Simondon, Gilbert. "The Genesis of the Individual." In Crary and Kwinter, eds., *ZONE 6—Incorporations*.

Stevens, William E. "An Eden in Ancient America? Not Really." *New York Times* (Science Section), March 30, 1993, C-1.

Stone, Alluquère Roseanne. "Virtual Systems." In Crary and Kwinter, eds., *ZONE 6—Incorporations*.

Turner, Frederick. "Biology and Beauty." In Crary and Kwinter, eds., *ZONE 6— Incorporations*.

Van Sertima, Ivan. "Dr. Lloyd Quarterman—Nuclear Scientist." In Van Sertima, ed., *Blacks in Science*.

Zaslavsky, Claudia. "The Yoruba Number System." In Van Sertima, ed., *Blacks in Science*.

Articles

Abelson, Philip H. "Testing for Carcinogens with Rodents." *Science*, September 21, 1990, 1357.

———. "Toxic Terror; Phantom Risks." *Science*, July 23, 1993, 407.

Adams, Carol J. "Ecofeminism and the Eating of Animals." *Hypatia—A Journal of Feminist Philosophy* 6, no. 1 (1991): 125–45.

Archer, M. S., H. W. Sheppard, W. Winkelstein, and E. Vittinghoff. "Does Drug Use Cause AIDS?" *Nature* 362 (March 11, 1993): 103–4.

Athanasiou, Tom. "Greenhouse Blues." *Socialist Review* 91, no. 2: 85–109.

Baum, Rudy M. "Biomedical Researchers Work to Counter Animal Rights Agenda." *Chemical Engineering News*, May 1991, 9–24.

Best, Steven. "Chaos and Entropy: Metaphors in Postmodern Science and Social Theory." *Science as Culture*, vol. 2, pt. 2, no. 11 (1991): 188–226.

Bethell, Tom. "AIDS Reporters Snooze." *Heterodoxy* 1, no. 9 (February 1993): 7.

Beyerchen, Alan. "What We Now Know about Nazism and Science." *Social Research* 59, no. 3 (1992): 615–41.

Blattner, W., R. C. Gallo, and H. M. Temin. "HIV Causes AIDS." *Science* 241 (July 1988): 515–17.

Brenner, Sydney. "That Lonesome Grail." *Nature* 358 (1992): 27–28.

Breo, Dennis L. "At Large." *Journal of the American Medical Association* 264 (November 21, 1990): 19.

Burros, Marian. "Environmental Politics Is Making the Kitchen Hotter." *New York Times*, September 30, 1992, C-1.

Chandler, Mark. "Not Just a Lot of Hot Air." *Nature* 363 (June 24, 1993): 673–74.

Darnovsky, Marcy. "Overhauling the Meaning Machines: An Interview with Dona Haraway." *Socialist Review* 21, no. 2 (1991): 65–84.

Dean, Tim. "The Psychoanalysis of AIDS." *October* 63 (1993): 83–116.

Derrida, Jacques. "L'Affaire Derrida: Another Exchange." *New York Review of Books*, March 23, 1993, 65–66.

———. Letter to the Editor, *New York Review of Books*, February 11, 1993, 44.

Diamond, Cora. "Eating Meat and Eating People." *Philosophy* 53 (1978): 465–79.

Diaz, Kay. "Are Gay Men Born That Way?" *Z Magazine*, December 1992, 42–46.

Donoghue, Denis. "Bewitched, Bothered, and Bewildered." *New York Review of Books*, March 25, 1993, 46–52.

Duesberg, Peter H. "HIV Is Not the Cause of AIDS." *Science* 241 (July 1988): 514–16.

Dunayer, Joan. "Censored: Faculty Who Oppose Vivisection." *Z Magazine*, April 1993, 57–60.

Durr, Detleff, Sheldon Goldstein, and Nino Zanghi. "Quantum Equilibrium and the Origin of Absolute Uncertainty." *Journal of Statistical Physics* 67 (1992): 843–907.

————. "Quantum Mechanics, Randomness, and Deterministic Reality." *Physics Letters A* 172 (1992): 6–12.

Easterbrook, Gregg. "Green Cassandras." *New Republic*, July 6, 1991, 23–25.

Ellis, Kate. "Stories without Endings: Deconstructive Theory and Political Practice." *Socialist Review* 89, no. 2 (1989): 37–52.

Embertson, Janet, et al. "Massive Covert Infection of Helper T Lymphocytes and Macrophages by HIV during the Incubation Period of AIDS." *Nature* 362 (1993): 359–62.

Epstein, Steven. "Democratic Science? AIDS Activism and the Contested Construction of Knowledge." *Socialist Review*, 91, no. 2 (1991): 35–64.

Fauci, A. S., et al. "Acquired Immunodeficiency Syndrome: Epidemiologic, Clinical, Immunologic, and Therapeutic Considerations." *Annals of Internal Medicine* 100 (January 1984): 92–106.

Fausto-Sterling, Anne. "Is Nature Really Red in Tooth and Claw?" *Discover*, April 24, 1993, 24–27.

Ferris, Timothy. "The Case against Science." Review of *Understanding the Present*, by Bryan Appleyard. *New York Review of Books*, May 13, 1993, 17–20.

Feyerabend, Paul. "Atoms and Consciousness." *Common Knowledge* 1, no. 1 (1992): 28–32.

Fox, Robin. "Anthropology and the 'Teddy-Bear' Picnic." *Society*, November/December 1992, 47–55.

Fromm, Harold. "Scholarship, Politics, and the MLA." *Hudson Review*, 46, no. 1 (1993): 157–68.

Gallo, Ernest. "Nature Faking in the Humanities." *Skeptical Inquirer* 15, no. 4 (1991): 371–75.

Gitlin, Todd. "The Rise of 'Identity Politics.'" *Dissent*, Spring 1993, 172–77.

Goldsmith, Edward. "Evolution, Neo-Darwinism, and the Paradigm of Science." *The Ecologist* 20, no. 2 (March/April 1990): 67–73.

Gori, G. B. Editorial. *Wall Street Journal*, August 27, 1992.

Gould, Stephen J. "The Confusion over Evolution." *New York Review of Books*, November 19, 1992, 47–54.

————. "Integrity and Mr. Rifkin." In *An Urchin in the Storm*. [Reprinted from "On the Origin of Specious Criticism"—review of *Algeny* by Jeremy Rifkin, in *Discover*, January, 1985, 34–35.]

Gross, Paul R. "Animals Still Crucial to Research." *Medical World News*, August 20, 1989, 87.

————. "On the 'Gendering' of Science." *Academic Questions* 5, no. 2 (Spring 1992): 10–23.

Grove, J. W. "The Intellectual Revolt against Science." *Skeptical Inquirer* 13, no. 1 (1988): 70–75.

Hacking, Ian. "His Father the Engineer." Review of *Understanding the Present* by Bryan Appleyard. *London Review of Books*, May 28, 1992, 5–6.

Harding, Sandra. "After the Neutrality Ideal: Science, Politics, and 'Strong Objectivity.'" *Social Research* 59, no. 3 (1992): 567–87.

Harris, Daniel. "AIDS and Theory." *Lingua Franca*, June 1991, 1 and 16–19.

Hayles, N. Katherine. "Gender Encoding in Fluid Mechanics—Masculine Channels and Feminine Flows." *Differences—A Journal of Feminist Cultural Studies* 4, no. 2 (1992): 16–44.

Hooks, Bell. "Columbus: Gone but Not Forgotten." *Z Magazine*, December 1992, 25–28.

Imbrie, J., A. C. Mix, and D. G. Martinson. "Milankovitch Theory Viewed from Devil's Hole." *Nature* 363 (1993), 531–33.

Jacobs, James R. "The Political Economy of Science in Seventeenth-Century England." *Social Research* 59, no. 3 (1992): 505–32.

Jay, Martin. The Academic Woman as Performance Artist. *Salmagundi* 98–99 (1993): 28–34.

Kamminga, Harmke. "What Is This Thing Called Chaos?" *New Left Review* 181 (1990): 49–59.

Kantrowicz, Barbara, et al. "Teenagers and AIDS." *Newsweek*, August 3, 1992, 44–49.

Kaus, Mickey. "TRB." *New Republic*, March 29, 1993, 4.

Keller, Evelyn Fox. "Feminism and Science." *Signs: Journal of Women in Culture and Society* 7, no. 3 (1982): 589–602. [Reprinted in Ann Garry and Marilyn Pearsall, eds., *Women, Knowledge and Reality: Explorations in Feminist Philosophy.*]

———. "Long Live the Difference between Men and Women Scientists." *The Scientist* 4 (October 15, 1990).

Kirp, David L. "R$_x$ Populi." *The Nation*, April 5, 1993, 458–61.

Laber, Jeri. "Bosnia: Questions about Rape." *New York Review of Books*, March 25, 1993, 3–6.

Lentricchia, Frank. "Reading Foucault—II." *Raritan Review*, Summer 1992, 41–70.

Levin, Margarita. "Caring New World." *American Scholar* 57 (Winter 1988): 100–106.

Loeb, Jerod M., et al. "Human vs. Animal Rights." *Journal of the American Medical Association* 262 (November 17, 1989): 2716–20.

Lukacs, John. "Atom Smasher Is Super Nonsense." *New York Times*, June 17, 1993, A-25.

MacDonald, Heather. "The Ascendancy of Theor-ese." *Hudson Review* 45, no. 3 (1992): 358–65.

———. "The Holocaust as Text." *Salmagundi* 92 (1991): 160–73.

Matthews, Anne. "Deciphering Victorian Underwear and Other Seminars." *New York Times Magazine*, February 10, 1991, 42ff.

Minogue, Kenneth. "The Goddess That Failed." *National Review*, November 18, 1991.

Nash, Gary B. "The Great Multicultural Debate." *Contention* 1, no. 3 (1992): 1–28.

Ortiz de Montellano, Bernard. "Afrocentric Creationism." *Creation/Evolution* 29 (1991): 1–8.

———. "Chariots of the [Black] Gods?" Forthcoming.

———. "A Critique of the Portland Schools Baseline Essay on African-American Science by Hunter Havelin Adams III." Paper presented at the meeting of the American Association for the Advancement of Science, Chicago, February 11, 1992.

———. "Magic Melanin: Spreading Scientific Illiteracy among Minorities: Part II." *Skeptical Inquirer* 16, no. 2 (1992): 163–66.

———. "Melanin, Afrocentricity, and Pseudoscience." *Yearbook of Physical Anthropology*, forthcoming.

———. "Multicultural Pseudoscience: Spreading Scientific Illiteracy among Minorities: Part I." *Skeptical Inquirer* 16, no. 2 (1991): 46–50.

Pantaleo, Giuseppe, et al. "HIV Infection Is Active and Progressive in Lymphoid Tissue during the Clinically Latent Stage of the Disease." *Nature* 362 (1993): 355–58.

Pardes, Herbert, et al. "Physicians and the Animal-Rights Movement." *New England Journal of Medicine*, 234 (1991): 1640ff.

Pecora, Vincent P. "What Was Deconstruction?" *Contentions* 1, no. 3 (1992): 59–79.

Plumwood, Val. "Nature, Self, and Gender: Feminism, Environmental Philosophy, and the Critique of Rationalism." *Hypatia—A Journal of Feminist Philosophy*, 6, no. 1 (1991): 3–27.

Polley, H. Wayne, et al. "Increase in C_3 Plant Water Use Efficiency and Biomass over Glacial to Present CO_2 Concentration." *Nature* 361 (1993): 61–64.

Potynen, Arthur. "Oedipus Wrecks: PC and Liberalism." *Measure* 113 (February 1993): 1–4.

Robinson, Jeffrey. "Lessons in Holistic Learning." *Rutgers Targum*, April 19, 1993, 6–7.

Rorty, Richard. "For a More Banal Politics." *Harper's*, May, 1992, 16–21.

———. "Love and Money." *Common Knowledge* 1, no. 1 (1992): 12–16.

Rosen, Charles. "The Mad Poets." *New York Review of Books*, October 22, 1992, 35.

Rosenthal, Michael. "What Was Postmodernism." *Socialist Review*, 92 no. 3 (1992): 83–105.

Ross, Andrew. "New Age Technologies." In L. Grossberg, C. Nelson, and P. Treichler, eds., *Cultural Studies*. New York: Routledge, 1992.

Rosser, Susan V. "The Gender Equation." *The Sciences*, September/October 1992, 42–47.

Ryan, Alan. "Foucault's Life and Hard Times." *New York Review of Books*, April 8, 1993, 12–17.

————. "Princeton Diary." *London Review of Books*, March 26, 1992, 21.

Scholz, Susanne. "The Mirror and the Womb: Conceptions of the Mind in Bacon's Discourse of the Natural Sciences." *Women: A Cultural Review* 3, no. 2 (1992): 159–66.

Seabrook, John. "Tremors in the Hothouse." *New Yorker*, July 19, 1993, 32–41.

Sheehan, Thomas. "A Normal Nazi." *New York Review of Books*, January 14, 1993, 30–35.

Slicer, Deborah. "Your Daughter or Your Dog? A Feminist Assessment of the Animal Research Issue." *Hypatia* 6, no. 1 (Spring 1991): 108–24.

Soper, Kate. "Postmodernism, Subjectivity, and the Question of Value." *New Left Review* 186 (1991): 120–28.

Sprinker, Michael. "The Royal Road: Marxism and the Philosophy of Science." *New Left Review* 191 (1992): 122–44.

Tarantola, Daniel, and Jonathan Mann. "Coming to Terms with the AIDS Pandemic." *Issues in Science and Technology* 9, no. 3 (1993): 41–48.

Taubes, Gary. "The Ozone Backlash" (News & Comment). *Science* 260 (June 11, 1993): 1580–83.

Wallerstein, Immanuel. "The TimeSpace of World-Systems Analysis: A Philosophical Essay." *Historical Geography* 23, nos. 1 and 2 (1993): 5–22.

Waters, J. W., et al. "Stratospheric C10 and Ozone from the Microwave Limb Sounder on the Upper Atmosphere Research Satellite." *Nature* 362 (1993): 597–602.

Weisberg, Jacob. "NAS: 'Who Are These Guys Anyway?'" *Lingua Franca*, April, 1991, 34–39.

Weiss, Robin A. "How Does HIV Cause AIDS?" *Science* 260 (May 28, 1993): 1273–79.

Wilson, Elizabeth. "The Postmodern Chameleon." *New Left Review* 180 (1990): 187.

Wittes, Benjamin, and Janet Wittes. "Group Therapy (Research by Quota)." *New Republic*, April 5, 1993, 15–16.

Young, Robert M. "Science, Ideology, and Donna Haraway." *Science as Culture*, vol. 3, pt. 2, no. 15 (1993): 165–207.

Zahler, R. S., and H. J. Sussmann. "Claims and Accomplishments of Applied Catastrophe Theory." *Nature* 269 (1977): 759–63.

Miscellaneous

African-American Baseline Essays. Portland: Multnomah School District 1J, Portland Public Schools, 1990.

Aronowitz, Stanley. "Science under Capitalism." Lecture. Plenary Session, Socialist Scholars Conference, Borough of Manhattan Community College, CUNY, New York, 1991.

————. "Science under Capitalism—The Ecological View." Lecture and panel discussion. Socialist Scholars Conference, Borough of Manhattan Community College, CUNY, New York, 1992 (tape available from J. Turney, 2214 Hey Road, Richmond, VA 23244).

Campbell, Mary Anne, and Randall K. Campbell-Wright. "Toward a Feminist Algebra." Paper read at a meeting of the Mathematical Association of America (San Antonio, 1993), MAA Abstract No. 878-00-1035.

Critical Review, vol. 5, no. 2 (1992).

Denitch, Bogdan. Address. Socialist Scholars Conference, Borough of Manhattan Community College CUNY, New York, 1992.

Galloway, Allison, et al. Exchange of letters in *Science* 250 (1990): 1319.

Gould, Stephen J. "Capitalism and the Environment." Lecture at Socialist Scholars Conference, Borough of Manhattan Community College, CUNY, New York, 1993.

Harding, Sandra. Seminar. Center for the Critical Analysis of Contemporary Culture, Rutgers University, 1992.

Hypatia—A Journal of Feminist Philosophy 6, no. 1 (Spring 1991). Special issue on ecological feminism.

Keller, Evelyn Fox. Broadcast interview by Bill Moyers, PBS.

Latour, Bruno. "Aramis." Lecture at the Center for Critical Analysis of Contemporary Culture, Rutgers University, 1992.

Merchant, Carolyn. Address. University of Virginia, 1992.

Models for Biomedical Research: A New Perspective. Washington, D.C.: National Academy Press, 1985.

Scott, Eugenie. Lecture and panel discussion, "Multiculturalism—The Good, the Bad, and the Ugly." Conference, Committee for the Scientific Investigation of Claims of the Paranormal, Dallas, 1992.

Spitzer, Matt. Letter to the Editor. Z, June 3, 1993.

Zmirak, John. Letter to the Editor. *New Republic*, July 12, 1993, 4–5.

Supplementary References

Cole, Stephen. *Making Science: Between Nature and Society.* Cambridge: Harvard University Press, 1992.

Cronin, Helena. "Oh, Those Bonobos!" *New York Times Book Review*, Sept. 5, 1993, 19.

Donovan, Arthur, Larry Laudan, and Rachel Laudan. *Scrutinizing Science: Empirical Studies of Scientific Change.* Baltimore: Johns Hopkins University Press, 1992. (Hingham, Mass: Kluwer Academic Publishers, 1988.)

Dyson, Freeman. "Science in Trouble." *American Scholar* 62, no. 4 (1993): 513–25.

Ronell, Avital. "AIDS and the Failure of Man's Custodianship: AIDS Update." In Marc de Guerre, ed., *Public—The Ethics of Enactment.* Toronto: Public Access, 1993.

Ross, Andrew. "The Chicago Gangster Theory of Life." *Social Text* 35 (1993): 93–112.

Small, Meredith F. *Female Choices: Sexual Behavior of Female Primates.* Ithaca: Cornell University Press, 1993.

Todd, Alexandra Dundas. *Intimate Adversaries: Cultural Conflict between Doctors and Women Patients.* Philadelphia: University of Pennsylvania Press, 1989.

Weinberg, Steven. *Dreams of a Final Theory: The Search for the Fundamental Laws of Nature.* New York: Pantheon Books, 1993.

Index

AAAS. *See* American Association for the Advancement of Science

Abelson, Philip, 279

Academic left, 8–10, 26, 72, 74, 151, 152, 159, 161, 166, 180, 199, 203, 255; and AIDS, 190; and animal rights, 197–98; careerism within, 237–41; definition of, 2–4, 9–10, 34–41, 260; hostility to science, 10–11, 14, 27, 46, 236, 237; threat to science education, 244–48

Acid rain, 172, 173–74

ACT-UP, 187

Adams, Carol J., 198–99

Adams, Hunter Havelin, III, 207, 208–9, 210, 212, 251, 285

Adler, Jonathan H., 279

Affirmative action, 111

Africans, pre-Columbian contacts with America, 211

Afrocentrism, 11, 14, 38, 225, 252, 289; and science, 203–14, 220, 241, 246–47, 251, 253–54, 284–86

Agent Orange, 163, 164

Agricultural "crisis," 172–75

Ahab, Captain, 179

AIDS, 14, 79, 145, 179, 180–96, 198, 214, 248, 280–83; in Africa, 184, 190, 192; in Asia, 280–81; black attitudes toward, 186, 189, 247; latency of, 281; placebo studies, 189–90; postmodern discourse on, 190–93, 194, 288; transmission of, 183–85, 187

Alar, 163

Algebra, 264, 265, 273; teaching of, 113–117

Alperovitz, Gar, 69, 265

American Anthropological Association, 213, 246–47

American Association for the Advancement of Science, 168

American Mathematical Society, 273

Amnesty International, 138

Ampère, André Marie, 22

Anarchism, 151

Anasazi, 165

Ancient aeronautics, 207

Andreski, Stanislav, 42, 87, 268

Androgenized females, 145–47

Animal research, 199–203, 242, 284

Animal rights, 14, 129, 180, 231, 242, 283–84; and science, 196–203

Antisemitism, 110, 129, 154, 274

Appleyard, Brian, 260, 261

Aramis project. *See* Latour, Bruno

Argyros, Alexander J., 80, 83, 99, 266–67, 268, 270, 272

Aristotle, 40, 117–18

Aronowitz, Stanley, 59, 70, 89, 91, 129, 171, 172, 212, 241, 249, 287; *Science as Power*, 50–55, 262, 279

Ascher, M. S., 282

Ashe, Arthur, 184

Association for Science and Literature, 103

Astrology, 225, 286

Athanasiou, Tom, 150–51, 276

Austen, Jane, 84

Ayer, Alfred Jules, 86, 101
AZT, 185; "Concorde" trial of effectiveness, 282

Bach, Johann Sebastian, 36
Bacon, Francis, 119, 123, 152, 153, 155, 171, 177–78, 277
Bailey, Sarah, 273
Balkanization, 177, 203, 204
Banneker, Benjamin, 209
Baudrillard, Jean, 25, 79, 85, 95, 266
Baum, Rudy M., 284
Beethoven, Ludwig van, 27, 85
Beldecos, Athena, 273
Bell, John, 101, 264, 278
Benin, 206
Berlin, Isaiah, 260
Berman, Morris, 153, 171, 177, 198, 276, 280, 284
Bernoulli, Jacob, 69
Best, Steven, 14, 95–98, 104, 170, 171, 172, 269–70, 279
Bethell, Tom, 282
Beyerchen, Alan, 269
Biehl, Janet, 176, 280, 286
Big Bang, 145
Biodiversity, 159, 167, 172, 174, 278, 288
Biology and Gender Study Group, 117–22, 124, 125, 131
Black studies, 203–214
Blake, William, 20–21, 105, 153, 155, 177, 223, 280
Blattner, W., 282
Bleier, Ruth, 143, 285
Bloch, Marc, 69, 265
Blood supply, protection of, 182
Bloor, David, 52–53
Blumenfeld-Kosinski, Renate, 284
Bohm, David, 261–62, 278–79
Bohr, Niels, 262, 270
Bonobo, 124–25, 288
Boyce, William E., 272
Boyle, William, 17, 63–65, 67–69, 264
Bramwell, Anna, 154–55, 161, 277
Branch, Taylor, 260
Brenner, Sydney, 273–74

Breo, Dennis L., 283
Brown, Lester R., 173, 279
Bryan, William Jennings, 28
Buchanan, Patrick, 260
Bunthorne, Reginald, 71, 106
Burke, Edmund, 20
Burros, Marian, 279
Butler, Judith, 265

Caesarian section, 206, 284
Cambodia, 32
Campbell, Maryanne, 113–17, 272–73
Campbell-Wright, Randall. See Campbell, Maryanne
Cantor, Georg, 100, 103, 263
Carcinogens, 163–64
Cargo cults, 40–41, 288
Carnap, Rudolf, 102, 271
Carroll, Lewis. See Dodgson, Charles Lutwidge
Castañeda, Carlos, 153, 222
Catastrophe theory, 267
CD_4^+ cells: defined, 181. See also T-cells
Cell biology, 117–20, 125, 181, 280, 282
Chalmers, Alan, 237, 261, 287
Chamberlain, Houston Stewart, 209
Chandler, Mark, 277
Chaos theory, 6, 92–105, 249–50, 262, 268–72
Charles II, 63
Cheng, Nien, 275
China, 20, 21, 31, 32, 128, 275
Church of England, 17, 66
Churchill, Winston, 29
Clark, Ronald, 262, 274
Clarke, John Henrik, 206
Clemens, Roger, 99
Climate change, 158
Clinton, President Bill, 53
Cokely, Steve, 281
Cole, Stephen, 288
Coleridge, Samuel Taylor, 20
Columbus, Christopher, 211
Combinatorics, 61
Common law, 87
Communist Party, U.S.A., 28–30, 73
Computational complexity, theory of, 94

Comte, Auguste, 22, 102
Copernicus, Nicolaus, 170
Corneille, Pierre, 87
Correlation coefficients, 62
Counterculture, 222–24
Cousineau, G. H., 274
Crary, J., 267
Creationism: afrocentric, 286; Institute
for Creation Research, 9, 129, 269
Critical Review, 268
Cronin, Helena, 288
Cuba, 32
Culler, Jonathan, 265
Cultural anthropology, 81–82, 213, 246–
47, 256
Cultural constructivism, 11, 13, 14, 56–
57, 59, 63, 69–70, 132–34, 162, 261,
264, 288; and AIDS, 193; Edinburgh
school, 25, 52; and Marxism, 226;
strong version, 44, 45–50; weak ver-
sion, 43–45
Cultural studies, 72, 85, 89–90

Dannemeyer, Rep. William, 125
Dante Alighieri, 85
Darnovsky, Marcy, 275
Darwin, Charles, 44, 132, 156, 171, 217;
social Darwinism, 23
Davidson, Eric, 121, 274
Day, David, 150, 276
Dean, Tim, 79, 192, 266, 283
De Beauvoir, Simone, 40
De Broglie, Louis, 52, 261, 285
Debs, Eugene V., 28–29
Deconstruction, 6, 75–78, 82, 85, 99,
120, 219, 255, 265–66. See also Der-
rida, Jacques
Deep ecology. See Environmentalism
De Grazia, Alfred, 268
De Landa, Manuel, 80, 266–67
Deleuze, Gilles, 80, 267
De Man, Paul, 76–77, 83, 99, 265
Democratic Socialists of America, 50, 83,
261
Denitch, Bogdan, 83, 234, 236–37, 268
Derrida, Jacques, 25, 39, 40, 75–78, 79,
83, 84–85, 88, 89, 95, 96, 99, 192,
265–66, 270, 283, 286

Descartes, René, 65, 68
Diaz, Kay, 283
Diderot, Denis, 18, 24, 165
Differential geometry, 94
Diggins, John Patrick, 72, 74, 265
Di Prima, Richard C., 272
Dirac, Paul M., 261
Discover Magazine, 122–26, 274, 279
Dittersdorf, Carl Ditters von, 128
DNA, 6, 118, 159, 171, 275, 279; as
"master molecule," 140–41
Dodgson, Charles Lutwidge (Lewis Car-
roll), 114, 273
Doell, Ruth, 145
Dogon, astronomical knowledge of the,
207–9, 285
Donne, John, 36
Donoghue, Denis, 179
Donovan, Arthur, 288
Doolittle, Doctor, 196
Duesberg, Peter H., 184–86
Dunayer, Joan, 284
Duplication of the cube, 67, 264
Dürr, Detlef, 261–62
Düttmann, Alexander García, 283
Dynamical systems, 92, 93, 99, 100
Dyson, Freeman, 274, 288

Earth First! 152, 169, 227
Easterbrook, Gregg, 278
Ecofeminism, 134, 162, 176, 231
Ecoradicalism, 150, 161, 163, 166, 169,
176
Edelman, Gerald, 147, 276
Edinburgh school. See Cultural construc-
tivism
Edison, Thomas Alva, 206
Egypt, 206–8; Afrocentrism and, 246–47;
mathematics in, 20; psychic science in,
208
Ehrlich, Paul, 169
Einstein, Albert, 24, 55, 85, 129;
Einstein-Podolsky-Rosen paradox, 264;
relativity theory, 46, 54, 79, 96, 101,
102, 128, 262–63, 274
Eisenman, Peter, 80, 267
Ekland, Ivar, 268–69, 270

Eliot, Thomas Stearns, 36
Ellis, Kate, 83–84, 268
Embretson, Janet, 281
Encyclopedists, 18
Engels, Friedrich, 129
Enlightenment, 3, 13, 25, 99, 148, 175, 197, 215, 220, egalitarian principles of, 19–20, 217, 250; and environmentalism, 154, 171; and postmodernism, 4, 38, 72, 85, 134, 288
Entropy, 95, 97
Environmentalism, 3–4, 8, 14, 15, 149–77, 250–51, 276–79; conflicts with scientific attitude, 231–32; deep ecology, 5, 162, 169, 227–28, 231
Environmental piety, 152
Epstein, Steven, 193–94, 283
Ethiopia, 173
Ethnic studies programs, 204
Euclid, 206–7
Eysenck, Hans, 263

Falwell, Jerry, 193
Fauci, Anthony, 188, 280
Fausto-Sterling, Ann, 125, 274
Feigenbaum, Mitchell, 99, 270
Felman, Shoshanna, 265
Feminism, 3, 6, 8, 25, 26, 32, 38, 39, 72, 151, 152, 198, 203, 204, 216, 221, 288; anti-essentialism, 229–31; eco-feminism, 10; essentialism, 229–30; "feminist standpoint," 135–36; hostility to science, 5, 11, 13, 14, 47, 100–101, 107–48, 227, 235–36, 238, 261, 276; science education and, 245, 251–52, 253–54; and women's studies, 287
Ferguson, Harvie, 46, 261
Ferris, Timothy, 24, 260
Fertilization, syngamic, 118, 120–21
Feuer, Lewis, 241–42, 287
Feyerabend, Paul, 49, 51, 52, 62, 102, 261, 270–71
Feynman, Richard, 123, 264, 276
Finch, Charles S., 284
Finney, Ross L., 272
Fish, Stanley, 265

Fomenko, Anatolii T., 273
Foner, Philip S., 260
Ford, Henry, 177
Foreman, Dave, 14, 152, 178, 276
Foucault, Michel, 25, 37, 39, 40, 75, 77–78, 82, 85, 89, 103, 194, 196, 221, 265, 268, 270, 271, 286
Fourier, Charles, 22
Fox, Robin, 81–82, 246, 248
France, 31, 75; French Revolution, 18, 19, 22
Franco, Francisco, 28
Franklin, Benjamin, 244
Freedman, Bill, 279
Freedman, David H., 125
Freemasonry, 18, 224
Frege, Gottlob, 103
Freud, Sigmund. See Psychoanalysis
Fromm, Harold, 260
Fumento, Michael, 156, 159, 184, 277, 287

Galilei, Galileo, 16, 24, 106, 128, 148, 218
Gallo, Ernest, 265
Gallo, Robert, 181, 280, 281
Galloway, Alison, 272
Garry, Ann, 134, 272, 275
Gates, Henry Louis, Jr., 212
Gauss, Carl Friedrich, 22
Gay, Peter, 260
Gay Men's Health Crisis, 187
Geddes, Patrick, 118
Gerdes, Paulus, 286
Germany, 31, 75
Gibbon, Edward, 18
Gilbert, Scott, 117–22, 273, 274
Gilbert, William Schwenk, 71
Giroux, Henry, 287
Gitlin, Todd, 284
Gleick, James, 94, 97, 104, 271
Global warming signal, 174
Glorious Revolution, 17
Gobineau, J. A., 209
Goddard Institute for Space Studies (NASA), 277

Gödel, Kurt, 80, 102; incompleteness theorem, 4, 78, 100, 102, 267
Goethe, Johann Wolfgang, 20–21, 142, 223, 275; as scientist, 137–38
Goffman, Erving, 57
Goldman, Emma, 35
Goldsmith, Edward, 155, 277
Goldstein, Sheldon, 261–62
Golub, Edward S., 280
Gore, Vice President Albert, Jr., 277
Gori, G. B., 279
Gornick, Vivian, 260
Gould, Stephen J., 44, 56, 69, 143, 171, 261, 269, 275, 279
Goya, Francisco, 215–16
Graff, Gerald, 265
Greek mathematics, 207
Green, Douglas R., 280
Greenhouse effect, 157–58, 160, 166, 167, 169, 172, 174, 232, 277–78, 279
Gross, Paul R., 274, 284
Grossman, Stanley, 273
Grove, J. W., 268
Guggenheim Fellowship, 103
Guth, Alan H., 97
Guzzo, Lou, 278

Hacking, Ian, 260, 261
Halley, Edmund, 16, 65–66
Handel, George Frederic, 19
Haraway, Donna, 14, 100, 132–34, 193, 227, 251, 275
Harding, Sandra, 14, 107, 109, 115, 126–32, 134–36, 143, 212, 227, 229, 241, 249, 251, 272, 274, 275, 285–86, 288
Harris, Daniel, 190–91, 283
Harvey, William, 16, 58
Hawking, Stephen W., 97, 274
Haydn, Franz Joseph, 99
Hayles, N. Katherine, 14; Chaos Bound, 98–105, 270–71
Heidegger, Martin, 23, 76–77, 234
Heisenberg, Werner, 262; uncertainty principle, 4, 51–52, 270
Helms, Sen. Jesse, 125–26
Herder, Johann Gottfried von, 18

Hertzsprung-Russell diagram, 240
Herzen, Alexander, 24
Hicks, Karen, 273
High-temperature superconductors, 43–44
Hilbert, David. See Hilbert space
Hilbert space, 261
Hitler, Adolf, 28, 260
HIV. See Human immunodeficiency virus
Hobbes, Thomas, 63–69, 264
Hodges, Andrew, 264
Hofstadter, Douglas, 276
Holton, Gerald, 262
Homer, 261
Homosexuality, 181, 182, 184, 187–88, 197, 203, 216, 266, 282, 283
Hooke, Robert, 63, 65
Hooks, Bell, 211, 285
Hrbacek, Karel, 263
Hughes, Robert, 259, 285
Human immunodeficiency virus (HIV), 181–82, 184–86, 187, 190, 191, 192, 247, 280–83
Hume, David, 18, 24
Huygens, Christiaan, 68
Hydroelectric power, 160–61
Hypatia, 198–99, 284

Imbrie, J., 279
Immunity, cellular basis of, 280
Institute for Creation Research. See Creationism
Italy, 75

Jacob, James R., 264
James, Henry, 36
Jameson, Fredric, 265
Japan, 5, 31
Jay, Martin, 287
Jech, Thomas, 263
Jefferson, Thomas, 18, 85, 244
Johnson, Earvin "Magic," 183

Kaku, Michio, 276
Kamminga, Harmke, 95, 97, 269
Kant, Emmanuel, 25, 85
Keeton, William R., 119

Keller, Evelyn Fox, 14, 115, 136, 137–42, 143, 171, 251, 273, 275
Kepler, Johannes, 16, 55
Khomeini, Ayatollah Ruhollah, 178
Kimball, Roger, 88, 220
King, Martin Luther, Jr., 26
Kinsey, Alfred, 125
Kirp, David, 194, 283
Klehr, Harvey, 260
Krafft-Ebbing, Richard, 125
Kramer, Larry, 179, 182, 186–89, 283
Kuhn, Thomas, 49, 56, 57, 94, 139, 180, 263, 280
Kwinter, S., 267

Laber, Jeri, 286
Lacan, Jacques, 193, 221, 266, 286
Lancet, The, 282
Laplace, Pierre Simon de, 20; Laplace's Demon, 262; Laplacian Operator, 262
Lamarck, Jean Baptiste, 156
Lather, Patti, 271
Latour, Bruno, 14, 57–62, 91, 263–64; Aramis project, 60–62; Science in Action, 57–60
Lattimer, Lewis, 206
Laudan, Larry, 275, 288
Laudan, Rachel, 288
Layzer, David, 274
Leary, Timothy, 222
LeGuin, Ursula K., 149–50, 152
Leibniz, Gottfried Wilhelm, 17, 69
Lenard, Philip, 129, 131, 274
Lenin (V. I. Ulyanov), 29, 129, 225–26, 252; Leninism, 73, 252
Lentricchia, Frank, 83, 268
Lessing, Gotthold, 24
Levin, Margarita, 131–32, 275
Lewis, Martin, 149, 164, 165, 177, 279, 280
Lillie, Frank R., 274
Linearity, 95, 97, 98, 99, 100, 104-5, 266–67, 269, 271–72
Lingua Franca, 39, 190, 260
Literary criticism, 12, 74, 75, 76, 81, 82, 84, 86, 89, 110, 115, 196, 239, 256, 266

Locke, John, 17, 24, 85, 177
Loeb, Jacques, 121
Loeb, Jerod M., 284
Logical positivism, 86, 101–3, 144
Longino, Helen, 14, 134, 136, 138, 142–48, 241, 251, 276
Lorentz, Hendrik A., 262
Lorenz, Edward N., 100
Loubere, Leo, 260
Lovelock, James, 169
Lukacs, John, 259
Lumpkin, Beatrice, 285
Lyotard, J. F., 25, 37, 80, 85, 88, 95, 266–67
Lysenko, Trofim, 156, 226, 252

McCarthy, Sen. Joseph, 30
McClintock, Barbara, 141–42
McClung, C. E., 118–19
MacDonald, Heather, 75, 265
McDonald's Restaurants, 278
Madonna, 118
Maistre, Joseph de, 21, 23
Malcolm X, 26
Malleus Maleficarum, 37
Mandelbrot, Benoit, 80; Mandelbrot set, 100
Mann, Jonathan, 280–81
Manning, Kenneth R., 285
Manson, Charles, 99
Mao Tse-tung, 31
Marine Biological Laboratory (Woods Hole), 274
Martin, Emily, 125
Marx, Karl. See Marxism
Marxism, 11, 32, 38, 40, 46, 55, 57, 72, 73, 82, 83, 84, 95, 151, 152, 162, 237, 252; and mathematics, 286; post-Marxism, 37, 158, 221; and science, 5, 10, 22, 225–26
Maternal messenger RNA, 121, 274
Mathematical Association of America, 272
Maxwell, James Clerk, 103, 262
Maya, 165, 246
Mayles, Peter, 134
Mead, A. D., 274

Melanin scholars, 209, 285
Melville, Herman, 84, 179, 280
Mendelssohn, Felix, 87
Merchant, Carolyn, 14, 152, 162, 171, 177, 249–50, 276, 278–79, 288
Messiha, Khalil, 207
Metabolism, 118
Metaphor: mongering, 116, 121; quibbles over incidental, 121
Michaels, Patrick J., 279
Michaelson-Morley experiment, 54, 263
Milankovitch theory, 170, 279
Mill, John Stuart, 24
Miller, J. Hillis, 265, 286
Minogue, Kenneth, 136, 275
Mittag-Leffler theorem, 108
Mobius strip, 80
Models, biomedical research and animal, 200–203
Modern Language Association, 103
Monoclonal antibodies, 280
Montagnier, Luc, 181, 280
Montaigne, Michel de, 165, 219; On Cannibals, 286
Montgomery, Thomas H., 119
Moralism, 220; of ecoradicals, 160–61
Morgan, Thomas Hunt, 274
Morse functions, 273
Mountcastle, Vernon, 176
Mozart, Wolfgang Amadeus, 85, 128, 218–19
Muir, John, 28
Multiculturalism, 5, 14, 110, 210, 213, 241, 246, 248, 259, 271, 287
Mussolini, Benedetto, 29

Napoleonic Code, 87
National Association of Scholars, 241, 287
National Institutes of Health, 284
Natural childbirth, 288
Naturphilosophie, 137–38
Navasky, Victor, 260
Nazism, 76, 105, 129, 155
Neuronal selection theory, 147
Newton, Isaac, 16–17, 65–66, 68–69, 85, 92, 98, 100, 105, 132, 137, 142, 153, 177, 218, 262; Principia Mathematica Philosophae Naturalis, 18, 65, 131
New York Academy of Sciences, 285–86
Nicolas, Gregoire, 270
Niemczyk, Nancy, 273
Nietzsche, Friedrich, 74, 234, 288
Nobel Prizes, 200–201, 274
North Korea, 19
Norway, 208
Nuclear power, 160
Nuclear winter, 167, 169

Occam's Razor, 262
October, 266
Oldenburg, Henry, 64, 264
Olmec, 246
O'Neill, John, 283
O'Neill, William, 260
Operations research, 61
Ortiz de Montellano, Bernard, 213, 246–47, 284, 285, 286, 287
Ottoman Empire, 22
Ovum, 118–22, 125
Ozone hole, 158–59, 162, 167, 172, 174, 278

Pagels, Heinz, 276
Paglia, Camille, 286
Pantaleo, Giuseppe, 281
Parthenogenesis, 121
Pathetic fallacy, 122, 274
Patton, Cindy, 190–92, 283
Pauli, Wolfgang, 261
Pax Romana, 20, 128
Peano, Giuseppe, 100, 263
Pearsall, Marilyn, 134, 272, 275
Peat, F. David, 278
Pecora, Vincent, 85, 260, 268
Penrose, Roger, 274
People for the Ethical Treatment of Animals, 138
Physics, worldview of, 127
Pi, value of, 207
Pinchot, Gifford, 28
Pius IX, 23
Planck, Max, 261
Plato, 40

Plumwood, Val, 176, 280, 286
Poincaré, Henri, 93–94, 115, 261; Poincaré conjecture, 115, 273
Political correctness, 8–9
Polley, H. Wayne, 279
Pope, Alexander, 19, 100
Popper, Karl, 89, 140
Population growth, 157, 169
Portland Baseline Essays, 208–9, 210, 246, 285–86
Postmodernism, 6, 8, 21, 175, 180, 204, 255, 269–70; and AIDS, 191–95, 288; and animal rights, 198–200; definition, 4–5, 71–78; and education, 247–48, 271, 287–88; and environmentalism, 151, 177; and feminism, 109, 135–36; and mathematics, 115, 267–68; obsolescence of, 272; and poststructuralism, 259; and science, 10–11, 13, 14, 47, 78–106, 235–36, 245, 274, 278–79
Poststructuralism. See Postmodernism
Potynen, Arthur, 287
Prigogine, Ilya, 96
Principia Mathematica (Russell and Whitehead). See Russell, Bertrand
Principia Mathematica Philosophae Naturalis. See Newton, Isaac
Pro-life movement, 198
Psychoanalysis, 56, 117, 140–41, 263, 266
Puck, Wolfgang, 279

Quantum mechanics, 49, 51–52, 59, 96, 98, 100, 255–56, 261, 264, 267; Copenhagen interpretation of, 262
Quarterman, Lloyd, 206

Racine, Jean, 87
Rauch, Jonathan, 272
Ray, Dixy Lee, 278
Reagan, President Ronald, 188, 278
Redondi, Pietro, 16
Regan, Tom, 197, 283, 284
Relativism, 26, 134, 225, 268; and anthropology, 246; and cultural constructivism, 44, 133, 162; and feminism, 109, 135; Kuhn's rejection of, 139; and

postmodernism, 84, 96, 199; and science, 49–50, 59–60, 62
Relativity. See Einstein, Albert
Rembrandt van Rijn, 27
Responsibility of scientists, 6–8, 14–15, 253–57
Reynolds number, 62, 263–64, 270
Richmond Times-Dispatch, 281
Riemann, Bernhard: Riemannian manifold, 267; Riemannian metric, 262
Rifkin, Jeremy, 14, 96–97, 154, 170–75, 178, 243, 249, 269, 276, 279, 288
Robeson, Paul, 26
Robinson, Jeffrey, 271
Rodents, carcinogen screening in, 163–64
Romania, 203
Romanticism, 20–21, 223
Ronell, Avital, 288
Roosevelt, President Franklin D., 29
Root-Bernstein, Robert, 185, 282
Rorty, Richard, 83, 102, 270
Rosen, Charles, 280
Rosenberg, Rebecca, 273
Rosenthal, Michael, 272
Ross, Andrew, 14, 89–92, 213, 249, 257, 288
Rosser, Sue V., 272, 285–86
Rousseau, Jean-Jacques, 19, 165
Royal Society, 63–69
Ruelle, David, 95, 268–69, 270
Russell, Bertrand, 2, 24, 102; Autobiography, 1, 270; Principia Mathematica, 102–3

Sagan, Carl, 169
Sahara Desert, 172–73
Schaffer, Simon, 14, 63–69, 70, 264
Schatten, G. and H., 120
Schiller, Friedrich von, 83
Schneider, Stephen, 167–68, 169, 279
Scholarship, craft of, 239
Scholz, Susanne, 277
Schrödinger, Erwin, 52, 98, 261; equation, 261; Schrödinger's Cat, 262
Schubert, Franz, 87
Schwinger, Julian, 264

Science: and humanism, 243–44; "works," 272
Science magazine, 111, 163, 272, 281
Sciences, The, 285–86
Science studies, 132
Scott, Eugenie, 246–47, 287
Scott, Joan Wallach, 265
Scrivenor, Patrick, 259
Seabrook, John, 279
Sex determination, 118–20; Y-chromosome in, 274
Sexist discrimination, 110–13
Sexuality, "spin" on, 122–26
Sexual selection, 44
Shakespeare, William, 27, 85
Shapin, Steven, 14, 63–69, 70, 264
Sheehan, Thomas, 265
Sheldrake, Rupert, 169
Sheppard, H. W., 282
Sherrington, Sir Charles, 137, 275
Simondon, Gilbert, 80, 267
Singer, Peter, 197, 283, 284
Slovakia, 203
Small, Meredith F., 124, 274, 288
Smith, Adam, 18, 19
Snow, C. P., 7, 244
Socialism, 19, 27–29, 229, 252
Society for Industrial and Applied Mathematics, 273
Sociobiology, 288
Solar power, 160
Somalia, 173
Soper, Kate, 83, 268
Space-filling curve. *See* Peano, Giuseppe
Spermatozoa, 118–22, 125, 126
Spinoza, Baruch, 24, 65, 264
Spitzer, Matt, 284
Sprinker, Michael, 55, 263
Squaring the circle, 67, 264
Stalin (Joseph V., b. Dzhugashvili), 25, 28, 29, 91, 156, 226, 256
Steiner, George, 79, 151, 215, 217, 220, 265, 276, 286
Stone, Alluquère Roseanne, 80, 267
Structuralism, 259
Stuart, House of, 64

Students for a Democratic Society (SDS), 31, 221
Sudan, 173, 203
Sussmann, Hector J., 267

Tanzania, 206
Tarantola, 280–81
Taubes, Gary, 278
T-cells, 181, 280
Temin, H. M., 282
Thermodynamics, 96, 97, 98
Thirty Years' War, 16
Thom, René, 80, 267
Thomas, George B., 272
Thomas, Lewis, 276
Thomas, Norman, 29
Thomson, John Arthur, 118
Three-body problem, 98
Todd, Alexandra Dundas, 288
Topoi, 266
Topology, 7, 59, 94, 100, 115, 265–66, 79; differential, 266; fundamental and higher homotopy groups, 273
Totalism, 225–26
Traveling salesman problem, 61
Trotsky (Lev Bronstein), 29
Turgenev, Ivan, 24
Turner, Frederick, 80, 267

Uganda, 206
Uncertainty principle. *See* Heisenberg, Werner
Union of Soviet Socialist Republics, 25, 29, 160, 171, 172

Values, constitutive and contextual, of science, 143–44
Van Sertima, Ivan, 206–9, 211–12, 246, 251, 285
Variables, intervening, 146
Velikovsky, Immanuel, 268
Vienna Circle, 101, 271
Vietnam, 19, 30, 32, 33, 163, 164
Vittinghoff, E., 282
Voltaire (F. M. Arouet), 24, 219
Von Neumann, John, 52, 261
Vulcanism, and sulfate aerosols, 277

Wallerstein, Immanuel, 269, 271
Wallis, John, 66–67, 264
Ward, Seth, 66, 264
Waters, J. W., 277
Wedel, Andrew, 273
Weinberg, Steven, 288
Weisberg, Jacob, 287
Weiss, Robin A., 281
West, Cornel, 212
Weyl, Hermann, 261
Whitehead, Alfred North, 104
Whitman, Walt, 142
Wilkins, John, 264
Wilson, E. B., 119
Wilson, Edward O., 159, 278, 288
Wilson, Elizabeth, 83, 268
Winkelstein, W., 282
Witten, Edward, 97

Wittes, Benjamin and Janet, 195, 283
Wittgenstein, Ludwig, 40
Women's careers, "spin" on, 123
Wordsworth, William, 20–21, 83, 142, 223
Worldwatch Institute, 173, 279

Yoruba, 206
Young, Frank, 188
Young, Robert M., 275
Yugoslavia, 203

Zahler, Raphael S., 267
Zanghi, Nino, 261–62
Zaslavsky, Claudia, 206
Zeeman, E. C., 267
Z Magazine, 211, 284
Zmirak, John, 260

LIBRARY OF CONGRESS CATALOGING-IN-PUBLICATION DATA

Gross, Paul R.
    Higher superstition : the academic left and its quarrels with science / Paul R. Gross,
Norman Levitt.
        p.   cm.
    Includes bibliographical references and index.
    ISBN 0-8018-4766-4 (alk. paper)
        1. Science—Social aspects.   2. Humanities.   I. Levitt, N. (Norman), 1943–
II. Title.
Q175.5.G757   1944
500—dc20   93-32914

    ISBN 0-8018-5707-4 (pbk.)